California's Fading Wildflowers

California's Fading Wildflowers

Lost Legacy and Biological Invasions

Richard A. Minnich

UNIVERSITY OF CALIFORNIA PRESS
Berkeley · Los Angeles · London

University of California Press, one of the most
distinguished university presses in the United States,
enriches lives around the world by advancing scholar-
ship in the humanities, social sciences, and natural
sciences. Its activities are supported by the UC Press
Foundation and by philanthropic contributions from
individuals and institutions. For more information,
visit www.ucpress.edu.

University of California Press
Berkeley and Los Angeles, California

University of California Press, Ltd.
London, England

Library of Congress Cataloging-in-Publication Data

Minnich, Richard A.
    California's fading wildflowers : lost legacy and
biological invasions / Richard A. Minnich.
        p.   cm.
    Includes bibliographical references and index.
    ISBN: 978-0-520-25353-7 (cloth : alk. paper)
        1. Biological invasions—California.   2. Plant
invasions—California.   3. Wildflowers—California.
    I. Title.

QH353.M56   2008
582.1309794—dc22   2007040978

Manufactured in the United States of America

17   16   15   14   13   12   11   10   09   08
10   9   8   7   6   5   4   3   2   1

The paper used in this publication meets the minimum
requirements of ANSI/NISO Z39.48–1992 (R 1997)
(*Permanence of Paper*).

*To my parents, my wife and foundation Maria,
and my daughters Victoria and Jenny, who
are taking on lives of their own*

and

In memory of my mother, Marion

Contents

Illustrations

Tables

Preface

The inspiration for this book may have been a glimpse down from a commercial jet of radiant golden hillsides of California poppies at Lake Elsinore in 2001 that will forever stick to memory. It was the finest outbreak of wildflowers in coastal southern California since the 1960s, and my first experience with hillside splashes of color, having grown up in Long Beach, California, in the 1950s when the city was surrounded by former pasturelands of European grasses and black mustard. My first instinct was to jump off the plane, but once I recognized my captivity I reminded myself that the flowers would await my return.

The topic of native herbaceous flora of California drew my attention to many sources of information: the important translation of Juan Crespí's original diaries by Alan K. Brown, numerous volumes of early explorers and settlers in California history, historical newspapers, and botanical collections. Gradually, a new coherent story emerged, different from the one I learned in university classes and in modern textbooks.

This volume may not have happened without a life-long experience of the California dream, and the inspiration of my father, J. Raymond Minnich, who was also born in California on a fruit ranch at Hemet in 1912. He was so resourceful that he built a cabin in the forests of Mount Baldy in the San Gabriel Mountains during the Depression. The wildlands of this remote place became my childhood "home away from home" on weekends and shaped my interest in ecology. As a graduate student at Stanford, Ray spent many trips hiking and skiing in the Sierra Nevada. When he wasn't looking I pulled out from his closet a roll of topographic maps of the Sierra and learned to visualize the contours. Ray also introduced me to the important natural history writings of John Muir, William Henry Brewer, and Clarence King, which prepared me for the literature to be tackled and the historical perspective needed to trace the history of California pasture. The story to follow is nearly beyond the living memory of my father, and virtually everyone else in the state.

I would like to acknowledge Michael Barbour, Brett Goforth, and three anonymous reviewers for their constructive reviews of the manuscript, and Kristin Hepper for her help in preparation of the maps. Ernesto Franco-Vizcaíno and his colleagues at the Centro de Investigación Científica y de Educación Superior (CICESE) in Ensenada and Andrea Kaus provided advice on the translation of Spanish plant names. Victoria Minnich accompanied me on weekend "flower expeditions" and gave me much fodder toward a philosophical basis of science. The intellectual foundation for this volume lies in graduate school decades ago with Jonathan D. Sauer, who views vegetation dynamically over broad spatial and temporal scales, and who believes that models have strong empirical foundation.

CHAPTER I

The Golden State

No poet has yet sung the full beauty of our poppy, no painter
has successfully portrayed the satiny sheen of it lustrous petals,
no scientist has satisfactorily diagnosed the vagaries of its
variations and adaptability. In its abundance, this colorful
plant should not be slighted: cherish it and be ever thankful
that so rare a plant is common.

—John Thomas Howell (1937)

California is historically and metaphorically symbolized as the "Golden
State" in tribute to the gold rush of 1849, but for many living in the state
gold is also a reminder of its sunny Mediterranean climate, or perhaps
the Golden Gate Bridge. The 'Washington' navel orange was "liquid gold"
from which fabulous wealth was created in the late nineteenth century.

The coastal plains and valleys were also once golden with fields of brilliant wildflowers, highlighted by the stunning California poppy (*Eschscholzia californica*), as well as goldfields (*Lasthenia* spp.) that created
bright yellow rugs. California hillsides also hosted a rainbow of other
colors from tidy tips, fiddlenecks, lupines, phacelias, owls clover, baby
blue eyes, penstemon, and many other genera. The splendor of California native wildflowers was early disseminated to many parts of the known
world by word of mouth and in the writings of those first explorers who
cruised along the California coast and saw long stretches of rolling hills
clothed in a mantle of gold. The first Spanish galleons sailing up and down
the coast in the eighteenth century called the region "a land of fire," noting the deep orange-colored hillsides of California poppies. Their spontaneous exclamation, "la tierra del fuego!" (the land of fire) became a
symbol of this rich, newfound land.[1] California's wildflower heritage was
appreciated by the generations of the late nineteenth century, was the topic
of books (e.g., Holder 1889; Saunders 1914, 1931), was institutional-
ized in floral societies, and was the primary inspiration of the New Year's
Day Rose Parade in Pasadena (the Tournament of Roses). Indeed, the

poppy was chosen as the state flower by the State Floral Society, an effort passed by legislation in 1903.[2]

In modern times, California's "gold" is advertised in tabletop books that show glossy photographs of yellow oak-dotted rolling hillsides. While the trees are native, the yellow undercarpet is an assemblage of bromes, oats, fescues, barleys, and mustards introduced to California from Mediterranean Europe and the Middle East. Over the past two centuries, European annual species have proliferated across the state, a process that began with the deliberate introduction of some of these invaders by Spanish Franciscan missionaries in 1769. This treasure deemed "golden" is in fact a biological transformation, but naïve Californians are oblivious to the immense change in annual herbaceous vegetation. Invasive grasses and forbs have diminished the diversity and abundance of the state's wildflower flora, degraded pasture, and have increased fuels that threaten urban areas with wildfire. The tragedy is that the poppy, California's state flower, is no longer common, to the point that reserves have recently been created to protect it.

Perhaps the most remarkable aspect of California's biological invasions is their furious pace, as herbaceous cover had already changed over extensive areas of the state before the arrival of the first botanists to California in the early nineteenth century, a topic of many scientific investigations (e.g., Mooney and Drake 1986; Bartolome et al. 1986; Huenneke 1989; Keeley 1989, 1993; Bossard et al. 2000). The rapid change in herbaceous vegetation has hampered investigation to a point that the pre-European herbaceous flora is enigmatic to most scholars in the modern scientific community (Barry 1972; Bartolome at al. 1986; Keeley 1989; Hamilton 1997; Mack 1989), leading to a plethora of hypotheses about the indigenous flora and time line and mechanisms of the transformation. The modern consensus is that California was carpeted not by wildflowers, but by perennial bunch grassland that became replaced by modern exotic annual grassland, encouraged by grazing and drought in the mid–nineteenth century (reviewed in Heady 1977; Keeley 1989, 1993; Sims and Risser 2000). This view appears to originate with the observations of William Henry Brewer, who led the first survey of the state's flora in the 1860s (Brewer and Watson 1876–80). The prominent early twentieth century ecologist Fredrick Clements (1934) formally proposed the bunch grassland model, which he deduced using his climax and "relict" theories. Modern grassland specialists have undertaken ecological and restoration studies on bunch grasslands (e.g., Nelson and Allen 1993; Dyer and Rice 1997; Hamilton et al. 1999).

Advocates of the bunch grassland theory write with certainty. Referring to purple needle grass, Heady (1977: 495) writes, "*Stipa (Nassella)
pulchra,* beyond all doubt, dominated the valley grassland. . . . Perennial
grasses associated with *Stipa* were *Aristida hamulosa (ternipes), Elymus
glaucus, E. triticoides, Festuca idahoensis, Koeleria cristata, Melica californica, M. imperfecta, and Poa scrabella (secunda).*" Burcham (1957:
90), referring to the pre-Hispanic landscape, writes that "the pristine
dominants were perennial bunchgrasses—purple needle grass (*Stipa pulchra*) and nodding needlegrass (*Stipa [Nassella] cernua*) [and] were outstanding in the Central Valley. Blue wild-rye, pine blue grass, and deer-
grass were important associates" (cf. Beetle 1947). In *North American
Terrestrial Vegetation,* Sims and Risser (2000: 342) state that the "original Pacific prairie was dominated by cool-season, perennial bunchgrasses
such as *Nassella pulchra, N. cernua, Elymus* spp., and *Poa secunda.*"

Annual and perennial wildflowers were important components of the
bunchgrass prairie (Barbour et al. 1993). Burcham (1957: 104) states that
"associated with the grasses, . . . were . . . broad-leaved herbs with brightly
colored flowers." Early studies describe "great masses of annuals" comprising several hundred species and more than 50 genera (Clements 1920,
1934; Weaver and Clements 1938; Jepson 1925; Beetle 1947; Barbour et
al. 1993). In "A California Flora," Munz and Keck (1959: 18) describes
a "subtropical type" of open treeless grassland, with a "rich display of
flowers in wet springs." Still other botanists and ecologists assert that
bunch grasslands were never a prominent component of the state's vegetation (Twisselmann 1967; Wester 1981; Hamilton 1997; Schiffman 2000).

A second theme is whether human disturbance was critical to the expansion of introduced annual grasses and forbs. Until the late nineteenth
century, the novel source of disturbance in California was the introduction of domestic livestock, largely cattle, but also horses for tending cattle and sheep. An important question is whether the expansion of introduced species was largely facilitated by livestock or was pushed ahead,
dispersing and colonizing based on the introduced species' ecological requirements, seed-dispersal capacity, and other life traits. As stated by
Blumler (1995), a major issue is the extent to which anthropogenic disturbance is necessary for alien species' invasions. On one hand species
may have been ruderal, strongly dependent on disturbance for success,
while other species may have transplanted vigorous adaptations from one
continent to another, outcompeting native species.

Advocates of the bunch grass hypothesis argue that perennial grassland was replaced by introduced annual forbs and introduced grasses pri-

marily during early American settlement in the mid–nineteenth century
due to overgrazing (Burcham 1957; Barry 1972; Keeley 1989). According to Burcham (1957: 192), a leading proponent, "With reservoirs of
seed of aggressive annuals widely distributed about the countryside . . .
recurrent [droughts] . . . combined with extremely heavy grazing during
the late 1850s and 1860s, struck the final blow at the once abundant
perennial grasses of the range lands. . . . As perennials were depleted the
burden of grazing fell upon the more palatable of the native and introduced annuals." The linkage between expanding exotics and displacement of bunch grasses was even made in top-flight botanical floras, used
by botanists for decades. For example, in *A California Flora*, Munz and
Keck (1959: 17) describe "Valley grassland" as "originally being covered with various bunch grasses such as *Stipa* (*Nassella*) *pulchra, S. cernua, Poa scabrella* (*secunda*) and *Aristida divericata;* now because of overgrazing largely replaced by annual species of *Bromus, Festuca, Avena,* etc."
    This view may even have deeper roots. Livestock grazing at scales of
millenia in Europe selected for weedy, aggressive annuals that expanded
across California (Zohary 1962, 1973; Baker and Stebbins 1965; Sauer
1988; cf. Blumler 1995 and Blumler and Byrne 1991). A corollary argument is that lightly grazed California pasture was vulnerable to invasion from grazing-adapted Mediterranean annuals (Mack and Thompson 1982), i.e., native California annuals were noncompetitive against
introduced species because they evolved without grazing pressure.
    Another hypothesis is that exotic species expanded across California
independently of human disturbance, i.e., they determined their own des-
tinies as invasive species (Mooney and Drake 1986; Huenneke and
Mooney 1989; D'Antonio and Vitousek 1992; Blumler 1995; Bossard
et al. 2000; Brooks et al. 2004). Introduced species were "preadapted"
in California wildlands because they came from similar climates in the
Mediterranean basin, invading almost exclusively preexisting herbaceous
landscapes (Heady 1977; Huenneke 1989). The invaders were also highly
productive, thereby using resources more efficiently than indigenous forbs
(Huenneke 1989; Blumler 1995), even strongly competing against native perennials (Biswell 1956; McNaughton 1968; Heady 1977; Bartolome and Gemmill 1981; D'Antonio and Vitousek 1992; Blumler
1995). Introduced Mediterranean annuals also came to California without their natural pathogens (Jackson 1985). Scholars have further
pointed out that not all of California was invaded at once, as exotic
species came at different times and spread at different rates into divergent habitats depending on their life traits (Heady 1977; Sauer 1988).
    As bunchgrass theory has come under scrutiny, some researchers have

posited another theory that annual forbs dominated California's prairies (Biswell 1956; Bartolome and Gemmill 1981; Wester 1981; Hamilton 1997; Schiffman 2000). Accounts of scarce plant cover in the Central Valley suggest that bunch grasses were not present in these areas and that vivid descriptions of wildflowers were made throughout California (Wester 1981).The remarkable success of exotic annual grasses and forbs cannot be denied. Modern exotic annual grasslands have extensive distribution on clay-rich soils and alluvium at lower elevations throughout California, including on the coastal plains and in interior valleys of central and southern California and inland across the Central Valley to the foothills of the Sierra Nevada (Figure 1.1) (Heady 1977; Sims and Risser 2000). Exotic annual grasslands are free of woody cover over extensive areas or grow beneath oak woodlands in the mountains and foothills. Exotic annual grassland reaches its limit where it interfaces with coastal sage scrub and chaparral on shallow, poorly developed soils on well-drained slopes of the Sierra Nevada, the central Coast Ranges, and the Transverse and Peninsular Ranges of southern California. California grasslands grow in a wide range of average annual precipitation, which mostly falls between November and April, ranging from 20 to 100 centimeters. Thin cover of exotic annual grassland even extends into the desert on the leeward side of the Sierra Nevada and southern California ranges. The dominant species of exotic annual grassland include *Bromus madritensis* (*rubens*), *B. diandrus*, *Avena fatua*, *A. barbata*, *Brassica nigra*, *B. geniculata*, *Hordeum murinum*, and *Festuca megalura* (*myuros*) (Heady 1977). Annuals forbs are diverse, but scarce, and include annual species in the genera *Eschscholzia*, *Phacelia*, *Cryptantha*, *Salvia*, *Nemophila*, *Viola*, *Chaenactis*, *Layia*, and perennials such as *Allium* and *Nassella* (Raven 1963; Ornduff 2003). The growing season is the winter rainy season, when temperatures are warm enough to maintain growth flushes and hard frosts are rare (Minnich 2006). Exotic and native annuals germinate soon after the first heavy rains, grow slowly in winter, and then grow rapidly to flower and seed in spring. Drying when soil moisture is depleted is followed by fires if there is sufficient growth from winter rains. The seed of natives and exotic forbs may survive for years to decades as soil "seed banks," whereas exotic grasses have short seed life of a year or two, but compensate through massive germination rates with the first rains.

While there is unanimous agreement in the scientific community that herbaceous vegetation has undergone enormous change since the late eighteenth century, little consensus has emerged on the nature of the transformation largely because existing hypotheses have limited empirical foundation. Thus far, the web of models and hypotheses on the history

Figure 1.1. Generalized distribution of exotic annual grassland in California (mapped from MODIS Rapid Response System, http://rapidfire.sci.gsfc .nasa.gov).

of California's herbaceous flora has little direction toward synthesis. For reasons of practicality, nearly all ecological research is based on local field studies covering a few years, which precludes the generation of realistic null hypotheses (Jackson et al. 2001). Studies have also resorted to deductive historical scenarios based on spatial evidence, often in relation to the modern population dynamics of already invasive-contaminated herbaceous ecosystems. Without historical perspective and baselines from which vegetation change can be reconstructed, the conclusions drawn may be ad hoc stories.

The goal of this volume is to assess pre-European herbaceous vegetation and its transformation to modern exotic grasslands. The approach here follows the perspective of Grove and Rackham (2001: 18), who assert that "landscape history is best arrived at from the records of identifiable sites which can be traced down the centuries." Hypotheses concerning biological invasions in California's herbaceous communities can be best "tested" by examining historical records of introduced species, native vegetation, and grazing of sufficient time scale to capture vegetation change.

The choice of baseline to reconstruct historical vegetation change also affects the outcome, a phenomenon that Jackson et al. (2001) call the "shifting baseline syndrome." Defenders of the bunch grassland model have built their case on historical evidence from the mid–nineteenth century, based on a longstanding view that only observations by trained botanists have scientific merit (Parish 1920; Burcham 1957). However, the first botanists saw already widespread grasslands of introduced European annual grasses and forbs. This book begins with the earliest historical baseline, the journals kept by Spanish missionary explorers during the Gaspar de Portolá and Juan Bautista de Anza land expeditions in the late eighteenth century. These documents are the only written record of indigenous herbaceous cover before invasive species began spreading across California. The Spanish account may possibly capture California's herbaceous vegetation at Holocene time scales. Until the arrival of the Spaniards, long-range seed dispersal was near background rates at geologic time scales because Native American hunting and gathering societies, the population in California possibly numbering 350,000 (Baumhoff 1963), were limited in mobility, precluding accelerated the anthropogenic dispersal of seed plants seen in recent centuries.

While Spanish botany is not at the level of modern scientific protocol, the journals of the Portolá, Anza, and other Spanish expeditions are a systematic survey of the state as required by mandate of the viceroy of Mexico, and they provide a baseline of aboriginal vegetation against which one can assess changes in California's vegetation. Moreover the recent publication of the original diaries of Juan Crespí (Brown 2001) has brought greater detail concerning California vegetation at the onset of European settlement. The Crespí diaries translated by Bolton (1927) were scribe copies that generalized the original manuscripts. To make effective use of the Spanish diaries requires two concessions from the reader: (1) the broadscale pattern does not require that the diaries be highly precise (Jackson et al. 2001); and (2) the diaries should be appreciated as originating from the mind-set of late–eighteenth century Spanish priests (Grove and Rackham 2001).

Indigenous forbfields similar to those described by the Spaniards were the object of discussion in the nineteenth and early twentieth centuries by botanists, naturalists, explorers, and book writers devoted to California's diverse landscapes, and in articles of the same localities in the *Los Angeles Times* well into the twentieth century. These sources and writings of explorers, naturalists, and settlers in the post-Hispanic era vividly capture the arrival and expansion of some European grasses and forbs, and related impacts such as the displacement of wildflowers and the proliferation of wildfires in the deserts.

This book evaluates the regions settled or explored during the Spanish mission period: California south of San Francisco and west of the Sierra Nevada, as well as the southeastern deserts (Figure 1.1). The following questions will be addressed: (1) the character of pre-European herbaceous vegetation; (2) patterns of aboriginal burning; (3) the expansion of introduced species and displacement of native herbaceous vegetation; and (4) where, when, and how many domestic livestock grazed, and what was the role of grazing in the transformation of California pasture.

The hope is that this study will encourage new studies and models of California's herbaceous vegetation that conform to the historical record. The central hypotheses of this book are the following: (1) California's pre-Hispanic vegetation consisted of vast carpets of wildflowers, not bunch grasslands; (2) the introduction of European species triggered a biological invasion without the help of man's activities such as grazing; (3) the transformation of herbaceous cover began along the coast and shifted inland, the pace of change being dependent on habitat, climate variability, and, most importantly, the time of arrival and adaptive modes of the invaders; and (4) the collapse of indigenous forblands over most of California happened right in front of our eyes with the invasion of bromes in the twentieth century.

CHAPTER 2

Pre-Hispanic Herbaceous Vegetation

All of the lands which they [Native Californians] occupy are
as fertile and beautiful as the regions beyond of this channel,
and the sight of them is certainly a pleasure, especially to
one who has witnessed the extreme sterility of the Gulf of
California, where neither trees nor useless herbs are to be
seen, while here, on the contrary, fields as verdant as they are
flower-covered touch the very waters of the sea.

—Juan Bautista de Anza, near Santa Barbara,
April 14, 1774 (Web de Anza Archives)

An investigation into the transformation of the herbaceous vegetation
of California must begin by asking a simple question: what was the aboriginal herbaceous cover, i.e., what was displaced by modern exotic annual grasslands? The Spanish expeditions are the only eyewitness written accounts of California before the expansion of introduced species
from Mediterranean Europe. The ocean voyages of Juan Rodríguez
Cabrillo in 1542 and Sebastián Vizcaíno 1602–03 (Bolton 1916) were
focused on discovery of a new continent and mapping coastal landmarks.
Since the vessels were mostly out at sea, the explorers provided only limited information about the vegetation. Hence, the vegetation baseline fundamentally begins with the Gaspar de Portolá and Juan Bautista de Anza
land expeditions of California from San Diego to San Francisco in 1769–
76, and our insights into California herb cover were vastly improved by
Alan K. Brown's recent translation of the original diaries of Juan Crespí
(Brown 2001).

Perhaps the greatest asset of the Spanish land expeditions is their remarkable geographic resolution. Vegetation was described almost daily,
and landmarks or place names permit the accurate mapping of the expeditions. The accounts highlight the vast extent of pasture in Califor-

nia, but the Spanish explorers also made clear that the quality of potential grazing lands changed with the seasons and from the coast to the interior. This chapter's epigraph by Anza illustrates another basic observation in Spanish texts: that wildflower fields were seen wherever the expeditions traversed California in the vernal season.

SPANISH EXPLORATIONS

European eyes first saw the California coast during the voyage of Cabrillo in 1542 (Bolton 1916), supported by Viceroy Mendoza of Mexico. Cabrillo sailed northward from Navidad, Mexico, and kept within sight of shore along Baja California and Alta California, as far north as latitude 40° N to a headland he called Cape Mendoza (Cape Mendocino). In 1602–03 Vizcaíno, under orders from Philip III of Spain, explored the coast of Baja California and Alta California, discovering the bay of San Diego and reaching the Point of Pines at Monterey, but he did not enter San Francisco Bay (Davis 1929; Bolton 1916). From the observations of Vizcaíno, the first pilot of the Philippines, Don José Gonzalez Cabrera Bueno, made several sea charts that, together with a theoretical treatise on navigation, were published in Manila in 1734. The maps and diaries of the Vizcaíno expedition provided latitude coordinates and place names that were used by the Portolá expedition, 167 years later, and that included the southern California Channel Islands, Point Conception, the Santa Lucia Mountains, "Monte Rey" (Monterey), Punta Año Nuevo (west of Santa Cruz), and the Farallon Islands. The pine forest that descended to the sea and a "great harbor" at Monterey were critical landmarks that defined the destination of the Franciscan expeditions of 1769–70. Vizcaíno's pilot, Cabrera Bueno, described the port as "a large bay until it comes out from a point of low land, very heavily forested to the very sea, to which was given the name of Punta de Pinos [Point of Pines]. . . . It is heavily grown with pine forest. . . . which they call Monte Rey . . . following the coast from the Point of Pines toward the south-southwest there is another fine harbor" [Carmel] (Cabrera Bueno 1734:303).

Spanish land explorations began with the 1769–70 Portolá expedition from northern Baja California to San Diego and San Francisco, two Anza expeditions (1774, 1775–76) from Tubac, Arizona, across the Sonoran Desert to Mission San Gabriel and San Francisco, and the 1774 Fernando Rivera y Moncada expedition from Monterey to San Francisco (accompanied by Francisco Palóu). Accounts were also made by Pedro Fages, Francisco Garcés, Father José Maria de Zalvidea, and Father Pedro

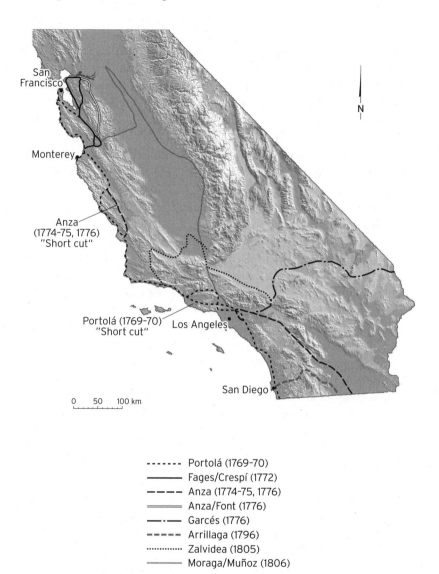

Figure 2.1. Routes of Spanish explorers, identified by the first explorer to travel each route.

Muñoz of the Central Valley and Mojave Desert (Figure 2.1; Brown 2001; Bolton 1916, 1927, 1930a,b, 1931, 1933; Priestly 1937; Coues 1900, Cook 1960, 1962; Simpson 1961; Montané Martí 1989; Brown 2001; Web de Anza Archives).

The Franciscan expeditions into northern Baja California and Alta

California resulted from a directive of the Spanish Crown to expand the sphere of influence along the Pacific coast of North America to impede foreign encroachment, in particular to resist the southward expansion of Russian settlements from Alaska (Bolton 1927). The Franciscans were thrust upon the North American scene with the expulsion of the Jesuits from the region in 1766 by the Spanish Crown. In 1768 a royal order was sent to Viceroy Carlos Francisco de Croix and José de Gálvez, the king's visitor-general for New Spain, to expand the mission system into California. While the Franciscans would have preferred to expand the missions contagiously northward, the threat of foreign encroachment made it necessary for the mission fathers, under orders of Gálvez, to organize a lengthy expedition northward under the command of Portolá, governor of the Baja California peninsula at the time, to establish garrisons at San Diego and Monterey (Figure 2.1). The Portolá expedition traveled northward from the last mission built by the Jesuits at San Fernando Velicatá (lat. 29° N) through the mountains and coastal valleys of northern Baja California, arriving at San Diego in July 1769 to join members that ported there by sea (Minnich and Franco-Vizcaíno 1998). The routes of all the Spanish land expeditions, including campsites, place names, latitude coordinates as well as references are given in Appendix 1.

After a month at San Diego, Portolá led the second segment of the expedition northward across coastal California, arriving at Monterey in October. The route hugged the coast, relying on guidance from the Vizcaíno expedition. They moved northward from San Diego to near Carlsbad, where the party moved inland 10 kilometers to San Luis Rey and San Juan Capistrano, and then they traced the eastern margin of the Orange County plain to the Santa Ana River near Yorba Linda, crossing the Puente Hills to the San Gabriel Valley. The expedition then moved west across Los Angeles and West Los Angeles, where they were forced northward by the cliff coast of the Santa Monica Mountains and across the San Fernando Valley to the upper Santa Clara River valley, where Native Americans warned them of nearby mountain barriers. The expedition descended the Santa Clara River and Oxnard Plain, and continued along the San Barbara Channel to the Vizcaíno landmark of Point Conception. The party then marched northward to San Luis Obispo, Morro Bay, and San Simeon, where the cliff coast of the Santa Lucia Mountains forced the expedition to traverse the range into the Nacimiento and San Antonio River drainages, reaching the Salinas River near King City. Reaching Monterey, they were disappointed with the harbor and continued northward around the bay to the Santa Cruz Mountains, where they followed the coast to the San Francisco Peninsula. The expedition explored the western bay southeastward to near San Jose and sent a scouting party to the eastern bay. The expedition returned using the same route to Monterey. In was only then, in retrospect, that the expedition realized that the large bay of Monterey described in the Vizcaíno voyage referred to the entire embayment from Punta Año Nuevo to Monterey, not just the Monterey harbor. The party returned to Mission San Diego in January 1770.

The most detailed information from the Portolá expedition comes from journals kept by Juan Crespí, which consist of two drafts: a field version and a first revision that cover most of the expedition (Brown 2001). Crespí described the landscape and topography in tedious detail that can easily be followed (Bolton 1927); he assigned place names, many of which survive to the present; and he recorded latitudes by transit, which were consistently about one-half degree too far north (Minnich and Franco-Vizcaíno 1998). Diaries were also maintained by Miguel Costansó, a civil engineer (Teggart 1911), and by Pedro Fages, a soldier and later military governor of New California in 1770–74 (Priestly 1937). In April–May 1770, Crespí returned to Monterey (on a second journey), during which he maintained a single journal (Brown 2001).

By 1772, missions were being established between San Diego and San Francisco but the Franciscan missionaries quickly realized they had insufficient manpower to establish missions in northern Baja California. In that year a "concordat" was made between the Franciscans and the Dominican Order, which also pressed the Spanish Crown to establish missions along the Pacific coast. In the agreement the Dominicans agreed to take over the mission system in the deserts of Baja California, left behind by the Jesuits, and to develop new missions in the Mediterranean lands of northern Baja California. The Franciscans had control of Alta California. The agreement eventually became the foundation for the modern international boundary dividing Alta and Baja California, Mexico (Meigs 1935).

In 1772, Juan Crespí and Pedro Fages left Monterey in an expedition around the arm of San Francisco Bay to Point Reyes, but they did not have the means to cross the bay to the north (see oration of General Vallejo, the top military leader during the Mexican period, in Davis 1929). However, the expedition did explore the east side of San Francisco Bay and did see the Central Valley for the first time. Leaving Monterey, the expedition closely traced modern Highway 101 through Gilroy and Hollister to San Jose. The party skirted the East Bay's shore to Richmond and then headed eastward through the San Pablo Bay/Suisin Bay strait

across Concord to Antioch, where they had their first view of the great
Central Valley and a "great river" they called Rio San Francisco (Sacra-
mento River). They returned by heading southwest from Antioch across
the Coast Range to Walnut Creek and Pleasanton, and across the East
Bay hills to their original route at Fremont, from which they returned to
Monterey.

Fages also left a brief account of an expedition across the San Joaquin
Valley to its south end in 1772 (Priestly 1937). From the Central Valley
he ascended Tejon Pass and followed the southern margin of the Antelope Valley and Mojave Desert to the San Bernardino Mountains, and
then returned to Mission San Gabriel.

The expeditions of Juan Bautista de Anza in 1774 and 1775–76 were
undertaken to bring settlers to San Francisco Bay and to develop a military presence there. Pedro Font explained the rationale for these expeditions his diary:

> It is necessary to remember that in virtue of the exploration of the port
> of San Francisco made by Captain Don Pedro Fages in company with
> Father Fray Juan Crespí in the year 1772, in the month of March, and
> of the report which was given of it, accompanied by a map in which they
> delineated a great river which they said they had found and called the Rio
> de San Francisco, an order came from Madrid to the effect that that port
> should be occupied and settled immediately. With this in view the Viceroy
> ordered Commander Ribera to go to examine the port and seek there a
> good site in which to found a presidio and settlement, to serve as a start
> or beginning for subsequent plans; and to this purpose the present expedi-
> tion for the escort of families by Commander Anxa [sic] was directed, as is
> evident from his Excellency's decree given in Mexico, November 28, 1774.
> In consequence of that order Commander Ribera, accompanied by Father
> Fray Francisco Palou, went to explore the port in the latter part of the same
> year of 1774 [but this expedition failed to find suitable locations for a port
> and Ribera, aka Rivera y Moncada, was an] avowed opponent to a new
> foundation. (Bolton 1930b: 224)

General Vallejo's oration records a similar rationale:

> The President of the Missions having become fully convinced of the impos-
> sibility of establishing that of San Francisco immediately at its own port,
> as he lacked the means to transportation by sea, and in order to proceed
> by land, additional exploring parties were deemed necessary. He reported
> the failure of the expedition of Fages to the viceroy of New Spain. The
> Viceroy gave orders to Captain don Fernando Rivera y Moncada, who
> had been appointed successor to Fages in command of the military posts
> (presidios) of New California and make a second examination for the
> purposes of discovering the most appropriate localites for the foundation
> of the Missions . . . He calls upon Father Junipero Serra to aid and assist

the new commander and to occupy and establish missions in the most convenient and suitable places. After having made necessary preparations, Captain Rivera started from Monterey on 23rd of Nov 1774 with Palou. (Davis 1929: 356).

With respect to the second Anza expedition, Vallejo stated, "The next attempt to found a religious and military establishment at San Francisco proved successful with expedition of Anza, by orders of the viceroy, who recruited settlers and soldiers from Sonora and give them all possible aid to facilitate their journey to their new homes in Upper California. This expedition was accompanied by chaplain Father Pedro Font" (357).

The route of the Anza expeditions originated at Tubac, Arizona, and went northwestward across the Mexican Sonoran Desert to Yuma, Arizona. The parties followed the Colorado River and its distributaries to near Mexicali, and then crossed the Salton Sea trough to the present-day Anza-Borrego Desert State Park. They crossed the San Jacinto Mountains, descending into the San Jacinto Valley, and moved west across the Riverside plains, Pomona Valley, and San Gabriel Valley to the Mission San Gabriel. From the mission, the Anza expeditions used the primary road to Monterey, termed El Camino Real, which was "straightened out" as Spanish explorers became familiar with easier routes of access. Two major shortcuts were used by Anza along modern Highway 101. One was from Los Angeles to Ventura via the Camarillo grade, avoiding the Santa Clara Valley traverse of the Portolá expedition. The other was a steep ascent from San Luis Obispo to Paso Robles and through the interior Coast Range to Mission San Antonio and Jolon where the expedition joined El Camino Real near King City in the Salinas Valley. This route avoided the Morro Bay–San Simeon coast and a rugged traverse of the Santa Lucia Mountains. The Camarillo shortcut was explained by Fages in 1772: "for at that time a more direct road, leading in the proper direction was chosen, for by that time [the explorers] had a terribly accurate idea of the country and did not advance so tentatively as on the first trip when . . . [there was a fear] of meeting some insuperable obstacle caused by the sea or by the asperity of the country" (Priestly 1937: 21). However, the Anza expeditions still followed the route of the Portolá expedition from Santa Barbara around Point Conception to San Luis Obispo, as the routes from Santa Barbara, over the Santa Ynez Mountains, and through San Marcos Pass or Gaviota Pass to the vicinity of present-day Solvang were yet to be discovered.

Francisco Palóu joined the first Anza expedition at Monterey and marched to San Francisco Bay via the San Juan Bautista Valley before turning north to join the primary road at Gilroy to San Jose (Bolton 1926). This expedition explored the west bay to the port of San Francisco and then returned to Monterey along the Pacific coast of the Santa Cruz Mountains. In the second Anza expedition, Anza and Font followed the same route through California, including the 1772 Crespí/Fages route to Antioch, but returned to Monterey farther inland in the Diablo Range through the east side of Livermore Valley, southward to a point 15 kilometers east of Mount Hamilton, eventually reaching the primary road at Gilroy. Anza and Font also made a trip from Mission San Gabriel to Mission San Diego, where they investigated an Indian revolt. In a separate journey, Garcés also crossed the Central Valley, the Mojave Desert, and the San Bernardino Mountains.

THE VICEROY MANDATE AND SPANISH "BOTANY"

The Spanish diaries are the only detailed written documents of herb cover before introduced species began spreading across California. The content of these documents is consequently invaluable, but must be interpreted in historical context (Grove and Rackham 2001). A mandate by the viceroy of Mexico required that a journal record be made of the local resources, including comments on the vegetation—such as pasture, timber, and fuelwood—to determine whether to finance a mission (Meigs 1935: 20). The requirement for daily observations is the reason for the systematic vegetation inventory. The geographic resolution of accounts depended on the pace of individual expeditions. Crespí's journal on the 1769–70 Portolá expedition had a resolution of 2–4 leagues (5–10 km) per day.[1] Later journeys of Anza and Fages proceeded faster in already familiar lands and have a resolution of 4–8 leagues (15–30 km). The expeditions of José Maria de Zalvidea (in 1805) and Gabriel Moraga (in 1806) into the Central Valley traveled as much as 50 kilometers per day. The spatial resolution of these accounts was not exceeded until U.S. surveys produced the first vegetation maps of California in the late nineteenth century.

José Longinos-Martínez, who was part of a Royal Scientific Expedition to survey medicinal plants in areas of modern-day Mexico (Engstrand 1981), visited California in 1792. Although he made a remarkable account of vegetation in the Sierra San Pedro Mártir in Baja California that year (Minnich and Franco-Vizcaíno 1998), he did not record journeys to specific localities in Alta California. He did write a general summary of the vegetation and ethnobotanical plants (Simpson 1938, 1961). California was also visited by the astute observer and soldier José Joaquin

Arrillaga, later governor of California, who made extensive descriptions of northern Baja California vegetation during four journeys in 1796 (Robinson 1969; Minnich and Franco-Vizcaíno 1998). On the California side, he maintained a journal only for a short stretch of San Diego and Imperial counties at the end of his third journey. In the review of vegetation below, the routes of Spanish land expeditions are traced from the locations of daily campsites as listed in Appendix 1.

Understanding the Spanish diaries requires the reader to get into the mind-set of the observers. The diarists had a mandate to look at California from a "barnyard/resource" point of view, with the intent to establish a mission system and to convert the indigenous population to Christianity. The quality of the landscape and vegetation descriptions is remarkable considering that priests and soldiers, not scientists, made them (Bolton 1927; Minnich and Franco-Vizcaíno 1998). But the clergy were the educated scholars of the period. Although the Linnaean taxonomy system had already developed in Europe, only Longinos-Martínez and Fages showed evidence of familiarity with the system. For example, Fages used scientific names for oaks including *Quercus suber* (cork oak), apparently for *Q. agrifolia,* and *Q. robur* (English oak) for *Q. lobata,* on the assumption that California oaks were conspecific with European species.

The diarists had remarkably good botanical vocabulary (see Appendix 2). Plants familiar to those in Europe or in lands already under Spanish control, including the deserts of Baja California and Sonora, Mexico, were identified unambiguously. New plants or vegetation brought confusion to both the writers and the readers of these texts. The diarists' level of observational skill is best seen in records of the central desert of Baja California, where the Franciscans identified the more conspicuous shrubs and trees to genus, often species, as the missionaries were informed by the previous experience of Jesuit missionaries in these lands during the early eighteenth century (Aschmann 1959b, 1966; Minnich and Franco-Vizcaíno 1998). Many names are now adopted by English speakers, including palo verde (*Cercidium floridum*), ocotillo (*Fouquieria splendens*), and mesquite (*Prosopis glandulosa*).

The quality of descriptions of vegetation and taxonomy becomes uneven once the Franciscans encountered unfamiliar Mediterranean vegetation beginning in the Sierra San Pedro Mártir of northern Baja California and northward through California (Minnich and Franco-Vizcaíno 1998). Most tree species, including oaks, conifers, and riparian species, were identified to genus because the Spanish missionaries grew up among congeners in Europe. Examples include *encino* (evergreen oak, *Quercus agrifolia*), *roble* (deciduous oak, *Q. lobata*), *álamo* (*Populus* spp.), *sauce* (*Salix* spp.), and *aliso* (sycamore, *Platanus racemosa*). Spanish taxonomy was also unequivocal among genera with bizarre morphologies such as palms, as well as for ethnobotanical plants such as *chia* (*Salvia columbariae*). Unfamiliar species were identified using morphological analogues. The chaparral shrub, chamise (*Adenostoma fasciculatum*), and possibly California buckwheat (*Eriogonum fasciculatum*), were termed "rosemary" due to the similarity of foliage. The writers also borrowed generic structural terms now applied to other plant assemblages in Mexico, such as *bosque chaparro, emboscado, ramajos, montuosa,* and *matoral* (Martínez 1947). Coastal sage scrub was identified as having "kitchen" plants useful for seasoning. The chaparral was viewed as a barrier to travel and as having inferior fuelwood and was largely avoided.

Each day's journal entry was typically a listing of the species encountered, just as in the diary records from Baja California (Minnich and Franco-Vizcaíno 1998). The vegetation structure can be interpreted from characteristic descriptions throughout the diaries. Areas dominated by herbaceous cover were described as being covered by *pasto, zacate,* or flowers. Areas without pasture were "sterile."

The quality of desert plant identifications in already missionized lands of central Baja California and of tree species in California affirm the excellence of the observations made by Spanish explorers. Comparable familiarity would have doubtless resulted in better descriptions of some elements of California vegetation. For example, the observations of Arrillaga in Baja California in 1796 (Robinson 1969) reveal the extensiveness of chaparral there, an insight that cannot be gleaned from the 1769–76 diaries of the Portolá and Anza expeditions (Minnich and Franco-Vizcaíno 1998). The barnyard view of vegetation is little different from that given by English speakers who described California during the gold rush (see Chapter 3).

The content of the vegetation descriptions also depends on the season of observation. Crespí's journal of the Portolá expedition captures California herbaceous vegetation in the dry season, but his return trip to Monterey the following year took place in the spring growing season. Both Anza expeditions took place in the winter–spring growing season.

The quality of journal entries varies with the observer. Crespí provided the best account of California, not only because of his insightful observations of vegetation and topography, but also because of the high resolution of entries resulting from a slow-moving expedition. Font and Anza also provided excellent insights into the vegetation. Unfortunately, Anza quit writing entries once he reached Mission San Gabriel because areas northward were already described, i.e., the viceroy's mandate was fulfilled. Anza did provide careful observations of new lands around San Francisco Bay. Fages provided short synopses of his expeditions, which were apparently extracted from field notes. Among the various translations of the Franciscan land expeditions, Brown (2001) and the Web de Anza Archives are preferred because English translations are paired with the original Spanish. The Brown (2001) translation of the original Crespí journal provides superior detail of the California landscape compared to the accounts translated by Bolton (1927). Costansó contributed little knowledge of the vegetation. According to Saunders (1914: 5), when Costansó departed the scurvy-racked ship at San Diego in May, his first entry on California raved about the "happy aspect" of the bay and the abundance of "rosemary, salvia and roses of Castile," but the source of greatest excitement was the riot of wild grapes along the river bottoms. The dirth of comments on the biota thereafter confirms Saunders conclusion that "Costansó . . . was a civil engineer, and, his first enthusiasm over the floral exuberance of the land expended, he turned to more practical matters" (5).

CALIFORNIA VEGETATION
IN THE SIXTEENTH AND SEVENTEENTH CENTURIES

Sparse descriptions during the voyages of Cabrillo in 1542 and Vizcaíno in 1602–03 affirm broad gradients in vegetation seen presently along the Pacific coast, with deserts spanning central Baja California giving place to "grassy" pastures dotted with oaks and riparian forests from Ensenada to Monterey (accounts of these expeditions are in Bolton 1916).

Most of the coastline received vague descriptions from a distance, except where the voyages ported to obtain water. Cabrillo described deserts and barren landscapes as the expedition moved north along the length of the Baja California peninsula. At Magdelena Bay (lat. 23° N), he saw "neither water or wood" (14). At Pequena Bay (lat. 26° N), they found that "the interior of the country is level, bare, and very dry" (14). Nearby "Zedros" (sic) Island was "high, rough, and bare" (16), but Vizcaíno saw "on the slope of a great mountain range . . . a large forest of pines" (67) in reference to Monterey pine (*Pinus radiata*) that survives on fog drip from the coastal low clouds in summer. Farther north, Cabrillo found that "the land was bare" (17) at La Playa Maria Bay (lat. 29° N). Cabrillo ported near the south end of the San Quintín plain (lat. 30° N), where "there are good valleys and some timber [riparian forests], the rest being bare" (18). Near San Quintín, where he may have viewed the high Sierra San Pedro Mártir at a distance, Cabrillo wrote, "The interior of the country consists of high and rugged land, but it has good valleys and appears to be good country, although bare" (19).

The first favorable description of Baja California was made by Cabrillo at Ensenada (lat. 32° N), where he wrote that "the land appears to be good; there are large savannas, and the grass is like that of Spain" (22). He also saw "groves of trees like silk-cotton trees, excepting they are of hard wood" (22). He apparently saw cottonwoods, which are now common in the arroyos of coastal Sierra Juárez (Minnich and Franco-Vizcaíno 1998). Along the coast from Ensenada to the Coronados Islands, near the international boundary, Cabrillo "saw very beautiful valleys and groves, the country both level and rough" (Bolton 1916: 22), very likely riparian forests of cottonwood, sycamore, and coast live oak (Minnich and Franco-Vizcaíno 1998).

The report of Father Ascensión of the Vizcaíno expedition gave a flattering account of San Diego Bay: "This realm of California . . . has a good climate, is very fertile, and abounds in many and various kinds of trees, the most of them like those in Spain, abundant pastures of good grazing land" (Bolton 1916: 109; Figure 2.1 in the present volume). Vizcaíno saw "many large, grassy pastures" as well as "Indians who were gathering their crops where they had made their *paresos* of seeds like flax" (82). He also indicated that Indians ate acorns, evidence of the coast live oak woodlands now seen near the bay. Vizcaíno wrote of "a great grove of trees on the east shore of the bay."

Similarly, Cabrillo saw "many savannas and groves [oak]" at Carpinteria and Santa Barbara (26), presumably the pasture and coast live oak now seen there. At Santa Cruz Island, Cabrillo observed that "the natives were eating acorns from the oaks . . . and raw plants from the field" (35). Vizcaíno saw fine plains and many groves and savannas near Santa Barbara. He also wrote that the Santa Barbara Channel "is fertile, for it has pine groves and oaks" (Bolton 1916: 90), very likely referring to the *Pinus coulteri* forests that grow in the Santa Ynez Mountains above Santa Barbara.

The account of Monterey by the pilot José González Cabrera Bueno (1734: 303) from the mid–seventeenth century states that the area is "a large bay until it comes out from a point of low land, very heavily forested to the very sea, to which was given the name of Punta de Pinos. . . . It is heavily grown with pine forest. . . . In this port which they call Monte Rey there are many pines. . . . Following the coast from the Point of Pines toward the south-southwest there is another fine harbor [Carmel] running from north to south . . . [which] has a river . . . whose banks are well grown with black poplars [*Populus trichocarpa*]." Vizcaíno wrote that the area "has many pines for masts and yards, as well as live oaks and white oaks" (Bolton 1916: 119). Ascención's report from the Vizcaíno expedition has a similar appraisal of Monterey. Ascención wrote that "it is a good harbor, well sheltered, and supplied with water, wood, and good timber, both for masts and ship building, such as pines, live oaks and great white oaks, large and frondose, and many black poplars on the banks of a river that near by entered the sea and was named the Carmelo. . . . The land of this country is very fertile and has good pastures and forests" (119).

The accounts of the Cabrillo and Vizcaíno voyages are consistent with the vegetation now seen along the coast. Desert scrub spanned the Baja California peninsula northward to near Ensenada, where it was replaced by Mediterranean vegetation, mostly herbaceous pastures and groves of stream trees and oaks. Pines were found at Monterey, Santa Barbara, and on Cedros Island. Pasture was seen at San Diego, Santa Barbara, and Monterey, but lands south of Ensenada were barren (*esteril*). Because the objectives of these voyages was to chart the continents, the journals leave little information on California pasture.

DESERT SCRUB *ESTERIL*

The expeditions of Juan Bautista de Anza (1774, 1775–76) and José Joaquin Arrillaga (1796) across the Salton Sea trough (Figure 2.2; Appendix 1), one of the driest regions of North America, record a dearth of pasture, sparseness of shubby vegetation, wide separation of water holes, and a story of human survival in a unforgiving desert (see Anza in Web de Anza Archives and Bolton 1930a,b; Arrillaga in Robinson 1969; see also Minnich and Franco-Vizcaíno 1998). Descriptions of the Sonoran Desert are reminiscent of Crespí (in 1769), Arrillaga (in 1796), and the Jesuit priest Wenceslaus Linck (in 1766) in the deserts of Baja California (Burrus 1966; Minnich and Franco-Vizcaíno 1998).

After crossing the Colorado River into California at Yuma, the first Anza expedition struck southwest along verdant pasture and cottonwood forests of the Colorado River to "Lake Olaya," a distributary of the river. This route avoided the Algodones dunes, a formidable barrier to livestock. Anza and Father Juan Díaz commented on extensive forests of cottonwoods and willows. To the west of Pilot Knob, Pedro Font saw desert scrublands with the cottonwood forests, where he stated that "the road was so thick with brush that in many places no more than a little trail

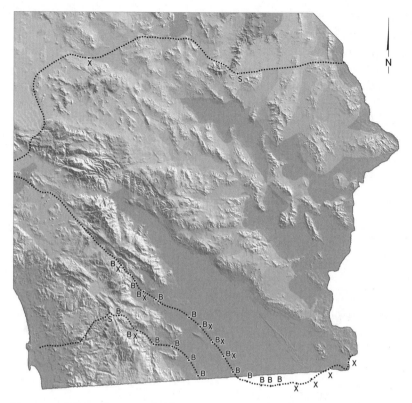

Figure 2.2. Vegetation recorded by Franciscan expeditions in the California desert (Web de Anza Archives; Bolton 1930a,b, 1933; Robinson 1969; Minnich and Franco-Vizcaíno 1998; and see Appendix 1).

was to be seen, the rest being densely grown with mesquite, *tornillo* [screwbean mesquite, *Prosopis pubescens*]" (December 4, 1775). Between Pilot Knob and Lake Olaya, Anza, Díaz, and Font described plenty of pasture for the cattle. Font found Lake Olaya to be "one of the many lagoons in the bottomlands which are left full by the river when it goes down," recognizing the relationship between the river flooding cycle and the vegetation. At the Lake Olaya camp he wrote that "the road is very thickly grown with trees [*emboscado*], although it has a great deal of *chamiso* [thicket, most likely *Prosopis juliflora*], while in places it is barren country [*tierra esteril en pedazos*]." Of the river, Font wrote, "It is a moist county with plenty of grass [*zacate*]." During the second expedition, Anza complained of the "hindrance of thickets to the driving of horned cattle."

When the first Anza expedition departed the lake for the open desert,

cattle and pack animals died in droves. The expedition was aborted part-
way across the basin, near Mexicali, and returned to the distributary in
order to build up provisions before undertaking a successful second attempt to face the desolation of the Salton Sea basin. Font perhaps said
it best: "This is a deadly place, with no pasturage except for a little *Carrizo* [cane grass near springs]. On the road about a league after starting
there is a salty lagoon without pasturage. . . . The road is level but over
bad, saline, and sterile country, which grows only *chamiso* [thicket],
hediondilla [little stinker, *Larrea tridentata*], the dominant species of the
California deserts, and another shrub which they call *parrilla* [possibly
Ambrosia dumosa], and other salty bushes [*Atriplex?*]" (December 9,
1775).

On February 13, 1776, about 30 kilometers west of Lake Olaya on
the Paredones (barren) River, Anza saw two pools of saltwater without
any pasturage. On his return trip, Font lamented that the region between
Imperial, California, and Lake Olaya was "without any *carrizo* for animals to eat" (May 9). Likewise, from San Felipe Creek in Borrego Valley to Imperial, Font wrote that the entire area was "a dry plain without
pasturage and water" (May 9). Near Signal Mountain, Díaz stated that
the region had "no pasturage whatsoever" (February 15). On the west
side of the Salton Sea basin, the only pasturage was found at springs, including Pinto Wash, Yuha Well (March 8), Coyote Wash near Plaster City,
along San Felipe Creek near the junction with Carrizo Creek, and near
Harpers Well, which Anza described as a "very large marsh with many
waters and much pasturage, but both very salty." Harpers Well, according
to Font (December 13–17) had *"carrizo"*—a large cane grass such as tule
(*Scirpus* spp.), or cattail (*Typha* spp.)—"and some grass, although it was
not very good because the soil is so saline in all this flat that in places the
salt whitens it like flour." In the second expedition, Anza wrote that the
"pasture made animal[s] purge because of the saltiness of the leaves."
Anza mentioned that the cattle ate the leaves of mesquite.

From Harpers Well, for "the whole march" to the east entrance of
Borrego Valley, Font wrote that "there is nothing more than now and
then a scrubby mesquite [*Prosopis glandulosa*] and the *hediondilla* [*Larrea tridentata*]" (December 18, 1775). Anza wrote on March 12–13,
1774, that the only pasturage along these days' route was at camp in
Borrego Valley, but it was "better pasturage than any which had been
seen since we left the Pimería [in Arizona]."

The barren terrain of the southern Santa Rosa Mountains caught
Font's attention, who wrote that the hills were "so dry that not a little
tree or even brush is seen in them, and only in the flats is there a little

hediondilla [Larrea tridentata]" (December 19, 1775). From Beatty's Ranch in Borrego Valley, Font wrote, "There is nothing but *chamizo* [possibly a dominant subshrub, such as *Ambrosia dumosa* or *Encelia farinosa*] and *hediondilla*, and in this place a little *galleta* [*Hilaria rigida*] and *choya* [*Opuntia* spp.], with which the cattle entertained themselves" (December 20–22). The second Anza expedition ascended the San Jacinto Mountains along Coyote Canyon, which Font described as "arid, fruitless, and without trees or any green thing. Of grass [*zacate*] in this place there is none, and on the way there are only a few small willows along the banks of the arroyo. The road in places is somewhat broken and grown with shrubs or brush and a little *hediondilla*, for since this is a shrub of evil augury, it not lacking in these salty and worthless lands" (December 23). Font ate some mescal (*Agave deserti*). On his return trip to Arizona, Font summarized his impressions of the Borrego Desert by stating, "All this country is bad, sterile and without grass [*sin zacate*]. . . . Indeed it is all sandy or sand dunes, as I said on going [on his trip to California]" (May 7, 1776). In retrospect, Font recalled the distributions of creosote bush and possible burrobush (*Ambrosia dumosa*), saying that "the most abundant weed or shrub" was what the Native Americans call *hediondilla* and the Spaniards call *la gobernadora* (the governor or dominant), along with "another shrubby and useless plant which, if horses eat it, burns their mouths." Font last saw creosote bush at its modern limit along the route just below the pass of San Carlos in the San Jacinto Mountains.

On his fourth journey of 1796, Arrillaga found little pasture when he crossed the Sonoran Desert from northern Baja California to San Diego County (Minnich and Franco-Vizcaíno 1998). Near the international boundary, he described the alluvial fans along the base of the Sierra Juárez as "gravel," clearly referring to the extensive desert pavements there. The only shrub noted was "Palo Adan" in reference to the ocotillo (*Fouquieria splendens*). On his way to Carrizo Wash near Fish Creek Mountain, Arrillaga described the region generically based on statements of his Native American guide: "Except near the river, the rest is arid and dry." His ascent of Carrizo Valley in San Diego County on October 25 gives a glimpse of the important species of that area. He wrote that the ground was covered with mescal plants (*Agave deserti*) and that the Native Californians have (as a food resource) "a lot of mescal; of other seeds, [but he] only saw the *tornillo* [screwbean mesquite, *Prosopis pubescens*] . . . in small quantity." He wrote that in general "all the route I traveled is bare hills, and those that are next to the arroyo are ravines of sterile earth." He managed to find "two seeps and next to a hill sufficient pasture for the horses." Arrillaga then entered San Felipe Valley, which had abundant pasture on the valley floor that he called *zacatón salado* (salt-tolerant grass, *Distichlis*, or less likely *Hilaria*). However, "the immediate hills, which are fairly high, are entirely without trees or pasture." He left the desert by ascending Banner Canyon to the Laguna Mountains near Julian. During his climb, Arrillaga wrote that he saw "pasture and abundant mescal, but I did not see any other seeds. The immediate hills are likewise very bare" (October 26).

In 1776, Francisco Garcés crossed the Mojave Desert from Needles to the Mojave River, and then to its source in the San Bernardino Mountains (Coues 1900). His brief diary provides little insight into the vegetation. At a location near the Providence or Granite mountains, Garcés stated that the "road was level and grassy" (March 6), perhaps an observation of *galleta* grass (*Hilaria rigida*) that is now abundant at higher elevations of this region. This location seems to be verified by his next day's observation of "hills of sand" that possibly identify the Kelso dunes. He wrote on March 8 that there was "sufficient grass" at a location near Soda Lake, no doubt a saline spring in the region. Pasture did not catch Garcés's attention again until he struck the Mojave River, possibly near Afton Canyon (March 9). In the following days along the river, he described pasture, riparian forests of mostly cottonwood, rushes, wild grape, mesquite, and "trees that grow the screw," believed to be *Prosopis pubescens* by Coues (1900: 240n14), but more likely *Prosopis glandulosa* found along the river from Afton Canyon to near Barstow. Garcés also described cottonwood forests at Victorville, where they are presently abundant.

Pedro Fages traversed the Mojave Desert north of Los Angeles from the San Joaquin Valley in 1772 (Bolton 1931). He found a "marsh" and "some lands covered with a little pasturage" in the western Antelope Valley (August 8). Near Palmdale, he was more infatuated with the grotesque Joshua trees than pasturage, stating that "the land both to the east and south have more and more palm groves [Joshua Tree, *Yucca brevifolia*]." On the north edge of the San Gabriel Mountains, Fages saw "a stream full of water," very likely Rock Creek, but said that the area was "without land for cultivation nor much pasturage in its vicinity." He found a "swamp full of water" as he entered the San Bernardino Mountains at the modern site of Silverwood Reservoir (on the West Fork of the Mojave River).

COASTAL CALIFORNIA *PASTO* AND *ZACATE*

Perhaps the most significant observation made by the Portolá and Anza expeditions was the abundance of pasture for cattle along the length of the California coast. Rarely was camp made along their routes when pasture was not mentioned, usually a cliff coast or a mountain pass covered with shrubland or oak forest.

The English versions of the diaries of Crespí, Anza, and Font (Brown 2001; Web de Anza Archives; Bolton 1930a,b, 1933) frequently translate herbaceous cover as "grassland," but is this a valid interpretation? While such a translation may impart a realistic description of green cover that is herbaceous, "grassland" also has a taxonomic implication that the cover is dominated by grasses. Resolving this inherently difficult incongruity requires examination of the original Spanish, and in the context of the late eighteenth century.

Most Spanish entries use *pasto* and *zacate* (*sacate*) to describe dry herbaceous cover, or variations of these words, such as *pastales* and *zacatón* (see Appendix 2). The Velázquez Spanish-English dictionary describes *zacate*, a Nahuatl word from central Mexico, as "grass, herbage, or hay." It can also refer to forbs (E. Franco-Vizcaíno, pers. comm.). *Pasto* is defined from an agrarian perspective as "pasture," and "the grass which serves for the feeding of cattle." The Velázquez dictionary also equates *pasto* with *hierba*, which means an "herb, a plant not possessing a woody stem, but dying down after flowering" (i.e., a dicot forb). *Pasto* also means "green food for cattle, grass (chiefly in plural), pasturage and grass." In view of these definitions, *pasto* and *zacate* doubtless have various meanings from one region to another. *Zacate* can refer to any form of grass, even a lawn. In the Chihuahuan Desert, *zacate* was used to referred to *Erodium cicutarium*, a dominant forb there, and other ephemeral ground cover (A. Kaus, pers. comm.). Other words used for "grass" consistently refer to plants in riparian settings, not broadscale cover. The word *pajon* (*panonales*), meaning "tall grass," was used only once in apparent reference to *Sporobolus* near a river. At one swampy locality, Garcés found a plant that looked like rye (*centano*), very likely *Elymus condensatus*, now called wild rye (the whole genus is called wild rye; Brewer 1883). Other words include *carrizo* (reeds) and tule, plants that both grow near streams or in swamps. *Prado* was used in the traditional usage, "wet meadow." The antonym for "pasture" is *esteril*, which means barren, sterile, or unproductive. Perhaps most significant is that words that traditionally refer to "bunch grass" such as *grama, sabaneta,* or *zacate amacollado,* never appear in the diaries. In addition, the word *grama* is Spanish and refers to a couple of species from Spain, but can also refer to pasturage (E. Franco-Vizcaíno, pers. comm.).

The next question is whether *pasto* and *zacate* have different meanings. This is addressed by examination of the field draft and the first re-

TABLE 2.1 USE OF *PASTO* AND *ZACATE*
IN THE JOURNALS OF CRESPÍ, FONT, AND COSTANSÓ

	Dry Herbage (summer)				Green Herbage (winter)		
	Salt grass	*Pasto*	*Zacate*	Both	*Pasto*	*Zacate*	Both
Crespí	2	26	24	26	9	5	4
Costansó	—	23	1	0	—	—	—
Font[a]	—	—	—	—	8	1	—

SOURCE: Brown 2001; Teggart 1911; Web de Anza Archives.
[a]. Font's explorations were entirely in the winter growing season.

vision of the original Crespí diaries (Brown 2001). Paired examination
of the texts for each day shows interchangeable use of the words (Table
2.1). On twenty-six occasions *pasto* was used in both diaries, and *zacate*
twenty-four times. However, one word in the field draft was substituted
for the other in the first revision on twenty-six occasions. Costansó and
Font used *pasto* almost exclusively. Since Font's accounts were largely
in the rainy season, when forbs were described, his use of *pasto* may de-
note the presence of forbs.

Clearly the words *pasto* and *zacate*, and their derivatives used to describe herbaceous cover across coastal California, represent overlapping,
ambiguous meaning. Brown (2001) translates all references to herbaceous
cover as "grassland," but does he have license to make such taxonomic
inference? These words mean pasture, i.e., herbaceous cover good for
livestock, and no more should be inferred from them.

The accounts of herbaceous cover in California also vary with the season of the expeditions. The Portolá expedition (journal of Crespí) and
the account of Palóu of San Francisco occur during the dry season. They
offer little botanical perspective because the vegetation was cured. Only
on his return to San Diego did Crespí see germinating green cover beginning near San Luis Obispo and Santa Barbara. The journals of Font,
Anza, and Crespí (on his second journey) were largely recorded in the
winter and spring growing season; they offer more detail on wildflowers and their botany as the diarists understood it.

The following synthesis of the Spanish journals focuses on the taxonomy and distribution of pasturelands, wildflowers—in particular the
ethnobotanical plant, *chia* (*Salvia columbariae*)—tule swamps, and Native American burning of pasture and harvest of seeds. To avoid taxonomic inference in English translation of Spanish words for herbaceous
cover, the summary below uses the word "pasture" for both *pasto* and

zacate. Words such as "grassland" and "prairie"—a mixture of grasses and forbs—are avoided. While the generic use of "pasture" evokes grazing, which ultimately transpired across California during the subsequent two centuries, the term does not contain taxonomic baggage. The reader is reminded that the Spanish explorers had no previous experience with California other than familiarity with species closely related to those in Spain, or information obtained by them from Jesuit missionaries who resided in Baja California before 1766.

Instead of tracing individual journeys, the sections that follow composite the Spanish expeditions regionally from south to north, simulating the actual route of the explorations. The first section briefly summarizes herbaceous vegetation in Baja California from Minnich and Franco-Vizcaíno (1998), and subsequent sections include regional descriptions and maps of California pasture from San Diego and San Francisco. For reference, the descriptions are identified by place names or calendar dates listed in Appendix 1, which also provides the journals' sources.

BAJA CALIFORNIA

The broad distribution of herbaceous vegetation in Baja California parallels the accounts of Cabrillo and Vizcaíno in the sixteenth and seventeenth centuries, during which pasture desirable for cattle grazing became more extensive as climates became moister with increasing latitude (Minnich and Franco-Vizcaíno 1998). The central desert of the Baja California peninsula was "sterile," and the first good pasture was described at Valle San Telmo (lat. 31° N). More extensive pasture was found at Valle San Vicente, the plains of Ensenada, Valle Guadalupe, and the inland basins of Ojos Negros. The outstanding descriptions of Arrillaga and Longinos-Martínez reveal that the surrounding foothills and mountains were covered extensively by coastal sage scrub and chaparral. The best summer pastures were mountain meadows in the pine forests along the crest of the Sierra Juárez and Sierra San Pedro Mártir.

Wildflowers are rarely mentioned in the Spanish journals. Forage was poor in the deserts and, although Crespí described northern Baja California in spring, the propensity of Portolá to guide the expedition into the brush-covered mountains for secure water kept the expedition from the coastal plains and valleys. Crespí's journey crossed the fertile northern valleys in early to late April through early May, possibly too late for the flowering season. Father Junipero Serra once described flowers in the open chaparral of the southern Sierra San Pedro Mártir, but this observation could have referred to blooming shrubs (Minnich and Franco-Vizcaíno 1998).

Wildland fires did not catch the attention of the Spaniards. The Portolá expedition traversed northern Baja California during the rainy season, and storms were reported in the diaries as late as May. During their descents of the Sierra San Pedro Mártir to Valle San Telmo, the expeditions saw Native Americans roasting mescal. In 1792, Longinos-Martínez ascended the Sierra San Pedro Mártir in July, after an unusually wet winter. He saw a stream bursting out along the western escarpment, and to dissuade the Native Americans of a myth he made them drink water out of a huge lake at San Rosa meadow that was drowning Jeffrey pines along its shore. Today this lake forms in years with twice normal rainfall (e.g., 1980 and 1993) (Minnich and Franco-Vizcaíno 1998). Apparently, the vegetation was not dry enough to burn, even in July. Arrillaga traversed the mountains in 1796 after a dry winter, and the normally reliable Laguna Hansen in the Sierra Juárez was dry. Still, he recorded only smoke columns at Indian encampments during his four journeys.

SOUTHERN CALIFORNIA

In southern California (Figure 2.3) the Portolá expedition saw mixed pasture and brushlands along the San Diego coast, and extensive flowered pastures from present-day Orange County to Ventura. The Anza expeditions of 1774 and 1775–76 from Arizona crossed pastures from Hemet to Riverside before joining the Portolá expedition route at Mission San Gabriel. Native Americans treated the Portolá expedition kindly by offering baskets of chia (*Salvia columbariae*) and other seeds, each group sending word northward to the next tribal leader to prepare for the expedition's arrival.

San Diego to San Juan Capistrano

Crespí's journal in 1769 (written on the Portolá expedition) traces the submerged coastline from San Diego to the south edge of the Orange County plain at San Juan Capistrano (Figure 2.3). The rivers draining coastward from the peninsular ranges produced a series of beach estuaries that gave way upstream to tule swamps mixed with sycamore forest and coast live oak woodland. The intervening mesas were covered by dry pasture in summer and flower fields in winter, interspersed with patches of coastal sage scrub and chaparral. The Portolá expedi-

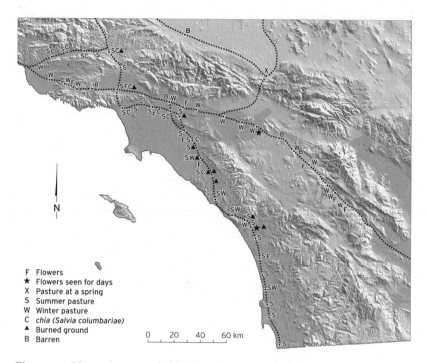

F Flowers
★ Flowers seen for days
X Pasture at a spring
S Summer pasture
W Winter pasture
C *chia (Salvia columbariae)*
▲ Burned ground
B Barren

0   20   40   60 km

Figure 2.3. Vegetation recorded by Franciscan expeditions in southern California (Brown 2001; Web de Anza Archives; Bolton 1927; Coues 1900;
Teggart 1911; Burrus 1967; Priestly 1937; and see Appendix 1).

tion encountered pasture burns beginning at Santa Margarita, offering
the first of numerous accounts of fires seen along the California coast
in 1769.

Crespí wrote of cured pasture at all his stops along the San Diego
County coast, including the San Diego mission site, Soledad Valley, San
Elijo Lagoon, Carlsbad, San Luis Rey, Santa Margarita, Camp Pendleton, San Onofre, and San Juan Capistrano. When the party passed
through the San Diego region in July, the dry pasture contrasted with
green cover along the river courses and estuaries. Several of Crespí's accounts of "green grass" at several localities near the coast refer to estuaries. This is unclear in Brown's (2001) translation of *estero* as "inlet" instead of the cognate "estuary." August Bernard Duhaut-Cilly
described in 1827–28 "a long sheet of seaweed, stretching more than a
league to the south-southwest" (Duhaut-Cilly 1929: 218). The estuary
was also described by him as a "floating field," most likely surf grass
(*Phyllospadix torreyii*) in quiet waters and eel grass (*Zostera* spp.), which
resides in the shallow water of bays and in muddy bottoms of the in-

tertidal zone (Munz 1974). Estuaries were also described at San Elijo Lagoon and Carlsbad.

Crespí's first glimpse of San Diego Bay took place at the end of the Baja California journey where, from near Tijuana, he saw a "long stretch of level shore, all the land being well covered with pasture" (March 13, 1769). In his letter to Palóu, he described San Diego as "a large, level place in the midst of great meadows and plains, with very good pasturage for all kinds of cattle." (Bolton 1927: 4). However, Font was less flattering after first seeing the pastures farther north, stating that there was "plenty of grass [*zacate*], although not as good or so abundant as in other places" (January 11, 1776). He further added, "This presidio has no advantages for raising crops nor, consequently, has the mission any."

Farther north at Soledad Valley, Crespí wrote about both dry pasture and riparian vegetation: "After a league and a half, we came to a vastly handsome valley or hollow that because of the greenness, viewed while coming down from the tablelands seems nothing other than a field of corn [*milperia*]. Having come down to the valley here, where was its greenness consisted of bitter wild gourds [*calabazas*], very large grass clumps [*grandes zacatónes*], a great many greens [*quelites*], many very lush wild grapevines [*parras, Vitis girdiana*], and a great many rose of Castile bushes [*Rosa californica*]. . . . This valley . . . has excellent pasture on its sides." San Elijo Lagoon had "pasture-covered rolling tablelands [*empastados*]" above the flood plain. The wild gourd was very likely *Cucurbita foetidissima* in San Diego County, which is congeneric with familiar gourds in Spain. Likewise, the wild grape and the rose of Castile are also European congeners. The "large grass clumps" (*zacatónes*) representing some perennial grass were the usual tule-rush swamps (*tular*). The valley was not cultivated as a field of corn, as agriculture is not known among California Indians except along the Colorado River (Kroeber 1953).

Similarly, lands around Carlsbad were "very low rolling knolls and tablelands with good grass [*pastos*]." On his way to the site of Mission San Luis Rey, Crespí "went up a grassy hill . . . and over very rolling tablelands and knolls all clad with good dry grass [*zacate seco*]. . . . We must have traveled a two short leagues . . . down to a vastly large, handsome, all very green valley. . . . Its greenness . . . is a great deal of wild grapevines, looking like planted vineyards [and] many tall grass clumps [*zacatónes*]. There are the usual tule-rush swamps [*tular*]." The expedition next encountered the Santa Margarita range. Crespí's account illustrates his acute observations of terrain: "We stopped here because a very high mountain range, ever since the day after leaving San Diego, has been con-

tinuing upon our right hand [to the east], and we now have it very close
by and seemingly running down to the sea."[2]

The arroyo where the party camped "bears north and south, [and]
there seems to be another hollow joining it upon the north-northwest,"
apparently identifying Talega Canyon. For the first time, they encountered evidence of recent pasture fires. Their entire route that day (July
20, 1769) was "among knolls and hills of sheer soil and everything very
overgrown with dry grass [*sacate*] burnt off here and there by the heathens." In Bolton's version of the journal (1927), Crespí explains that
the lands were "burned by the heathen for the purpose of hunting hares
and rabbits."

On the east side of Camp Pendleton, Crespí "climbed a high [point on
the terrain] with a great amount of rock. After this we came onto a very
open, rolling knolls and tablelands of sheer soil, everything very overgrown
with dry grass [*sacate*]."[3] The following day at San Onofre, Crespí wrote
that the expedition "went amid the range of very rolling hills and tablelands very covered with dry *zacate,* going up and down through hollows
and dry creeks . . . with grand, very fine *zacate.*" During the march to
San Juan Capistrano, the expedition "came out on tablelands, low
rolling knolls, hollows, and dry creeks . . . all of it well grass-grown [*empastados* = well pastured] as before."[4] As in most previous camps
through the San Diego region, the Portolá expedition camped among
"very tall green grasses" (*grandes pastos*), sustained by stream water at
San Juan Capistrano. Crespí also wrote that they camped at a "tule-rush
marsh."

Crespí recorded several large burns north of Santa Margarita. On the
east side of Camp Pendleton, he wrote that through "most of this march
we found it burnt off by the heathens." Even at a "very green hollow"
there was "a great deal of grass [*sacate*] though burnt off in places." Farther west, at Camp Pendleton, he described "the very open, rolling knolls
and tablelands of . . . dry grass, though over most of this march burnt
off by the heathens" (July 21, 1769). Again just to the west, "a tableland with very grand pasture . . . and all sorts of other very lush green
plants [*yerbajes*] . . . had been burnt off by the heathens not long ago."[5]
At the original mission site of San Juan Capistrano, about 1 league north
of the modern foundation, Font wrote on January 8, 1776, that the site
was called La Quema because Crespí's party saw a burn in the "grass
patches."

The journal of Crespí's second journey (in 1770) and Font's journal
of the second Anza expedition, six years later, give a strikingly different
picture of the San Diego coast in the rainy season when the annual veg-

etation was in full growth. In his march from San Luis Rey to Soledad Valley, Font wrote that "the country is green with a great deal of *zacate*" (January 10). On April 17, Crespí wrote that the vernal herb cover near San Elijo Lagoon was being harvested by Native Americans, similar to what Father Ascención reported nearly two centuries before at San Diego Bay. Crespí also recorded that the Indians showed displeasure that their harvest was being disturbed by livestock:

> All the country is very grass-grown with green grass in seed [*empastada de zacate*], from which the heathens were now plucking their seed crops. And in many spots along the way we came upon a great many heaps of them. Some heathens came over from the village and gave us to understand that, since they were plucking [their seeds] at present and we had stopped in the hollow at a place where they had not yet plucked, we should move onward, and the attempt was made to keep our mount away so as not to [do] them damage, because they gave it to be understood that they were unhappy with their [livestock] eating the seeds.

The next day north near Carlsbad, Crespí was more specific that the harvested fields had abundant wildflowers, stating that his party "met the villages at all of the spots, all of them engaged in plucking the seeds of their grasses [*sacates*]. It is a pleasure to see how the fields are abloom everywhere, and how fine their *zacate* is." As the journey approached San Luis Rey the following day, Crespí in hindsight suggested that wild-flower fields were common, possibly since his departure from San Diego. He wrote that the expedition "crossed a handsome valley. The entire way [on this day] has been, like the preceding ones, very flowery," i.e., the expedition had encountered wildflowers along the entire journey.

San Jacinto Mountains and interior valleys of Riverside

Pre-Hispanic herbaceous cover in the interior valleys and mountains of southern California was described in the rainy season by the Anza expeditions of 1774 and 1775–76 (Figure 2.3). The diaries of Anza, Garcés, Díaz, and Font (Web de Anza Archives; Bolton 1930a,b, 1933) describe a remarkable transition from the vast desert scrub crossed in Sonora, Arizona, and southeast California to Mediterranean scrub and flower fields in the San Jacinto Mountains at Terwillinger Valley. Pastures continued from the San Jacinto Valley and Riverside and Mission San Gabriel.

The Anza expeditions left the desert at the "pass of San Carlos" opposite Terwillinger Valley. On the second expedition, Anza wrote, "We came out to level country with an abundance of the best pasturage, trees and grass, that we have seen thus far" since Arizona (December 22, 1775). Anza also wrote that "from the pass of San Carlos are seen the most beautiful green and flower-strewn prairies [*llanada hermosisimas mui verdes y floridas*]." In Anza Valley he wrote that "camp is surrounded by flower-strewn and pleasant valleys [*esta circundada de floridos*] and several snow-covered mountains [Thomas Mountain, Coahuilla Mountain, San Jacinto Mountains]" (March 15, 1774). Font wrote, "Here the country is better than the foregoing, for after leaving the Pass of San Carlos the country completely changes its aspect, in contrast with that left behind on the other side . . . as if the scenery of the theater were changed, one beholds the Sierra Madre de California now totally different—green and leafy, with good grass [*yervas*] and trees . . . where as in the distance looking toward the California sea [Gulf of California] it is dry, unfruitful and arid" (December 27, 1775).

The descent from Anza Valley through Bautista Canyon to the San Jacinto Valley traversed chaparral and riparian forests of cottonwood, sycamore, and oak, after which Anza (on the second expedition) wrote that the party "came out on a level country with an abundance of the best pasturage, trees and herbs [*yervas*] we had seen thus far" (December 27, 1775). Font appears to have captured the germination of annuals with the first rains, stating, "The valley of San Joseph [San Jacinto Valley] is very large and beautiful [*hermoso*]. Its lands are very good and moist, so that although this was winter time, we saw the grass [*zacatito*] sprouting almost everywhere in the valley" (December 30, 1775). During the first expedition, Anza saw the valley in the full glory of spring: "All its plains are full of flowers, fertile pastures [*pasto fertiles*], and other herbs useful for the raising of cattle" (March 18, 1774). Impressed by the San Jacinto River and the line of cottonwoods along its banks, Anza wrote that the expedition came "to the banks of a large and pleasing lake, several leagues in circumference and as full of white geese as of water" (March 19), clearly Mystic Lake. The poorly drained parts of the valley had tule swamps like those along the San Diego coast. On this day, Garcés noted that "this is a very miry road. . . . The [cottonwood] groves are thickly sown with *sacate,* one species of which bears a seed much like rye [*Zacate*]. I have no doubt this is the grain the Gilenos call wheat [*trigo*], for they [Native Americans] told me there was . . . wheat which they harvested without planting it. There is a . . . very large marsh with much pasturage."

Pasture continued west into the Riverside plains. According to Díaz, the party "traveled over plains well grown with pasturage, bearing on the right a snow-covered mountain [Mount San Gorgonio]. Having ascended a small ridge through a very easy pass [western Moreno Valley], we entered a most beautiful and broad valley [Riverside]" (March 19). After crossing the Santa Ana River, Font made the most poignant account of the vegetation on New Year's day of 1776:

> In a word, all this country from the Puerto de San Carlos forward is a region which does not produce thorns or cactus. In fact I did not see in all the district which I traveled as far as the port of San Francisco any spinous trees [*matoralles espinosos*] or shrubs such as there are in the interior [deserts], except some prickly pears [*nopalis*] and some nettles [*hortigas*] which I saw near the port of San Diego. . . . Finally, this country is entirely distinct from the rest of America which I have seen; and in the herbs and the flowers of the fields [*las hierbas y lo florido de los campos*] and also in the fact that the rainy season is in winter, it is very similar to Spain.

Farther west, Anza wrote during both of his expeditions that the plains between the Santa Ana River to San Antonio Creek in Pomona Valley had good pasturage. In the San Gabriel Valley, Díaz wrote that the expeditions traveled "7 leagues to the northwest over very fertile plains with much pasturage" (March 21, 1774). On the second expedition, Font wrote that from "Rio San Antonio to a tributary of the San Gabriel River . . . one enters a country very level in all directions, which we found very green in places, the flowers already bursting into bloom [*y rebendado y a las flores*]" (January 3, 1776).

Los Angeles plains

The two Anza expeditions joined the route of the Portolá expedition at Mission San Gabriel. As a consequence, this region's vegetation received detailed accounting at seventy-three camps from Orange County to Los Angeles, and from San Fernando Valley to the Santa Clara River and west to the Ventura plain. The journals reveal that both the coastal plains and interior valleys of southern California were carpeted with pasture and flowers (Figure 2.3). Specific localities where *pasto* or *zacate* was described, from south to north, include El Toro, Tustin, Yorba Linda near the Santa Ana River, La Brea, Puente Hills, San Gabriel Valley, Whittier Narrows, Los Angeles River, west Los Angeles, San Fernando Valley, Newhall, Santa Clara River valley, and the Ventura plain. Burned pasture was frequently recorded and apparently extensive. Green cover and wildflowers were recorded in spring. One annual herb—*chia (Salvia columbariae)*—came to prominence in the Portolá expedition, as Native Americans offered this high-energy food to the Spaniards nearly every day. The journals also describe gallery forests of sycamore and cottonwood along the major rivers in otherwise treeless plains. Coast live oak woodland grew in the Puente Hills and the Santa Monica Mountains.

At San Juan Capistrano, the Portolá expedition left the hilly coastal terrain of San Diego County and, in the words of Crespí, "came down to a spacious plain whose limit was lost to the eye." From near Irvine, Crespí noted, "There is not a tree toward this quarter, nothing but ground and grass [pasto]" (July 24, 1769). The ravine "had all been burned off by the heathens." East of Tustin, he wrote of extensive partially burned pasture: "All of this canyon, hollows and knolls [are] very grass-grown with dry grass [empastada de sacate seco]. . . . We traveled over very open country of very low rolling knolls and tablelands, all very grass-grown. We came onto a large plain, all sacate, although [it was] burnt off by the heathens" (July 26). He made clear the vastness of the pasture the following day, stating, "The size of this flat is vastly great in leagues, all grass grown with dry sacate." The expedition crossed burned pasture in the plains immediately south of the Santa Ana River. The San Gabriel Valley was described as having "very grass-grown soil [empastada] and which must be at least six leagues in length from east to west" (July 30). Crespí also wrote that the area was an "exceedingly spacious valley of dark soil, all burnt off by the heathens."

The Portolá expedition crossed a swampy region and identified a different riparian grass at the San Gabriel River. According to Crespí, the party "went WNW through fields of dry grass [pajonales] and thickets, which detained us for a long time as it was necessary to clear a path at every step" (July 31).[6] Crespí contrasted the dry fields with the lushness of the Porciúncula (Los Angeles) River, which contained "very large, very green bottomlands, seeming from afar to be cornfields [milpas]" (August 2). West Los Angeles was clearly identified by the La Brea tar pits, described as "large marshes of a certain substance like pitch; they were boiling and bubbling, and the pitch came out mixed with an abundance of water" (August 3). To the west the party traveled "across level tablelands of . . . very good dry zacate" (August 4). Near Ballena Creek, Fages "passed by two leagues of well grassed fields which skirt the range," i.e., the herbaceous cover reached the base of the nearby Santa Monica Mountains.

The San Fernando Valley, as seen from the crest of the Santa Monica Mountains, was also covered with extensive burned pasture. Crespí wrote in his journal, "This is a large valley . . . very grass-grown soil [empastada], though most of it had been burnt off; many patches however had not been, where pasto still showed" (August 5). In his field draft he wrote, "The valley has all been burnt off, but to us from the height looked to be fallowed cornfields." Pasture and burns continued to Newhall, where Crespí offered additional speculation on Native American burning, stating that the expedition crossed a "high pass, clad with dry sacate wherever it had not been burnt, as almost all of it had been, by the heathens, who perhaps burn it so that the grasses and plants [*yerbajes*] they subsist upon will give them a better yield of seeds after the rains" (August 8).

Descending the Santa Clara River, Crespí wrote that the flood plain consisted of "very good soil, very grass grown [*empastada*] with very tall broad grass," again in riparian habitat. Crespí's field draft states that the Ventura plain was "very grass-grown flat land, widest in extent from east to west [*empastada de sacate*]." In his first revision, Crespí indicated the extensiveness of the pasture there: "From this point onward, the land makes a very large embayment of flat, very grass-grown soil [*llana mui empastada*] some five or six leagues in extent from east to west" (August 13). Fages described the area as a "spacious plain which stretches southward and westward to the sea; it is well grassed" (Priestly 1937: 46). He even proposed a mission site along the Santa Clara River (Mission Buenaventura), in part because of the "grassy fields" there. This section of the coast was also extensively burned. Fernando Rivera y Moncada, the military captain for the Portolá expedition, wrote on April 24, 1776, that his boat, passing outside the Santa Barbara Channel, landed at the Santa Clara River where "the gentiles destroy and consume the pastures with their burnings" (Burrus 1967: 310).

The writings of Crespí and Font record green pasture and flowers across the Los Angeles plains in winter and spring. In early winter, the Portolá expedition returned to San Diego and the second Anza expedition returned to Mission San Gabriel, and diaries from both capture pastureland in the early growing season. In southern Orange County, Crespí wrote, "It is a pleasure to see what good green grass [*sacates verdes*] there is at all places" (January 19, 1770). He also wrote that "the whole country [is] now greatly pastured [*empastada*] on all sides with very good sacate." Mission San Gabriel had "fine pastures for cattle and horses" (January 5, 1776). Along the new "short-cut" near modern Highway 101 from Ventura to Los Angeles, Crespí saw a "a very sightly valley covered with pasturage" at Thousand Oaks (January 13, 1770). El Triunfo were "plains of considerable extent and much beauty . . . with much pasture" (January 14).

Font and Crespí also saw the Los Angeles plains in vernal flowering season. Font first suggested the presence of forbs on January 7, 1776, at the Santa Ana River, where "the road is almost entirely level, except for some hills about halfway, and all very green and covered with *zacate* and various herbs [*hierbas*], among which is found a species of very small wild onion [*cebollin*]," very like a species of *Allium*. A month later into the growing season (February 11), Font gave a detailed account of wildflower fields in Orange County, flavored with naïve European taxonomy:

> Today I counted twenty-seven hills which we ascended and descended. Among the infinite variety of flowers, such as tulips and others of very diverse colors and very pretty, with which from now on the fields, groves, and valleys of those lands begin to be clothed, I saw several like those in Spain. Among them are some very pretty and small ones with five petals, which look like a face, exactly like those which I saw in some gardens in Cataluña and which are there called *pensamientos* [heartsease] with only the difference that those are yellow and somewhat brown on the edges of the petals, while these are entirely yellow and have no odor.[7]

The tulips may be the California poppy (*Eschscholzia californica*), which is shaped like the European tulip. The five-petaled species may be a species of *Layia*. Font's language implies that the flower fields covered the whole landscape, i.e., the fields, groves, and valleys.

Flower fields were also seen in the direction of Los Angeles. At the Porciúncula River (the Los Angeles River near the modern Civic Center), Font stated that "the land was very green and flower-strewn" (February 21). Crespí portrayed the San Fernando Valley similarly: "The valley here is very much grown over with green grass [*sacate verde*] and various kinds of flowers; as at the rivers behind us, there is a great deal of *chia*, which is all in bloom" (April 27, 1770).

Crespí's account of the *chia* is one of many observations of this annual wildflower throughout southern California, because this herb was an important food source among Native Americans (Timbrook et al. 1982). It was clearly an important member of the wildflower fields. Brown (2001) accurately translates the Spanish word *chia* to "sage," because it is a member of the genus *Salvia* (*columbariae*). However, his English translation leads to confusion with shrubby members of *Salvia* in coastal sage scrub, called *salvia* in Spanish. Hence, the original word *chia* is used here.

When the Portolá expedition crossed the northern plains of Orange County from Santiago Creek to the Santa Ana River, the same area of flower fields described by Font seven years later, Crespí gave the distribution of *chia* in relation to the local landmarks. The party went north-northwest along an extensive plain to the Santa Ana River and in view of a peak "full of snow in the mountain chain to the north [Mount San Antonio of the San Gabriel Mountains]." He wrote, "There is a great deal of *chia* good for refreshment here at the river. . . . The entire countryside, all over this great plain, is full of *chia* that is very good for refreshment, so much of it that I thought it impossible for the heathen folk, a great many of them though there are, to gather even half of it. It was in bloom at present; purple-colored bloom" (April 23, 1770).[8]

From San Juan Capistrano to Ventura, the Indians sent word to the next *rancheria* to prepare baskets of *chia* for the Portolá expedition at their next camp. At the Los Angeles River, Crespí wrote that the "heathens brought *chia*" and he realized in retrospect that "*chia* is plentiful at all these places ever since San Francisco Solano [i.e., since San Juan Capistrano]" (August 2, 1769). At Tustin the heathens gave them "*chia* gruel (*pinole*), and other very good, well flavored sorts of *zacate* seed" (July 28). The chief told the expedition to remain in the area because "they would sustain us on very good refreshing *chia*. Once he recognized the daily routine, Crespí wrote to Palóu that at the Santa Ana River the Indians "would sustain us with sage and with other seeds" (February 6, 1770). He also wrote, "From here onward [Santa Ana River northward], at all of the towns and villages that we encountered, they brought out a vast number of bowls of every good gruel and mush and pies (tamales) for us, not merely once but three times a day, morning, noon, and evening. There was one town were they carried over a good-sized double-hundredweight of *chia*" (Brown 2001: 787–88nn.).

At Encino, the "heathens brought *pinole* and *chia*. There are prickly pear patches (*nopaleras*) not very far from the southern pond and *chia* from which they make very good gruel, and very good it is for drinking, abounds here" (August 5, 1769). The next day the expedition's generous Native American hosts "brought baskets full of gruel and *chia*." Crespí also wrote from this site that "the party spent the day there while scouts were looking for a route [and] . . . to our astonishment they have repeatedly brought us a vast number of baskets full of gruels and *chia*, at morning, noon, and at evenfall."

Other localities where *chia* was offered to the expedition included a location several leagues north of San Juan Capistrano, the San Gabriel River at Whittier Narrows, east Los Angeles, west Los Angeles, Castaic, and two camps along the Santa Clara River. In 1792, José Longinos-Martínez, botanist of the Royal Scientific Expedition, wrote that "there were three species of *chia* that were abundant and could be used in commerce" (Simpson 1961: 85). He also stated that the "seeds of sages (Salviae)" were most commonly used in *pinoles* (34).

Other foods that were offered were *pinoles,* gruels, and seeds of "*zacate.*" At Castaic, "the heathens brought honey dew scraped from reed grasses [*carrizales*]," as well as "raisins [*pasitas*]" (August 8, 1769). This is a very tiny fruit, yielded by some trees that are very plentiful in this hollow; many of them I saw were laden with this little fruit, which is like so many grape seeds, very small and turning black when ripe," possibly elderberry (*Sambucus mexicana*).

SANTA BARBARA TO MONTEREY

From southern California, the Portolá and Anza expeditions traced closely along modern Highway 101, or the coastline (Figure 2.4). After crossing the pastures of the Santa Barbara Channel, the expeditions crossed brilliant flower fields at Point Conception, a labyrinth of swamps near San Luis Obispo, the rugged Santa Lucia Mountains, and heavily burned pastures along the Salinas River to Monterey.

Santa Barbara Channel

Leaving the Los Angeles plains, the Portolá and Anza expeditions followed the coast westward along the Santa Barbara Channel, then turned the corner of Point Conception mapped by the Vizcaíno voyage, and thereafter headed northwestward toward their destination of Monterey. The Santa Barbara Channel, in the eyes of the Spaniards, was bounded by two mountain chains, the northern Channel Islands (Anacapa, Santa Cruz, Santa Rosa, and San Miguel) to the south, and the Santa Ynez Mountains to the north. Between these ranges lies a "topographic low" of the sea and their route along a narrow coastal plain, which that skirts the Santa Ynez Mountains. While the cliff coast west of Ventura was "barren" (free of useful vegetation), pasture, flower fields, and local tule swamps spanned the Santa Barbara plain. Summer pasture burns were seen at many places, even in the tule swamps (cf. Timbrook et al. 1982).

Pasture was nearly absent along the coast between Ventura and Carpinteria, where cliffs descend to the shore. Crespí stated that the expedition passed "steep bare mountains coming down to the sea" (August 15, 1769). They spent another day along the cliff coast, sometimes walking the beach at low tide, before reaching Carpinteria. Here Crespí saw "a plain that must measure, from north to south, about a league of good black soil, well covered with grass [*empastada*]. From east to west it is four leagues long . . . [and] lies between the shore and the mountains running upon the north." Near Santa Barbara, Crespí viewed a "point of land that reaches out to sea [Santa Barbara Point]" where "there are good size tablelands of good very grass-grown pasture [*empastadas*]." Pasture

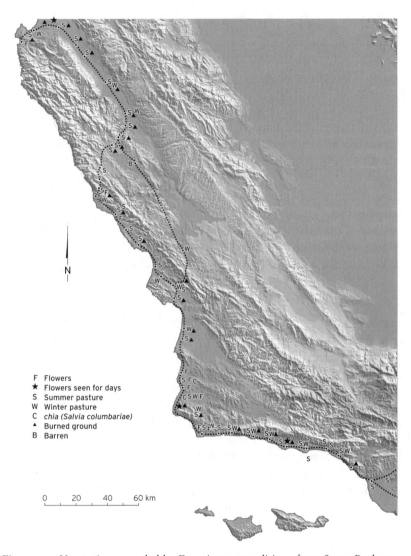

Figure 2.4. Vegetation recorded by Franciscan expeditions from Santa Barbara to Monterey (Brown 2001; Web de Anza Archives; Bolton 1927, 1930a,b, 1933; Teggart 1911; Burrus 1967; Priestly 1937; and see Appendix 1).

was also reported at Montecito and Dos Pueblos at the west end of the plain (August 17).[9]

Pasture formed a narrow strip on marine terraces from Gaviota to San Guido and continued to Goleta, where the Portolá expedition again saw evidence of Native American burning, even in the tule swamps adjoining an estuary. Crespí observed "level soil, very much clad with very fine pasture and very large clumps of very tall broad grasses [*grandes zacatónes de sacates mui altos y anchos*], burnt off in some spots and not in others; the unburnt [grasses] were so tall that they topped us on horseback" (Timbrook et al. 1982), most likely the giant rye grass *Elymus condensatus*. Local pasture and burns seen at Dos Pueblos were followed again by cliff coast without pasture, which Crespí described as "dissected and difficult to travel." At Gaviota, Crespí described Eocene sandstones that outcrop on the south front of the Santa Ynez Mountains: "The mountains that have been alongside us during the last two days are bold, rough and steep; white-colored here and there, as though from white earth or stone." Burned pasture was seen at Dos Pueblos, Gaviota, and near Tajiguas Canyon.

As the expeditions passed modern-day Santa Barbara, nearly all the diarists described the Coulter pine forests (*Pinus coulteri*) in the nearby Santa Ynez Mountains, possibly the same grove seen by Vizcaíno in 1602–03, attesting to the quality of their observations. Crespí wrote that the party halted in a "valley which had running water. Not far from its source is the valley, covered with live oaks and alders (*Platanus racemosa*), and on the summits there are some pines" (August 19, 1769). Likewise, Costansó wrote that "pines grew on some hill tops" and Fages stated, "In the mountains there are seen many pines like those of Spain" (Priestly 1937: 35). During the second Anza expedition, Font marched from Santa Barbara to Tajiguas Creek and saw that "a quite high sierra . . . ran along our right, where there are seen many pines which bear good and large nuts" (February 26, 1776).

The Santa Barbara coastal plain was green, verdant, and in bloom during the rainy season. On their return to San Diego in January 1770, the Portolá expedition suffered from extreme cold and want for pasture. Conditions improved only once they left the San Luis Obispo plains for the Santa Barbara Channel, where Crespí stated, "It is a pleasure to see the good soil here along the Channel and how everything is well-pastured" (January 7). During his second journey, Crespí in early May wrote, "All the spots on the channel are a joy to the sight . . . a great deal of *pasto* for stock." The western edge of the Santa Barbara plain consisted of "good-sized tablelands of very good dark level soil, having a great deal of pasture all the way out to the sea" (May 3).

Native American seed harvests were recorded at two locations along the channel. Near Carpinteria, Crespí wrote in the spring that "men must be gathering their seeds in the fields" (May 1). From near Ventura, the account of Rivera y Moncada on April 4, 1776, describes the seed harvest in relation to Native Californian burning practices: "In the country side [there has been] extreme need of pasture for the animals, which in some areas has caused me difficulty in staying overnight and even in stopping at midday, due to the horses and mules not having grass, all occasioned by the great fires of the gentiles, who, not having to care for more than their own bellies, burn the fields as soon as they gather up the seeds, and that [burning] is universal, although on some occasions it happens that it may be greater or less, according to the winds and calm" (Burrus 1967: 310).

Juan Bautista Anza, who wrote precious few landscape descriptions in this portion of his expedition, was clearly impressed with the Santa Barbara Channel, and in retrospect contrasted the area with coastal southern California and the Sonoran Desert: "All of the lands which they [Native Californians] occupy are as fertile and beautiful as the regions beyond this channel, and the sight of them is certainly a pleasure, especially to one who has witnessed the extreme sterility of the Gulf of California, where neither trees nor useless herbs are to be seen, while here, on the contrary, fields as verdant as they are flower-covered touch the very waters of the sea" (April 14, 1774).[10] In his view, the lands since leaving the desert in the San Jacinto Mountains were verdant, well-pastured, and flowery.

Point Conception

Point Conception lacked fuelwood and trees, but hosted spectacular flower fields when the Portolá and Anza expeditions passed through this section of the coast in late spring. Although the explorers described the California coast late in the growing season, the prevailing cold winds and fog at Point Conception maintained a late bloom that captured the imagination of the Spaniards. In summer they were less impressed with the pasture, much of which was burned (Figure 2.4).

Approaching Point Conception, Constansó reported a change in climate as the expedition "began to experience to cold and violent north winds" (August 25, 1769). The narrow belt of pasture leaving Santa Barbara expanded as the expedition turned the corner of Point Conception and was so at all the camps, including Canada del Coja, Point Conception, Los Pedernales, Santa Ynez River, and San Antonio Creek. Crespí gave a bleak but accurate view of present-day Point Conception: "The mountains that have been accompanying us to the right are bare [*pelonas*] as all the others, sheer soil and *zacate,* with some live oaks seen here and there on the summits and slopes. There is very little wood near this spot" (August 27). Beginning at Point Arguello, Crespí described salt grass (*marisco*) at several localities, no doubt related to the salt spray from the violent ocean and winds. He saw salt grass at the Santa Ynez River and north to San Antonio Creek. Pasture burns were reported by Crespí at Point Conception, near Jalama Beach, and near Point Arguello, including on lands covered with salt grass. Burning was extensive enough to make it difficult to find pasture. At Point Arguello, he wrote, "The soldiers had scouted up to this point, and it was not a full day's march, nor was there grass [*pasto*] for the animals, as it had all been burned off, after about three hours in which we must have made about two leagues and a half, we came to a hollow, . . . and although it had been burned off, there were spots that had not been and where there was good grass [*zacate*] for the animals" (August 29; Timbrook et al. 1982).

Crespí and Font were delighted with the pasture of Point Conception in the growing season. On their return to San Diego, Crespí found the area "grown over with new *zacate*" (January 4, 1770). Even the engineer Costansó wrote that "the country, covered with beautiful green grass [*ierva*], offered excellent pasture for the horses." Winter pasture had not germinated farther north, as Crespí observed that "the mounts have now been greatly restored by good pasture" (January 2).

Crespí's and Anza's journal entries of the Point Conception region during the spring season offer the most detailed account of California wildflowers of all the Spanish expeditions, and even offer a glimpse of the important species in California, at least from a European perspective. Crespí's journal entry from Los Pedernales to the Santa Ynez River also states that he had seen wildflower displays well back toward Santa Barbara:

> We . . . [kept] on a northwesterly course over very grassy level land [*empastada*] near the shore. We shortly passed the point here, and in the distance made out another that formed still another bight [between Point Conception and Point Arguello]. At once after setting out, we commenced to find the fields all abloom with different kinds of wildflowers of all colors, so that, as many as were the flowers we had been meeting all along the way and on the Channel, it was not in such plenty as here, for it is all one mass blossom, great quantities of white, yellow, red, purple, and blue ones; many yellow violets or gilly flowers of the sort that are planted in gardens, a great deal of larkspur, poppy and *chia* [*Salvia columbariae*] in bloom, and what graced the fields most of all was the sight of all the different sorts of colors together. (May 7, 1770)[11]

Among the species seen by Crespí (according to botanical interpretations in Timbrook et al. 1982), the violets are likely *Viola pedunculata*. The Spanish word *aleli* may refer to *Dianthus*. Larkspurs (*Delphinium*) are familiar to the European eye, but may also refer to a lupine (*Lupinus*). The poppy (*cardosanto*) may refer to prickly poppy (*Argemone*), which is common in northern Mexico but is not found at Point Arguello. It may also be the California poppy (*Eschscholzia californica*). Sage is the already familiar ethnobotanical food *chia* (*Salvia columbariae*) which the Portolá expedition had accepted almost daily from Native Americans since San Juan Capistrano.

The wildflower fields continued up the coast. Reaching the Santa Ynez River, Crespí wrote, "We went to hollow with great deal of pasture. On going another league and a half we went down to a stream deep down in a hollow with a great deal of *pasto*. We went up a knoll at the very edge of the shore and came into very large sand dunes, with the flowers still continuing wherever there was not dune. On going about another league and a half beyond the aforesaid stream, now through dunes, now over the blooming fields, we arrived close to another good-sized embayment."

On the following day the Portolá expedition reached San Antonio Creek, where Crespí recorded in his journal that they went "north over salt grasses [*mariscos*] and dunes . . . past this embayment of sand dunes—traveling over level ground and rolling knolls, with the inviting flowers continuing still almost as plentiful and varied as on the previous march."

Font recorded a similar picture of the Point Conception region. On February 28, 1776, the Anza party traversed the coast from El Cojo to the Santa Ynez River at "Punta la Concepcion the range which we have had on our right ends [Santa Ynez Mountains], and from this point the country sharply changes it appearance. All the land is thickly covered with flowers, and green with a great variety of *zacate*, good pasturage, and fragrant and useful plants [*hierbas olorasas*]. Today I saw much samphire. Today it was in full bloom, with yellow blossoms like small sunflowers, which were so abundant that they provided a very pleasing sight along the way."[12]

Font's samphire (*hinojo*) may be *Crithmum maritimum* (*hinojo*), which grows among the rocks and cliffs along the sea coast. His account of Point Conception is similar to that of Crespí's six years before: "All the country was thickly strewn with flowers, . . . I saw many larkspurs [*Delphinium* spp.], and some little red and very pretty marigolds."

San Luis Obispo coast

The Portolá expedition described the coastal plain west of Lompoc and the Santa Maria Valley as a "labyrinth" of estuaries, tule, and lakes behind beach sand dunes. Toward the interior were treeless pastures. Once the expedition reached hard ground again at Pismo Beach, Crespí's account reveals extensive, partially burned pasture from there to San Simeon. A detailed description of the Monterey pine forest at Cambria, and at Pico Creek 1 kilometer to the north, prematurely excited the soldiers, who thought that the expedition had reached its destination of Monterey (Figure 2.4).

North of San Antonio Creek, Crespí wrote on September 1, 1769, that the party "descended to a beautiful valley . . . well grown with pasturage, about three leagues wide and more than seven long," clearly identifying the plains of Santa Maria. A canyon with dry *zacate* was almost entirely burnt off by the heathens. They "camp[ed] [where] a large patch of very good *sacate* for the animals had been left unburnt" (September 4). Most of the Indian village, called El Buchón by the soldiers—in reference to an Indian chief with a large tumor on his neck—"was out plucking their seeds." The expedition then became mired in swamps and sand dunes. North of Guadalupe Lake, Crespí described the area as "covered with rushes and grass and the ground is very wet and swampy." The swamps brought engineering dilemmas for Costansó, who wrote that "there were immense sand-dunes along the shore, and on the plain there were creeks, estuaries [*estero*], and marshes [*pantáños*] which formed a labyrinth" (September 3).

Crespí wrote that the expedition proceeded "northward through very large sand dunes [Montaña de Oro State Park]" and came out onto "hard ground" at Price Canyon with "good soils and dry *zacate*, burnt off everywhere, except for a few patches" (September 4). From here north, pasture was described at all the camps to the cliff coast of the San Lucia Mountains, beginning at San Carpoforo Creek and including Valle los Osos, Morro Bay, and the coastal plains north to San Simeon. On the second Anza expedition, Font was impressed with the "fertile lands" and "pretty fields" at the new mission foundation of San Luis Obispo (March 2, 1776). Valle Los Osos, according to Crespí, had "very good dark soil and very tall dry *sacates*, all truly a marvel." In his second journey, Crespí wrote that this valley had "all sorts of plants [*yerbajes*] so tall and lush that one could scarcely go forward, and they usually topped our mules." Indians were also gathering their seeds. Eventually, the Portolá expedition reached a "high, round island rock . . . in the shape of a head [Morro Rock]" (September 8, 1769) where the lands surrounding the harbor had "rolling knolls and tablelands of very good soil and *sacate*" (September 9). Summer pasture north of Morro Bay consisted of "dry *sacate* that had been burned off in some spots but not in others" (September 9).

On September 10 the expedition reached the pines at Cambria, the present-day forest of Monterey pine (*Pinus radiata*) and a few bishop pines (*Pinus muricata*). Crespí's sharp eye even recorded the second smaller pine forest at Arroyo Pico, 1 kilometer north of the main grove. "Two short leagues" north of the Cambria pines, the expedition "stopped at a burnt-off tableland" (September 11). From there, Crespí wrote at Point Piedras Blancas that in the pasture along the coast a "great many live oaks show up upon the knolls and skirts of the mountains, which are bald [*pelonas*], but a few pine trees are visible on the summits of the nearest mountains" (September 12), mostly ghost pine (*Pinus sabiniana*) and Coulter pine (*P. coulteri*). The following day Crespí arrived at Ragged Point, where the impassable coastline forced the expedition to move inland cross the Santa Lucia Mountains. Approaching camp they found "rolling knolls and tablelands, close to sea with very good soil and *sacate* but also more burned pasture." They made camp in a small area of good pasture "that had not been burnt."

Crespí's return trip to San Diego traversed this portion of the coast during the early winter rainy season, and he recorded the mass germination of annuals that had not yet taken place in the colder weather farther north. He wrote, "As we had had rain a while back up country, we have been finding thin shoots of it during these last days' marches, but already there is good *sacate* here" (December 22).

On his second journey, Crespí saw flower fields along the base of the Santa Lucia Mountains. Near San Simeon, he wrote that he "saw a half dozen women gathering seed during the day's march; but the way the fields are abloom on all sides in a splendid sight." He also saw wildflower fields ascending Arroyo San Carpoforo at the western base of the range: "It is a pleasure to see how flowery and well-pastured the mountains are" (May 16–17, 1770). This was Crespí's last observation of flower fields, as the remainder of his second journey to Monterey took place in the warmer interior of Salinas Valley, after wildflowers had completed their annual growth cycles.

Santa Lucia Mountains and Salinas Valley

The Portolá journey tried to stay within eyeshot of the Pacific coast at all cost, relying on the navigational accounts of the Vizcaíno expedition.

However, the cliff coast of the Santa Lucia Mountains forced the expedition inland over the range at Arroyo San Carpoforo, where they descended into the San Antonio and Nacimiento drainages before reaching the Salinas Valley near King City (Figure 2.4). For the second time in the journey, the expedition briefly departed the California pasture along the crest of the mountains. Crespí described "treacherous slopes" with dense oak woodlands and pine (September 16–20, 1769). Arroyo San Carpoforo had "no soil, but vast amount of stone." On his second journey, Crespí saw a local patch of cypress (*cipres*) in this area, one of the rare groves of Sargent cypress (*Cupressus sargentii*) in the southern Coast Range (May 18, 1770). This Crespí expedition missed the southernmost grove of coast redwood (*Sequoia sempervirens*) by a single canyon (ca. 5 km), although the Anza expedition seven years later heard rumors of it in the Santa Lucia Mountains from Native Californians. Once over the first crest of the range, Crespí wrote that "this high range, once climbed, was not toilsome to get down from, as it was rolling descents, with sheer soil and plentiful dry *sacate*" (September 17, 1769). The Portolá party was not disappointed: the interior Santa Lucia Mountains hosted magnificent stands of valley oak, coast live oak, and ghost pine, as well as extensive pasture, "like parks in Europe," with extensive understory of partially burned pasture. At Jolon Valley, Crespí wrote the "terrain gets easier, . . . with dry grass [*pasto de sacate*] getting ever better, wherever it was not burnt" (September 24). They camped in partially burned pasture at the San Antonio River near Jolon, and about 10 kilometers north of Jolon.

By the time of the second Anza expedition in 1775–76, the Spaniards had already discovered another shortcut—a well-pastured route through Cuesta Pass, Atascadero, and Mission San Antonio to Crespí's route in the Jolon Valley. Although Font saw the region in early March, he wrote that the area was mostly dry because there had been little rain. Between Cuesta Pass and Paso Robles, Font saw "very green meadows [*prados*]" with their arroyos, which formed the Santa Margarina (Salinas) River (March 4, 1776). Font appears to have been describing drying pasture, with green forage following the river. From Mission San Antonio to King City, Font stated that "all the country is good and well grown with pasturage" (March 8).

The Salinas Valley, described as a "spacious and beautiful plain" by Anza in spring (March 9), brought only frustration to the Portolá expedition in summer because much of the valley was burned, leaving little forage for the horses. Indeed, the Spaniards frequently lamented the scarcity of unburned forage. On their arrival in King City, Crespí wrote of "the whole way being over level ground, the area grown with dry *sacate* that had all been burnt" (September 26, 1769). His journal records that camp at modern-day Metz was located at a location where "pasture . . . could not be obtained anywhere else in the valley." At Camphora, "Everything is burnt off by the heathens, so that there is hardly enough grass [*pasto*] for the mounts" (September 28). They found "unburned pasture" at camp near Chualar, and near Salinas there was "very good soil but with everything very much burnt off by the heathens" (September 29–30).

The Portolá expedition finally saw the destination of Monterey and the pine forests descending to the sea. Crespí wrote on September 30 that the Salinas River "empties into an estuary which enters the sea through a valley. That the beach can be seen to the north and the south surrounded with sand dunes, and the coast forms an immense bay; and that to the south is seen a ridge which terminates in a point in the sea, and is covered with trees which look like pines." They camped the following day at the beach, but again, according to Crespí, "All the pasture had been burnt off." The expedition was relieved to see the Point of Pines and Punta Año Nuevo on the north side of Monterey Bay, landmarks recorded by the Vizcaíno expedition.

When Font marched northward across the northern Salinas Valley in spring a little more than six years later, Crespí's burned pastures consisted of flower fields. Font wrote, "The road like all the rest is through pretty country, green, shady, flower strewn, fertile, beautiful, and splendid" (March 10, 1776).[13] The phrase "like all the rest" suggests that flower fields were encountered on the second Anza expedition not only this day, but on previous days along the valley.

MONTEREY TO SAN FRANCISCO

The lands from Monterey to San Francisco were almost continually covered with pasture, in Anza's view the best in California (Figure 2.5). Observations in spring document an abundance of flowers near Monterey and on both sides of San Francisco Bay. The Spaniards were impressed with the great size of the Central Valley, but were disappointed by the poverty of the pasture.

Monterey Bay, San Benito Valley, and Santa Cruz coast

The Monterey pine forest grew to the sea, as described in 1602–03 by the Vizcaíno expedition, but the stature and utility of the trees—good

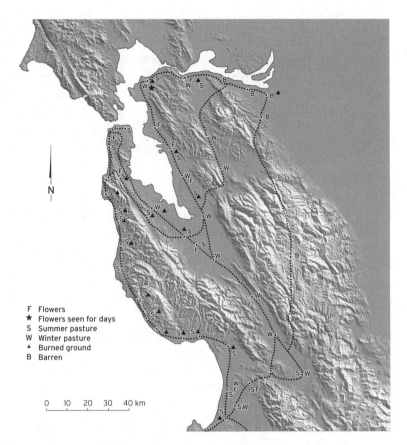

F   Flowers
★   Flowers seen for days
S   Summer pasture
W   Winter pasture
▲   Burned ground
B   Barren

0   10   20   30   40 km

Figure 2.5. Vegetation recorded by Franciscan expeditions from Monterey to San Francisco (Brown 2001; Web de Anza Archives; Bolton 1926, 1927, 1930a,b, 1933; Teggart 1911; Burrus 1967; and see Appendix 1).

enough for "ship masts"—was found to be an exaggeration by Crespí and Font. The Portolá expedition was so disappointed with the harbor at Monterey—it was exposed to the open ocean to the northwest—that they decided to continue moving north in search for "Monterey Bay," ultimately reaching the superior embayment of San Francisco. Crespí revisited San Francisco Bay in 1772, and Palóu visited two years later, accompanied by Anza.

At Monterey, Crespí "went up the hill of pines" and after "two leagues came down to [a] small harbor where the point of pines begins [downtown Monterey; Figure 2.5]. . . . One side of the river with tall grasses [*sacatones*] and other plants [*yerbajes*]—it had been burnt off by the heathens." He saw "very many hills close to here with very good pasture"

(November 28, 1769). When Rivera y Moncada was in Monterey, he speculated that the heathens burned "so that new weeds may grow to produce more seeds" (Clar 1959: 5). After making permanent residence at the Carmel mission, Crespí wrote after the first rain that "all the seeds of the *sacate* have sprouted" (October 21, 1770), clearly indicating the dominance of annual herbs. In the spring of 1776, Font went to Mission Carmel, which was "a most beautiful site and pleasing to the view, because it is so near the sea and in a country so charming and flower covered that it is a marvel" (March 11). Crespí saw what he believed were either cypresses or junipers, in clear reference to the Monterey cypresses near Carmel (*Cupressus macrocarpa,* cf. Brown 2001: 797n).

The Monterey Bay coastal plain was covered by pasture and flowers but lacked trees. Moving northward along the coast, Crespí recorded that "the road for the next two leagues was well covered with pasturage" (October 7, 1769). He then came to "a swamp . . . that was very green with very lush mallows [*malvas*] and other plants [*yerbajes*]" (October 7). This is a rare observation of wildflowers in late summer, perhaps reflecting the cold coastal summer climate. The mallow, very likely the California poppy, has a more perennial habit in coastal central California than in the south or the interior. Above the *cienega,* the surrounding "rolling hills . . . were very much grown over with very good zacate."

In his second expedition, Anza described this area in March as "a good plain of beautiful pastures" (March 23, 1775). Likewise, Font wrote that day, "We entered some hills when a very peculiar odor was perceived. I dismounted to smell some of the many and various flowers which there are in these fields." On his return trip from San Francisco, Palóu described the coastal lands of Monterey Bay as "hills of earth covered with grass" (December 13, 1774). Palóu (November 24, 1774) and Crespí (March 21, 1772) saw pasture-covered lands throughout Monterey Bay.

At the north end of the bay, Crespí entered swampy lands that he described as "a very large plain with grass clumps [*sacatones*] and other lush herbs" (October 8, 1769). Crespí also saw "two leagues of *zacate*-grown plains" to College Lake and the Pajaro River. It was here that scurvy affected nearly everyone of the Portolá expedition, the party camping here for recuperation for several days. Perhaps the diet of corn tortillas and beans could not provide the vitamin C of *chia* in southern California, where scurvy was not reported. From College Lake, the Portolá expedition also discovered the coast redwood as well as "hazelnuts."[14] The land of redwoods and hazelnuts near Santa Cruz had burned pasture at Corralitos Creek (October 15). The Pajaro River, according to

Palóu, was "level land with mellow soil and with beautiful pasturage" (December 11, 1774). He concluded that "this great plain [is] suitable for a large mission, with all the advantages of land, water, [and] pasture. "At present-day Watsonville, Palóu also wrote on this day that while he was "in sight of the beach, . . . [he saw] a wide plain which skirts the range of hills, all good arable land with fine pasture."

The journals of the expeditions of Crespí and Fages in 1772, Palóu in 1774, and of Font on Anza's second expedition in 1775–76 record pasture extending from Monterey Bay into San Benito Valley and north along modern Highway 101 to Gilroy and San Jose. Palóu wrote that San Benito Valley was "entirely of earth grown with *pasto*," and Hollister Valley was "level as a palm of the hand . . . and [had] good pastures" (November 24–25, 1774). From Hollister Valley to Gilroy, Fages wrote, "We found the whole country well covered with *pasto*" (March 22, 1772). Near Gilroy, "The land is very good, with abundant pasturage." Font wrote that the area from San Juan Bautista past Gilroy to Llagas Creek consisted of "extensive and good lands" (March 24, 1776).

The Portolá (1769) and Palóu (1774) expeditions described the coastline of the Santa Cruz Mountains and frequently commented on redwood forests, California buckeye, and madrone in the mountains, as well as on a narrow belt of pastures that skirted the range, the San Lorenzo River, Scott Creek, Whitehouse Creek, Waddell Creek, Pascadero Creek, Purisima Creek, and Half Moon Bay. In Crespí's account, nearly all of the pasture was burned, making it difficult for the mounts. Flower fields were not recorded because neither expedition passed through the coast in spring (Figure 2.5).

At Soquel Creek near Santa Cruz, Crespí saw "very pastured hills [*empastadas*], most of them burnt off by the heathens" (October 16, 1769). At the San Lorenzo River, the expedition went over "very grass-grown soil [*empastada*]—in sight of the sea about a league off, . . . almost all of it having been burnt off" (October 17). To the west of Santa Cruz, both Crespí and Palóu suggested that the pastures, largely burned, formed a narrow belt between the foothills and the coast. At Coja Creek, near present-day Santa Cruz, Crespí wrote that "tablelands must be a league in breadth up as far as [the] bare ranges of knolls that had been burnt off by the heathens, lying at the foot of the mountains" (October 18). Palóu described Scott Creek as "a wide plain of good land and plentiful pasture, which skirts the hills" (December 9, 1774). Punta Año Nuevo, according to Crespí, was a "tableland of pasture soil half league in width" (October 23, 1769). From Arroyo de los Frijoles to Pescadero Creek, pastures on a large range of foothills were "all burnt off" (October 24). Widespread burned pasture continued to Pescadero Creek, Purisima Creek, and Half Moon Bay.

San Francisco Bay and Concord

The pasture on the west side of San Francisco Bay was viewed as the best in California (Figure 2.5). Much of it grew with open woodlands of coast live oak (*encino, Quercus agrifolia*) and valley oak (*robles, Q. lobata*), and was extensively burned in summer. From Gilroy to Santa Clara, Palóu wrote that "the valley widens again, with good land . . . of pure earth and grass. . . . We followed the spacious plain west by northwest, and we found that the valley continues with good pastures" (November 27, 1774). From Palo Alto to San Francisco, Anza's journal states that the "pasture for cattle [was] without equal in quality and abundance" (March 29, 1776). When the Portolá expedition reached Palo Alto, Crespí lamented the chronic problem of burned pasture. He saw "a large plain that must be five or six leagues in extent, entirely grown over with white oaks—large ones, and ones of all sizes—and a few live oaks. . . . The soil of this plain is the finest we have seen, will seemingly yield well, or so the *sacate* that have not been burned indicate" (November 6, 1769). Palóu reported on November 29, 1774, that the pasture on the chain of hills near Belmont did not extend into the brushy slopes of Crystal Springs Reservoir. From Palo Alto to present-day San Bruno the journals indicate that, similar to what was seen along the Santa Cruz Pacific coast, pasture on the west bay stopped below the woodlands and forests.

On March 29, 1776, Font took note of wildflowers in his march between Arroyo Dolores, Palo Alto, and Laguna Merced (San Andreas Lake). Passing San Mateo Creek, he "passed through wooded hills and over flats with good lands . . . plentiful *zacate*, fennel [*hinojo*] and other useful herbs [*hiervas*]." He also saw "manzanita [*manzanilla*], . . . other plants [*yervas*] and many wild violets [*violeta*]" at the San Andreas Rift Valley, very likely *Viola pedunculata* or perhaps *Nemophila* spp. Font also wrote of "many wild violets" on the banks of Arroyo Dolores. San Vicente Creek in the hills south of San Francisco was covered with burned pasture.

Crespí's account supports the Rivera y Moncada perspective that Native American burning improved the food resource. On October 30, 1769, he wrote from Moss Beach that there were "tablelands and rolling knolls, very good pasture, though the latter [was] all burnt, since the heathens burn it all off in order for a better yield of the *sacate* seeds that they eat." At Fort Point in San Francisco, Font camped near a lake, where "there

are beautiful herbs [*hierba buena*] and so many lilies [*lirios*] that I had
them almost inside my tent" (March 27, 1776). At the San Francisco presidio, Font wrote, "On leaving we ascended a small hill and then entered
upon a mesa that was very green and flower-covered, with an abundance
of wild violets" (March 28).

The East Bay from San Jose to Richmond and eastward along the Carquinez Strait to Concord was covered with pasture and flower fields. In
summer, soldiers scouting for the Portolá expedition who were camped
near Palo Alto complained about burned pastures as their horses had little to eat. The scouts also reported the discovery of a full-flowing river
(Sacramento River) that flowed into an estuary. The scouts lamented that
"pasture upon the other side [of the bay] had all been burnt and was toilsome for the mounts," i.e., most of the east bay plains had been burned
by Native Americans that year. The scouts also saw another immense estuary to the northeast (Suisun Bay) (November 10, 1769).

The accounts of the spring expeditions of Crespí and Fages (1772)
and Anza (1776) draw a picture of extensive pasture and flower fields
across the East Bay. Near Fremont, Crespí wrote that the party "continued
NW toward east side . . . the land is level, . . . and well covered with
sacate" (March 24, 1772). Farther north the valley floor is "all black . . .
and well covered with several sorts of herbs and grass. . . . We soon found
the valley widening again [approaching San Jose]; all the land is level,
good, and well covered with *sacate*" (March 25). Font described the land
from San Jose to near Hayward thus: "All the rest of the road is through
very level country, green and flower-covered [*florida*] to the estuary." The
Indians "eat *zacate* and herbs [*ervas*] and some roots like medium size
onions [*raizes*], which they call *amole*, and greatly abound in these plains"
(March 31, 1776). From camp at San Lorenzo Creek, Crespí wrote that
"all the land is level, black, and every well covered with good grass, mallows [*Eschscholzia californica*], and other herbs" (March 25). The next
day he continued northwestward over "level land of the same nature as
the preceding," i.e., *sacate* and mallows continued through Fruitvale and
south Oakland. Fages wrote from near Lake Merritt in Oakland, "On
the plain we saw many lilies and an abundance of very leafy sweet mar-
joram [some kind of shrubby perennial mint] with herbal scent rather
than sweet scent" (March 27, 1772). From a point opposite San Mateo,
Font wrote that "all of the rest of the road is through very level country, green and flower-covered all the way to the estuary" (March 31,
1776). The following day, Font recorded that the lands near Oakland
were "very green and flower-strewn, with an abundance of lilies [*lirios*]."
Farther north, in view of Carquinez Strait, the Indians gave him many

cacomites (a species of *Iris*). Anza wrote that day that "the fields are as green with herbs [*yervas*] and as thickly covered with various wildflowers [*diversas flores*], as those farther back."[15] He suggested that he had seen wildflowers for several days over more extensive lands than he had seen that day. Near Richmond, Anza saw "ten heathen . . . adorned with plumes and garlands of flowers [*guiraldas de flores*]" (April 2, 1776). At Suisin Bay, Font contrasted the lands of Concord with the other side of the bay: "The hills which form this channel [between Carquinez Straight and Suisun Bay] are without trees, but those on this side have plentiful pasturage while those on the other side appeared somewhat bald, with little *zacate*, the earth being reddish in color" (April 2). Crespí found the Concord region to be "all level land covered with grass" (March 30, 1772).

SUMMER BARRENS IN THE INTERIOR

The Spanish expeditions never examined herbaceous cover of interior California in the winter growing season, but the Crespí and Fages expedition of 1772 and the second Anza expedition of 1775–76 reached the Central Valley at Antioch just after herbaceous cover had cured. Garcés and Fages also traversed the Central Valley, both journeys in summer, but their brief journals left little record of the vegetation. All these expeditions leave the same impression: a spectacular barrenness of the interior in the dry season (Wester 1981), in striking contrast with coastal pastures. In contrast with the coast, no burns were described in the interior except in tule swamps.

Southern California interior valleys

A few accounts by the Franciscans suggest that the interior valleys of southern California lacked good forage for livestock in summer. On his return to Sonora in 1776, Font crossed present-day Thousand Oaks in late spring (April 27), and found abandoned villages (Figure 2.3). He wrote that the Indians had moved, "since it did not rain much this year their watering places gave out and the country was very dry and cracked." Not only had the winter herbage dried out, but so little residue was left that the cracks of the earth and the soil were the most conspicuous aspects of the landscape. In the area from Riverside to San Jacinto Valley, rainy-season flower fields were found on May 4 to be "dry and with very little *zacate* [*toda esta tierra hallamos sea, y con mui poco zacate*]" (Figure 2.3). Of course, Font had no experience with California to determine whether 1775–76 was a drought year. In fact the number of rain days in his journal is near normal for the southern California climate. In addition, heavy rains caused storm runoff at times between December and February that stranded the Anza party at Mission San Diego for two weeks. Most watercourses had high runoff between San Diego and the Santa Ana River.[16]

Central Valley

The Central Valley was first crossed by Fages in 1772 and later by Garcés in 1776 (Figure 2.6). Fages described the valley as "part of the mountain range which slopes down toward Monterey [southern Coast Range], is bare of trees, but has plenty of seeds [herbaceous forbs and grasses]" (Priestly 1937; Bolton 1931). Fages also wrote that "the range on the other side of the river [Sierra Nevada] . . . is very high and the peaks are always covered with snow." In the southern San Joaquin Valley from the village of Buena Vista, he wrote that "the plain continues seven leagues more, over good lands with some water. And at the end of these seven leagues, one goes toward the south through a pass [Tejon Pass] . . . very thickly grown with groves of live oaks, and [which] are also on the hills and sierras which form these valleys." Tejon Pass has an abundance of *Quercus lobata* and *Q. douglasii*.

The San Joaquin Valley, which he called the "Llano de San Francisco," was a "labyrinth of lakes and tulares, and the river of San Francisco. All this plain is thickly settled with many large villages." The "labyrinth" implies a sharp contrast between watered and unwatered lands. Garcés's brief diary suggests that the best lands and largest Native American populations were found along the rivers (Coues 1900). When he traveled the plains near Bakersfield, he wrote that he was "passing by dry lagunas . . . and a level plain much undermined by pocket gophers [*tusas*]" (301). His observations of a landscape of rodent tailings suggest that vegetation was too sparse to obscure them. Near the top of Tejon Pass he was regaled with *chia* (*Salvia columbariae*), one of the dominant herbs in that area (Twisselmann 1967).

Crespí, Font, and Anza all recorded rapid change in herbaceous cover when they marched from Concord across Mount Diablo to witness the vast Central Valley (Figure 2.5). Crespí "saw that the land opened into a great plain as level as the palm of the hand, the valley opening about half the quadrant, sixteen quarters from the northwest to the southeast, all level land as far as the eye could reach" (March 30, 1772). Descending the east side of Mount Diablo to Antioch, Font wrote, "The land over

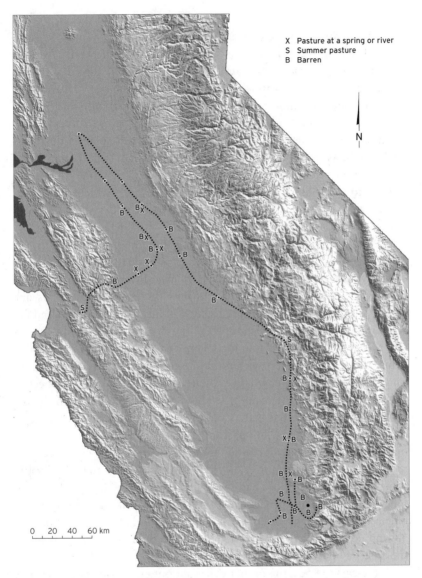

Figure 2.6. Vegetation recorded by Franciscan expeditions in the Central Valley (Cook 1960; Coues 1900; Priestly 1937; Bolton 1927; Web de Anza Archives; and see Appendix 1).

which we traveled was very dry . . . and for this reason the *zacate* was quite dry" (April 3, 1776). Anza had the same opinion. While the north flank of Mount Diablo hosted a "good *zacate*-covered plain," Antioch was "a very sterile and dry plain [*llano bastantemente esteril, y seco*]." Three leagues south of Antioch, Font described the plain as "an arid salty land, all water and mud flats" (April 4). He reported only one burn in the tule swamps; in his short journal he stated that "we trod over spongy decayed ground covered with dried scum, with such an acrid dust arising from the ashes of burnt reeds." In his long journal, Font wrote, "Travelling south from San Joaquin River, the party traveled more than three leagues to the southeast in the midst of the *tulares,* which for a good stretch were dry, soft, mellow ground, covered with dry slime and with a dust which the wind raised from the ashes of the burned tule."

Having given up their goal of crossing the "immense" valley to the Sierra Nevada, Font and the second Anza expedition returned to the Coast Range, which was described as "bare hills" near Livermore Valley (Patterson Grade). Font was emphatic about how disagreeable this landscape was, stating, "In all the journey today we did not see a single Indian, finding only human tracks stamped in the dry mud. It appeared to me that the country is so bare that it could not easily be inhabited by human beings. At least I was left with no desire to return to travel through it, for . . . I had never seen an uglier country" (April 4). Anza's perspective was similar: "All of the country we have traversed today, with the exception of that which has water, and the stretch which we have traveled the last hills is barren of any pasturage [*esteril de todo pasto*], or brush or trees, and apparently it continues that way toward the east [in the direction of the San Joaquin Valley]." Anza also wrote that the Diablo Range south of Livermore had "very little pasturage" (April 5).

These accounts of the Central Valley reveal a striking gradient in the phenology of herbaceous cover. Summer heat had already dried out the ground cover, while the same writers crossed extensive flower fields of the East Bay only three days before.

In the early 1800s, the Franciscans sent out expeditions to determine whether to expand settlement and mission activity inland into the Central Valley and Sierra Nevada, first seen by Fages and Crespí in 1772 and by Font and Anza in 1776. Two key expeditions were undertaken by Jose Maria de Zalvidea in 1805 and by Gabriel Moraga the following year, from which we have the journal of Pedro Muñoz (Cook 1960a,b). The Zalvidea and Moraga expeditions were followed by a number of smaller excursions into interior California before such explorations ceased with the Mexican Revolution and the inevitable decline of the mission system beginning in the 1820s (Figure 2.6). The journals of these expeditions do not have the same level of insight as those of Font or Crespí, but the bleak images of barren lands in the heat of summer did not demand sophisticated descriptions.

In 1805, Zalvidea explored the southern San Joaquin Valley (Cook 1960: 245–51). The party left Mission Santa Barbara and crossed the Santa Ynez Mountains to Mission Santa Ynez. From there, the expedition traversed the Sierra Madre to Cuyama Valley on to Lake Buena Vista, Kern Lake, and the Kern River in the San Joaquin Valley.

The Zalvidea expedition provided an unwelcoming assessment of the southern San Joaquin Valley. The western and central Cuyama Valley was "arid and saline [with] . . . no grass" (July 22–23, 1806). The eastern Cuyama Valley, was "arid, without herbage . . . [but] pasture grew on the nearby hills" (July 24). The southern San Joaquin Valley and Lake Buena Vista was a labyrinth of tules surrounded by barren land. According to Zalvidea, the area "consisted of extensive plains. In quality the land is alkaline. The shore of the lake is completely covered with a great deal of tule. Elsewhere, and in the hills bordering the plains, I saw neither pasturage nor watering places." Areas northeast of the lake "have little *zacate*" (July 26).

At Tejon Creek, Zalvidea "traveled about four leagues over arid, slightly grassy plains." He made an unflattering comparison between the San Joaquin Valley and lands around an interior Southern California mission: "All this territory is similar in character to that around Mission San Gabriel," suggesting that the forage at San Gabriel was also of low quality (July 27). He then took a foray toward present-day Arvin, where he described the valley generally and even identified some summer herbs. He wrote that the southern San Joaquin Valley "from north to south . . . is surrounded by hills which make a semicircle. All this territory is covered with a species of herb which has a little stem with a yellow flower, the stalk being no more than a quarter [of a yard] high. All the hills which encircle this area have also a little herbage that . . . is not very dense" (July 28–30). The flower appears to be a tarweed (*Hemizonia* spp. or *Madia* spp.).

Zalvidea estimated the carrying capacity for a future livestock economy: "All the hills which encircles this area have also a little herbage such that, although the vegetation is not dense, the great extent of the plains will make it possible to maintain twelve thousand head of cattle." The expedition then traveled "four leagues north over pure plains with a little grass" and reached the Kern River, where "all the territory is alkaline with some grass. To the north were bare hills" (August 1). Kern Lake was an immense plain with little pasturage. Zalvidea estimated that "from the end of the lake to the rivers eight thousand head of cattle could be maintained." The estimate covers an area of about 90,000 hectares. Hence, the area could support a density of about one head per 10 hectares. Zalvidea left the valley following Grapevine Canyon to Tejon Pass and traversed the southern edge of the Mojave Desert to near Silverwood Reservoir of the San Bernardino Mountains, before returning to Mission San Gabriel.

The Moraga expedition of 1806, from which we have Muñoz's journal, recorded barren lands and poor pasture along the length of the Central Valley (Cook 1960; Wester 1981). Tule swamps, meadows, valley oak woodlands, and riparian forests grew along the floodplains (Figure 2.6). The expedition departed Mission San Juan Bautista on the coastal slope of the Diablo Range, where nearby San Benito Creek was "a great plain well covered with forage" (September 21). The party crossed Pecheco Pass into "saline" lands of the Tulare Plain at San Luis Creek (September 23) and then entered a dense array of distributaries and swamp of the Merced River, which was "somewhat saline and very heavily covered with green vegetation at this season" (September 23–24). The San Joaquin River itself had "fine meadows of good land and excellent pasture" (September 25–26). Near present-day Merced, the party pushed through "a league of high, thick tules, in the midst of which could be seen a few clearings well covered with grass." They camped in the tule swamps and concluded that "the land is really miserable. Salt flats and alkali patches, with the innumerable ground squirrel burrows are all that one can see." The forage was "extremely scanty and that the country appeared to have been burned over by the Indians did not conceal the fact that the land is very poor. Consequently there is little pasturage" (September 27).

From the Merced River, the expedition traveled northward along the east flank of the Central Valley across sparsely covered plains with green pasture and tule swamps along the rivers. Muñoz wrote that the narrow Tuolumne River flood plain "provides only small meadows and a shortage of pasture" (October 1). The grass was sparse on the way to the Stanislaus River, but the Cosumnes River near Lodi "had excellent land for agriculture and grazing" (October 2–5). The Moraga expedition returned southward and again found "little pasturage at the Tuolumne River." Muñoz was most impressed when they returned to Merced River, which has "wide meadows with land perfect for raising crops [and] grazing cattle." However, the foothills to the east had "restricted pasturage" (October 7–9).

Pasture became increasingly sparse farther south in the semiarid southern San Joaquin Valley. From Mariposa Creek to the Chowchilla River, Muñoz recorded that "all the country traversed today has very poor *sacate*" (October 10). At the San Joaquin River, east of Fresno, "All the country we observed between the Tecolate [Chowchilla River] . . . and the Santa Ana [Fresno River] is worse than bad . . . there is little pasturage, although it is sparse and spread out widely" (October 11–12). Forays down the Fresno River led to "nothing but bad lands, with little *sacate*." The bleakest pasture was found at the Kern River, where "all the country which we have seen today is the most miserable noted in the entire expedition. There is no green grass . . . even at the river . . . there is a great scarcity of pasturage" (October 28). Leaving the valley at Grapevine Canyon, there was a "lack of *zacate* for the horses at Camp" (October 31). The expedition concluded by following a route along modern-day Interstate 5 to Mission San Fernando.

Later expeditions into the San Joaquin Valley were likewise disparaging of the pasture (Cook 1960). The Father Luis Martínez expedition of 1816 saw the same region in May and even reported a bunchgrass growth form, very likely *Sporobolus airoides,* a plant tolerant of high salinity and common in the marshes of the valley even now. Otherwise, the herbaceous vegetation was very poor. Martínez wrote that "in all our trip we did not see a good tree, nor wood enough to cook a meal, nor a stone, nor even grass enough for horses, more than bunchgrass, or what grows in the swamps" (Cook 1960: 271; cf. Wester 1981). A letter from Fray Antonio Ripoll to Father President Vicente Francisco de Sarria records that in the southern San Joaquin "there was a great dust storm on May 5, 1824" (Cook 1962: 153).

FLOWERS AND BARRENS

The pre-Hispanic vegetation of California was not what is envisioned by modern scientific consensus. The record in Spanish journals reveals coastal plains covered by extensive pastures and fields of wildflowers. While the taxonomy of wildflowers is unclear in these accounts, what is most important is that the dominant life-forms seen in California were cured pasture in summer and forbs in winter. The extensiveness of flower fields is suggested by the manner of observations in the Spanish journals. Flowers described for a location were also observed in days past, or in a broader region. From Santa Barbara, Anza wrote that all of southern

California had flowers. Font wrote that flowers were important along the entire route from the pass of San Carlos to Riverside. He also saw flowers for many days along the Salinas Valley and along eastern San Francisco Bay. Flower fields were not reported on the San Cruz Mountains coast or in the Central Valley because the expeditions did not travel through these regions in spring. Crespí captured the wildflowers at their phenological peak at Point Conception, where annuals have a late phenology linked to cold summers. Except for a report of "mallows" in Monterey Bay, Crespí did not report flowers north of the Santa Lucia Mountains because his journal of the first Portolá expedition took place in summer and fall.

Wildflowers were integrated into Native American traditions. They were recorded in Indian myths (Blackburn 1975), preserved in Indian burials (Timbrook et al. 1982), and the Franciscans saw men wearing flowers as garlands. Timbrook et al. (1982) report *Hemizonia ramosissima* and *Calandrinia* spp. (red maids) in a burial carbon-dated to 600 B.P. These plants, which are prepared as a *pinole*, were also found in cemeteries on the mainland and on the Channel Islands.

Wildflowers were an important food resource. Timbrook et al. (1982) list plants from which seeds or leaves were used, including species in the genera *Amsinckia, Aster, Astragalus, Calandrinia, Camissonia, Chaenactis, Cryptantha, Eschscholzia, Hemizonia, Heterotheca, Layia, Lepidium, Lotus, Lupinus, Malva, Phacelia, Salvia,* and *Senecio.* The Harrington notes on ethnobotany of the Chumash Indians of the Santa Barbara region reveal that the small seeds of flowering annuals were still remembered by Chumash after 1900 (Timbrook et al. 1982). Chia (*Salvia columbariae*) was a staple food that was stored in baskets, toasted and ground into flower, and eaten dry or mixed with water to form a gruel. Baskets of *chia* were offered to the Portolá expedition virtually every day from San Juan Capistrano to Santa Barbara. Indeed, Crespí (writing to Palóu) stated that all the towns and villages provided the party with "trays of very good *pinoles, atoles* and tamales as much as three times a day," and that they "gave us a good supply of *chia* for lunch, for there is good *chia* in most places" (Bolton 1927: 33). Longinos-Martínez proposed that *chia* could be grown as a crop (Simpson 1961).

There was fierce competition for seed among Indian groups for the native plant food resource. Longinos-Martínez in 1792 wrote that "in this part of the Santa Barbara Channel . . . their wars are frequent and always originate over rights to seed-gathering grounds" (Simpson 1961: 117). He also stated that "the gentiles living between San Diego and San Buenaventura store up against the winter the plants that bear the most seeds. . . . These nations continually keep on hand small baskets of seeds and other footstuffs" (Timbrook et al. 1982).

Inspired by the extensive burning in coastal California during early land explorations, the Spaniards proposed a positive feedback theory in which burning increased the production of desired foot plants, a topic of several studies (e.g., Lewis 1973; Timbrook et al. 1982). On the San Diego coast, Crespí proposed that Native Americans burned for the hunting of "rabbits and hares" (cf. Longinos-Martínez's account in 1792, Simpson 1961: 58–59). Near Santa Barbara, Rivera y Moncada wrote that they "burn the fields as soon as they gather up the seeds, and that [burning] is universal" (Timbrook et al. 1982). At Monterey, he wrote that the heathens burned "so that new weeds may grow to produce more seeds" (Clar 1959: 5). Timbrook et al. (1982) suggest that most annual wildflowers ripen in late spring and that seed shatter of native grasses in June or July explains Crespí's reports of burned grasslands in August, consistent with seasonal burning practices of other Indian groups elsewhere in California (Bean and Lawton 1973; Anderson 2005). The first burns were seen in late July along the San Diego coast. Indians through burning may have selected for food crops such as *chia* and nursemaids. While fire-stimulated forbs grow in early postfire successions in chaparral, the germination of most flowers of California pasture are not dependent on fire. The hypothesis of Parsons and Stohlgren (1989) that burning was probably rare before exotics can be refuted based on Crespí's account of California in 1769. Native American burning continued well into the Spanish mission period. Timbrook et al. (1982) and others cite Arrillaga's proclamation of 1793 to stop Indians from burning the land. There was also a mission questionaire in 1798 that discussed whether punishment would be meted out to Native Americans when there was transgression against the common good, like killing cattle and sheep or firing pasture.

Timbrook et al. (1982) propose that the cessation of burning during the European period led to a decline of *chia* and other seed species. Alternatively, one can conclude that native forbs found across California were not dependent on fire (cf. Keeley and Fotheringham 1997), with most germinating under cue of winter rains. Most forbs also grow in interior California valleys and southeastern deserts, where burning was seldom described, and wildflowers respond to precipitation cues. While the winter growth flush left cured pasture in summer along the coast, the interior valleys were barren in the dry season. Most herbs, having low lignin silica content, disarticulate upon desiccation. Some forb species have shown the capacity to cure into flammable biomass, including *Amsinckia*,

Phacelia and *Cryptantha, Chaenactis, Lepidium,* and *Salvia columbariae,*
species common in Indian burials. The Franciscans never saw interior
California in the growing season, but a dominance of forbs is consistent
with the barrenness of these lands described by the Fages, Anza, Zalvidea,
and Moraga expeditions.

Remarkably, the Harrington notes for Chumash territory had no reference to Native American burning to promote the growth of plants for
human consumption (Timbrook et al. 1982). In view of the extensive
burns seen by Crespí, this finding is enigmatic unless explained by the
tradition of burning having been forgotten.

Spanish words for bunch grasses were not used to describe the herbaceous vegetation of California except in tule swamps. Instead, the expedition diarists used the virtually synonymous *pasto* and *zacate* in reference to dry herbaceous cover. The newly discovered Crespí field diary
and first field revision published by Brown (2001), show that both words
were used interchangeably in daily journal entries. Hence, whether *pasto*
and *zacate* represent grasses or forbs is ambiguous. Neither word con-
veys botanical information. From the barnyard perspective of the journal mandate, both words depict the potential pasture resource for livestock. Spanish words for bunch grasses, e.g., *grama,* also a livestock term,
were not used in California even though grazing economy was part of
the Franciscan mission mandate.

Is this finding an artifact of Spanish botanical knowledge, or do these
accounts indicate that bunch grasslands were actually rare at European
contact? Timbrook et al. (1982) write that grass seeds were barely discussed in the Harrington notes and conclude that this was an indication
that native grasses had been substantially reduced. Alternatively, the
Spanish journals suggest that grasses may not have been important to
the vegetation. Longinos-Martínez, the only Spanish scientist to visit California in the eighteenth century, does not discuss bunch grassland in his
brief summary of the state's vegetation. Besides, perennial bunch grasses
would have drawn accounts of green pasture in summer, even in the interior barrens. Although the Zalvidea and Muñoz journals of the interior valleys were written three decades into the Franciscan mission period,
their explorations took place long before there was any settlement or grazing. Both expeditions record extensive barrens across the Central Valley
in summer. An exotic wild oat pasture would certainly have attracted
their interest, yet their bleak assessment of the Central Valley would appear to preclude its presence. The common Spanish plant name *aveno*
was not used.

Spanish colonization in the decades to follow brought a new biota from

Europe, as well as livestock. The intercontinental transfer of new species that by chance would occur at scales of near millions of years was instead compacted into two centuries, with several species reducing or displacing the indigenous flora throughout the state. The findings here opens the door to the prospect that introduced annual grasses and forbs entered flower fields, not perennial bunch grasslands.

Invasion of Franciscan Annuals, Grazing, and California Pasture in the Nineteenth Century

The Great Central Plain of California, during the months
of March, April, and May, was one smooth, continuous bed
of honey-bloom, so marvelously rich that, in walking from
one end of it to the other, a distance of 400 miles, your foot
would press a hundred flowers at every step.

—John Muir in 1868 (1904: 339)

We encountered more than one forest of mustard, whose
tall stalks were above the rider's heads, and made, as it were,
two thick walls on the two sides of the way. This plant has
become, for some years, a terrible scourge for part of California. It invades the finest pasture lands, and threatens to
spread over the entire country.

—August Bernard Duhaut-Cilly in 1827 (1929: 246)

The deliberate introduction of European annual grasses and forbs by the
Franciscan missionaries began an extraordinary transformation of the California herbaceous flora, which is an ongoing process. In explaining the
transformation to modern exotic annual grassland, the scientific community is still at the first step: detailing the history of invasions and associated change in California pastures. As it stands, there is disagreement on
when individual species arrived and how they expanded geographically
across the state. Differences in the invaders' habitat preferences are seldom
appreciated. The observations of John Muir and August Bernard Duhaut-
Cilly illustrate the complexities of species invasions. Muir saw the Central
Valley still covered with indigenous wildflowers forty years after introduced
mustards had expanded across the Los Angeles plains in the 1820s.

Conceptual models on the history of California pasture are frustrated by the sparse written record from the Spanish and Mexican periods. Scenarios have been deduced from ecological models based largely on theories of Fredrick Clements from the early twentieth century, which use botanical data and field observations that became abundant after the gold rush and California statehood (Hamilton 1997), not the Spanish journals. Because the early expansion of exotic species coincided with the introduction of livestock, whose collective numbers increased to millions of animals by the nineteenth century, ecologists have also addressed the role of role of grazing in the early expansion of European invasives.

THE BUNCHGRASS-GRAZING HYPOTHESIS

The bunchgrass-grazing hypothesis has its origin with the eminent early twentieth-century ecologist Fredrick Clements (1934) and was perpetuated by Burcham (1957), Heady (1977), and Sims and Risser (2000). In the model, wild oats are thought to have replaced bunch grasses dominated by species of *Nassella* (*Stipa*) by the mid–nineteenth century. Native perennial grasses were selectively eliminated because of high palatability. Wild oats and other introduced species were vectored by livestock. Growth of perennial grasses is slow and the germinable seed production was limited compared to annual plants (Major and Pyott 1966). Annuals are characterized by heavy seeding, self-pollination, and rapid maturation and so replaced perennial cover. The transformation was facilitated by extreme drought (1857, 1862–64) and overgrazing by livestock.

Clements (1934) hypothesized that the destruction of the grass prairie by overgrazing and fire was based on "relict" vegetation, using his "plant indicators" method to determine potential "climax communities" (Clements 1916). Clements noted the presence of *Nassella pulchra* along fenced railways, which he believed to be undisturbed. From these observations he concluded that grasslands dominated by *N. pulchra* were the "original great climax" of California (Clements 1934; Clements and Shelford 1939: 285; Beetle 1947; Burcham 1957: 90, 189, 192; Frenkel 1970). Climax theory, a supraorganismic concept of vegetation dynamics, is thought to permit inference of mature vegetation assemblages. A fundamental assumption is that disturbance by livestock is not natural and that, without it, California succession would inevitably proceed to bunch grassland, the climax, the closest affinity being the bunchgrass community, e.g., the Palouse of Washington State.

Defenders of the bunchgrass-grazing hypothesis deny the Hispanic record by asserting that reliance should be on botanical collections exclusively, i.e., a credible body of botanical records developed so late that pre-Hispanic vegetation is an enigma. In the process the vegetation baseline is brought forward to the mid–nineteenth century, when pre-Hispanic vegetation was already contaminated by European annual grasses and forbs. In Heady's view (1977), history has not recorded the vegetational dynamics of the pristine California prairie. Most collecting began with William Henry Brewer's state survey in 1860. By raising the bar of scientific scrutiny, the early transformation before the arrival of scientists makes difficult the evaluation of pre-Hispanic composition of herbaceous communities (Parish 1920; Burcham 1957: 193; Frenkel 1970; Bartolome et al. 1986; Hamilton 1997).

But this is a two-edge sword. While the Spanish diaries have taxonomic limitations, the expeditions provide a systematic spatial sampling between San Diego and San Francisco, unparalleled by any survey of California until the late nineteenth century, and a protocol that has never been undertaken by American botanists to this day. Adding rigor to observations in poorly understood, complex ecosystems raises more questions than answers and diminishes the prospect of synthesis of broadscale landscape patterns. The shifting baseline syndrome from 1769 to the mid–nineteenth century also changes the story (Jackson et al. 2001). Brewer (1966) also had the misfortune of basing his synthesis of California pastures on experiences during catastrophic drought and overgrazing of the early 1860s. What conclusions would have emerged from his experience if he had first visited California with John C. Frémont in the 1840s, or John Muir in the late 1860s, or with Spanish missionaries in 1769?

Proponents of the bunchgrass-grazing hypothesis have also proffered the hypothesis that the abundance of wildflowers was an "optical illusion" (Heady 1977). Aboriginal herbaceous cover was bunch grass, only people just notice the flowers. Quoting Muir's account of the Santa Clara Valley on April 1, 1868, "the hills were so covered with flowers that they seem to be painted." On the same trip, Muir referred to the Central Valley as a "garden of yellow compositae" (Heady 1977: 492–93). According to Heady, "but that was early April, when *Baeria chrysostoma, Amsinckia intermedia, Eschscholzia californica,* and others make their brief flowering season. . . . Six weeks later the descriptions could well have suggested a sea of mixed grasses gently moving in the morning breeze. One could argue from these observations that the vegetation was either annual or perennial, or a mixture of both" (492–93).

The central question is how European land use contributed to the conversion of California pastures. Livestock grazing and primitive agriculture of the Franciscan mission system have been viewed as a transformation from the Native American hunting and gathering economy (Kahrl et al. 1979). The Spaniards came to Alta California in 1769 with a different culture, which included introducing a grazing economy devoted primarily to cattle, and with the intent to establish permanent settlements. Sparse information in the first century of European colonization until the gold rush of 1849 suggested that settlement of California was primitive, as if the entire state existed as a national park. In terms of wild landscapes, European colonization just meant the addition of a few more animal species to join the wildlife, the extension of a primitive frontier mission system, and growth of tiny coastal settlements near the best harbors.

When John Bidwell crossed the Sierra Nevada and Central Valley to the central California coast in 1839, his first impressions were that cattle and other domestic stock lived as wild animals in pristine wildlands with an abundance of other herbivores such as elk, deer, and antelope. Domestic animals were in effect "wildstock" more than livestock. Cattle fended for themselves, horses became feral, and pigs hid in the tule swamps. It is perhaps ironic that Zenas Leonard of the Joseph Walker party of 1833–34 was perplexed that another cow, (i.e. the buffalo), a primary food source on their journey from the Great Plains, was last seen near the Great Salt Lake, given the verdant green pastures they first saw in the Sierra Nevada foothills (Leonard 1959: 83–84). Leonard recalled in 1833–34 that "hunters complained there was no buffalo as the grass was equally as good and plenty, and the prairies and forests as extensive as those of the region of the Rocky Mountain" (84). These early Americans had not yet experienced the summer barrens already familiar to the Spaniards and doubtless hostile to the animals' survival.

Conventional wisdom is that Spaniards rapidly destroyed Indian culture and changed the California landscape, but historical records show that the transformation was a gradual process, as Native Californians in interior California continued to practice their ways and adapted to Spanish culture well into the mid–nineteenth century, with a dietary shift to horses (Phillips 1993). Native Americans along the coast, of course, were subjugated by the mission system as neophytes for the purpose of integrating them into this new economy and instilling in them the social norms of Roman Catholicism. The Spanish Crown and the Mexican government after 1822 had no political control except along the coast. Neophytes escaped to the interior Central Valley after the collapse of the mission system in the 1830s. While cattle numbers had already built up along the coast, Bidwell, the Walker party, Frémont, and others in the 1830s and 1840s did not see animals in the Central Valley except near the Sacramento delta. Agriculture was limited to a few small fields next

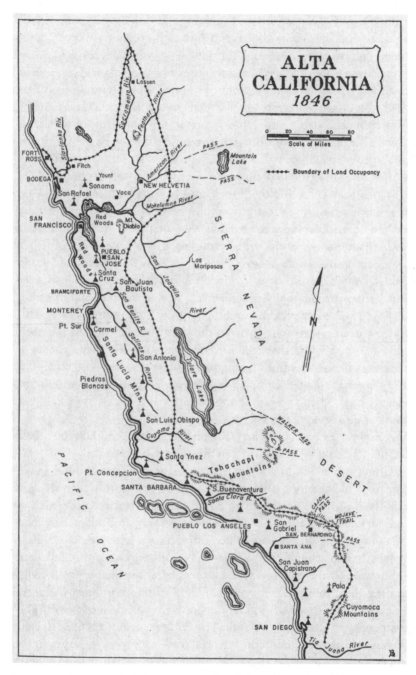

Figure 3.1. Lands under Mexican control and domestic livestock in 1846 (from Clar 1959).

to settlements, and the numerous ranchos operated as small feudal estates. Irrigation was limited to ditches, dams were not built, and roads were ruts, if not unnecessary. Small numbers of cattle, sheep, horses, and other livestock brought to California in 1770 gradually built up large herds approaching carrying capacity within about thirty to forty years. In the end, the livestock economy was restricted to the coast, first with the establishment of the Franciscan mission system and then with ranchos (Figure 3.1). The California interior had no livestock except for feral horses that escaped the ranchos.

The open-range grazing economy continued to dominate the state well after the gold rush and California statehood, until about 1880, but American settlement of the California interior after statehood virtually obliterated Native American culture by that time. While the Spanish and Mexican periods resulted in embryonic transformation of the natural landscape, the Americans began a process of converting much of the state to agriculture, mining, irrigation/hydraulic engineering, and development of aqueducts (Kahrl et al. 1979).

The goals of this chapter are to evaluate grazing by domestic and native herbivores, to establish a time line of the expansion of grazing by domestic livestock, to develop a time line for the expansion of exotic annuals, to describe California pasture in the nineteenth century, and to assess the importance of grazing in the transformation of pasture.

THE WRITTEN RECORD, SPANISH LAND TENURE, AND *DISEÑOS*

The Spanish Crown never established a vigorous commerce in eighteenth and early nineteenth centuries (Cronise 1868: 59–60), even though they "had possession of the entire Pacific coast of the New World, several of the best harbors in the world, and had resources afforded by markets in Europe and Asia, and no scarcity of materials or labor for ship-building." Trade was nonexistent. Among the earliest settlers, John Gilroy, who had resided in California since 1814, said that for several years after his arrival the entire trade and commerce of California consisted of the shipment of a cargo of tallow once a year in return for a "a few cotton goods and miscellaneous articles for the missionaries" (Cronise 1868: 61). The economy was self-sufficient and remote from any central politics of Mexico City and New Spain (Bancroft 1888: 300). California was particularly isolated during 1810–22 because of revolutions and collapse of the Spanish Empire in the New World.

There are few writings of the California frontier after the initial Spanish explorations. The Franciscans were no longer required to maintain diaries, and with their efforts focused upon the conversion of indigenous Californians to Christianity, they wrote little of the California landscape. Most records were baptisms and burials, number of livestock, and accounts of battles with interior Indians who incessantly raided missions for their horses, especially after about 1810 (Aschmann 1959b; Phillips 1993). The Franciscans sent a few expeditions to the California interior to make peace with the Indians, including the Zalvidea and Moraga expeditions of 1805–06. Few visitors came to California either by land or ship until the 1830s (Davis 1929).

Information on the number and distribution of animals can be understood in relation to Spanish land tenure and rancho practices. Livestock grazing developed hand-in-hand with large land holdings coupled with the Royal Spanish land tenure system, almost exclusively along the Pacific coast from San Diego to San Francisco and Sonoma where year-round pasture was secure. The decision to colonize the coast was doubtless shaped by the observations of Crespí, Font, Fages, Palóu, and later by the Zalvidea and Moraga expeditions, which clearly showed this region's superiority of pasture compared to the California interior. Open-range grazing continued into the American period until around 1880.

The history of California livestock grazing is detailed in Cleland's *Cattle on a Thousand Hills* (1964). After the Franciscan expeditions, California lands were partitioned by grants under Spanish law that recognized the king as owner of all the colonial possessions in the Americas. The Spanish colonial system used three institutions—the presidio, mission, and pueblo—to expand the frontier, in which control of the land was dependent on royal land grants (concessions). The presidio was a military post both for the defense of the province and the maintenance of internal order. It also possessed large land units to supply the garrison with food and to furnish pasture for the king's cattle and other livestock. The pueblo (town) was founded by civilian colonists with land holdings covering areas of typically 3 square leagues in communal pasture. The missions had the responsibility of converting the indigenous population to Christianity and also converting them from hunting and gathering to agriculture and grazing activities to support the mission. As emphasized by Cleland, the mission was a "frontier institution." It had no place in a settled, well-ordered society over the long run, its function ceasing whenever the wilderness was adequately civilized. Mission land grants were never made in perpetuity but had enormous size. Most California lands were under mission control until secularization in the 1820s.

From the period of Governor Fages in 1782 to the Mexican overthrow of Spanish rule in 1822, there were only twenty private rancho concessions in California, half of them in Los Angeles. *Diseños*, or land claim sketch maps, are a limited source of information on the vegetation (Becker 1964; Hornbeck 1983). According to Cleland, the possession of a land grant required a concession from the Spanish Crown. The recipient was required to build a stone house, stock the ranch with at least two thousand head of cattle, and provide enough *vaqueros* (cowboys) and sheepherders to prevent the stock from wandering. The grant could not intrude into water, pastures, wood, or timber allotted to pueblos, or into mission holdings, nor into Indian *rancherías*. The Mexican Congress in 1824 passed the Law of Colonization in response to threats by England, France, Russia, and expansion of U.S. settlement into the Great Plains. The law was passed to develop stable, well-armed populations of foreigners and Mexicans in sparsely populated areas of Mexico, including California. It also provided a legal structure for the establishment of land grants in California, but Mexican governors made little effort to break up mission lands into concessions. The overthrow of the Spanish dominion in Mexico in 1822 was the death blow to the mission system, although it had been decaying for years (Cronise 1868: 15). No new missions were founded after 1823. The Secularization Act of 1833 led to the collapse of the mission system, a self-fulfilling legal prophesy under Spanish law because the mission was a frontier institution that never owned land (Cleland 1964). The California provincial government distributed national domain in hundreds of land grants, especially after 1840.

To obtain title of a land grant, a petitioner produced a formal petition that included a *diseño* showing the area, location, natural boundaries, and landmarks of the grant, as well as information on the vegetation. The lack of *diseños* until the 1830s is explained by Beechey (1831: 11):

> At that time soldiers entered for a term of ten years, at the expiration of which they were allowed to retire to the Pueblos—villages erected for this purpose, and attached to the missions, where the men have a portion of ground allotted to them for the support of their families. This afforded a competency to many; and while it benefited them, it was of service to the government, as the country by that means became settled, and its security increased. But this privilege has lately been withheld, and the applicants have been allowed only to possess the land and feed their cattle upon it. . . . The real cause, however, was not explained to the soldiers; they merely heard that they would not have the land ceded to them for life as usual, and they were consequently more dissatisfied.

Bidwell (1937: 39), who obtained the Chico land grant (Gillis and Magliari 2003: 73), said that 1 Spanish league (6.5 square miles, 10 km²) was considered a farm, the smallest grant conceded by the Spanish government. Land grants were as large as 11 leagues and could "take in the shape of the valley or tillable land, and not include the mountains which bound the valleys." Bidwell also wrote that to qualify for a Mexican land grant, a grantee had to reside in Mexico for two years, become a naturalized citizen, and convert to Roman Catholicism (Gillis and Magliari 2003: 73).

The precautions the fathers had taken to prevent free emigrants settling in the territory ultimately worked against their interest, because it deprived them of self-defense. According to Cronise (1868: 15), "Upper California was denied representation as a state in Mexico, and was declared a Territory. The great riches possessed by the California missions had become a subject of the much solicitude to the Mexican congress and in 1826 a law was passed to deprive the Fathers of their lands, and the labor of the Indians, stopping their salaries."

The oration of General Vallejo records a "political tornado" (secularization) that "burst upon the mission system," which led the priests to convert their livestock wealth into currency and also led to the mission slaughter of the early 1830s, from which they only saved the hides. Vallejo wrote that "the pecuniary wealth of the Missions . . . was sent out of the country to Spain, Mexico, or Italy. . . . Neither the governors nor the Californians ever partook of that wealth" (Davis 1929: 367). According to Cronise (1868: 16), "One of the missionaries at San Luis Obispo took $100,000 with him when he left for Spain in 1828. Clearly, the allegiance of the men of cloth was to the church of Spain and ultimately the Vatican, not to California." Cronise (1868: 59–60) continued, "In 1834, this branch of trade was increased by the missionaries killing immense numbers of their cattle, possibly 100,000, to obtain hides and tallow, an anticipation of the movement for secularizing the missions." It was for this reason that the missionaries made an unparalleled detailed inventory of livestock that year. In more prosperous days, the Franciscan mission system was a country unto itself: "One mission would assist another with hides and tallow, or with fur, skins, or money, in payment for goods which it had purchased. . . . These numerous Missions were in reality one institution, with a common interest" (Davis 1929: 204).

Beginning in the 1820s, the accounts of vegetation by non-Hispanic visitors, resident Californios, and expeditions were sufficiently comprehensive to permit general description of the herbaceous landscape and wildlife.[1] Fur trappers from the eastern United States and Canada began visiting the region to harvest sea otter and beaver, but they left few writings of the California landscape, mostly only commenting on the wildlife. Two important early accounts were the diaries of Zenas Leonard of the Walker party of 1833–34 (Leonard 1959) and John Bidwell in 1839 (Bidwell 1928, 1937, 1948; Gillis and Magliari 2003). In the late Mexican period, foreign threats brought many explorers who took stock of California as potential spoils for their respective countries, with detailed descriptions of these lands during the Mexican period. Important early writings include those of Captain John C. Frémont, August Bernard Duhaut-Cilly, Eugene Duflot de Mofras, Charles Wilkes, and writings from the voyage of the *Blossom* (Beechey 1831). Speaking for the U.S. perspective, Leonard (1959: 94) wrote that

> the Spaniards are making inroads on the south—the Russians are encroaching with impunity along the seashore to the north, and further northeast the British are pushing their stations into the very heart of our territory (the west), which even at this day, more resemble military forts to resist invasion than trading stations. Our government should be vigilant. She should assert her claim by taking possession of the whole territory as soon as possible—for we have good reason to suppose that the territory west of the [Rocky] mountain will some day be equally as important to a nation as that on the east.

In Cronise's account (1868: 67), "The enunciation of the Monroe doctrine caused France and England, who were deeply interested in the Pacific coast to use all possible means to prevent the expansion of the United States into this region. Duflot de Mofras made his explorations of California to prepare the way for France to acquire possession of the country. In 1841 Maschal Soult, Minister of War under Louis Phillipe, appointed Duflot de Mofras, an eminent French savant and diplomat, to make a thorough exploration of California. The U.S. Government dispatched Commodore Charles Wilkes, with a squadron, consisting of five vessels of war at San Francisco Bay." According to Cronise, Wilkes thoroughly surveyed the bay of San Francisco and the Sacramento River as far as Sutter's Fort and made the following assessment: "De Mofras, in page 68, vol ii, of his report states that he was satisfied, from information he gathered on board of the English and United States vessels, that both parties expected to obtain possession of the country; while his own book was written to instruct the French officers how best to accomplish the same subject" (67). A few early non-Spanish obtained land grants, such as Captain John Sutter and John Bidwell, the latter of whom became the "right-hand man" at Sutter's Fort and built his own rancho at Chico. Sutter also closed down the Russian settlement at Fort Ross by international agreement, including the purchase of two thousand stock, and was surprised that the Indians there spoke Russian (Gillis and Magliari 2003: 93).

The earliest baseline of the California herbaceous flora were the observations of Captain Frémont, an excellent botanist and naturalist, who provided superb detail of the herbaceous flora of California in two journeys (1844 and 1846), journeys that were directly linked to the U.S. policy of Manifest Destiny (that the United States should stretch from the Atlantic to the Pacific coast). This policy was pushed hard by his future father-in-law, Senator Thomas H. Benson of Missouri, during the administration of President James K. Polk. In William Davis's view (1929), Washington DC wanted to know what this country was like, and was curious how Boston merchants grew rich on California's resources in a few years. Frémont's mandates were to inventory the state and win over the Californians by peaceful means. His first trip in 1844 crossed the Central Valley from north to south and exited coastal California via Tehachapi Pass. His second trip in 1846 followed the same route but also entered into Mexican settled lands along coast to the capital at Monterey. In the words of Davis (1929: xxix), from June to the end of July was a "recital of lawless acts committed against a people we hoped to win by kindness, including the Bear Flag revolt, murder of the de Haro boys and old man Berryessa, and kidnappings." In his diary Frémont recorded killings and botany on the same day. The end of the second expedition was a march to Los Angeles, where he met up with Robert F. Stockton's brigade, and a brief battle resulted in the capitulation of the Californios.

With the gold rush came a people with a laissez-faire belief that lands were open for the strong minded and the enterprising to seize and use in whatever way that would profit them (Kahrl et al. 1979). California was rapidly overwhelmed by Americans, dismantling both Mexican and Native American cultures. Botanical collections are an inadequate source of data before the gold rush (Frenkel 1970). Brewer and Watson (1876–80: 553) of the California Geological Survey state survey of the flora, report that were only twenty-six known collections between the Malaspina expedition in 1791 and U.S.-Mexican boundary survey of 1849. Frenkel (1970: 40) states that "anyone searching the earliest botanical collections is bound to be disappointed by the dirth of common exotic plants." However, about forty collecters were active between 1849 and the state survey in 1860. Information increased with the U.S.-Mexican boundary survey of 1849 (Whipple 1961; Parry 1859), the Pacific Railroad survey of the 1850s (U.S. Department of War 1855–61), and the recording of the state's botany under William Henry Brewer in the 1860s (Brewer and

Watson 1876–80). A wealth of information was contributed by explorers, naturalists, and environmental historians such as John Muir (1904), Clarence King (1915), John Hittell (1874), Archduke Ludwig Louis Salvator (1929), and Titus Fey Cronise (1868). John Bidwell (1928) had many famous guests at his Rancho Chico, including Asa Gray, Sir Joseph Hooker, Muir, and David Starr Jordan. The gold rush also brought a flurry of books and articles on the "new" California.

GRAZING BY NATIVE HERBIVORES

An important question with respect to early invasions of European annual grasses and forbs is the intensity of livestock grazing. The effects of domestic herbivores were not novel. The Hispanic grazing economy was established in landscapes where native herbivores had been abundant for millennia during aboriginal times and where there existed diverse megafauna in the last glacial maximum (see, e.g., Martin 2005). At European contact in the late eighteenth century, the most important herbivores were pronghorn antelope (*Antilocapra americana*), elk (*Cervus canadensis*), and deer (*Odocoileus hemionus*). Their broad distributions at the beginning of European colonization are known from the Spanish journals (Brown 2001) and the late Mexican period (Figure 3.2).

The Spanish missionaries observed antelope in flocks of perhaps half a dozen to fifty, mostly in the plains throughout coastal California (see Appendix 1 for locations of observations). In their march northward along the Pacific coast, the Portolá expedition first recorded antelope in Valle San Telmo in northern Baja California (lat. 31° N; Crespí, April 14, 1769, in Brown 2001). In Alta California, Juan Crespí first observed antelope north of San Diego and at La Jolla (July 16, 1769), Camp Pendleton (July 22), and the Irvine Ranch (July 26). In the San Gabriel Valley he wrote that "there are vast numbers of antelopes on this plain travelling in large bands just as though they were goats. We have seen twelve of them running, from very near by" (July 31). Antelopes were again encountered near the Los Angeles River and the west Los Angeles plains, where Crespí wrote, "All over the plains here are to be seen a great many antelopes." In his second journey, he saw "a great many antelope bands of five, eight, or ten apiece, there being vast numbers of them here on the San Miguel and Porciúncula Rivers [San Gabriel and Los Angeles rivers]" (August 2). When the second Anza expedition crossed the Ventura plain in 1776, Pedro Font recorded in his journal, "We saw in the plain a very large drove of antelopes which, as soon as they saw us, fled like the wind, looking like a cloud skimming along the

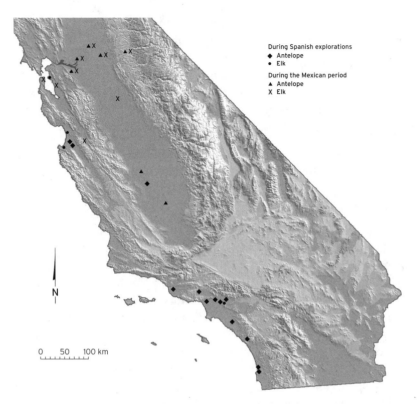

Figure 3.2. Distribution of antelope and elk during Spanish explorations and the Mexican period (Brown 2001; Web de Anza Archives; Bolton 1926, 1927, 1930a,b, 1933; Bidwell 1948; Brewer 1966; Frémont 1845, 1848; Gillis and Magliari 2003; Muir 1974; Cronise 1868).

earth" (February 23). In the northern Salinas Valley, Crespí saw "two bands of twenty-odd antelope," and near Salinas "there are a great many antelopes and bears found throughout these lands" (September 30, 1769). At San Delphina, on the Salinas plain east of Monterey, Crespí wrote that "the plain here has a vast number of large antelopes; they looked like flocks of goats when seen the other time, and various bands of six or eight of them have also been seen this time" (May 21, 1770). Pedro Fages stated that the antelope was abundant in the San Joaquin valley in 1772 (Bolton 1931).

Antelopes were abundant in the Mexican period. Zenas Leonard (1959) recalled that antelope were "remarkably plenty" when he crossed the southern San Joaquin Valley in 1833. When John Bidwell first saw the San Joaquin Valley in 1841, "The antelope could be seen in those days by the thousands. When frightened they would gather into immense herds, preparatory to taking flight" (Gillis and Magliari 2003: 87). Bidwell (1937: 28) had "hundreds of antelope in view at the Stanislaus River in the Sacramento Valley." As soon as they reached the bottom of the Stanislaus River "they saw an abundance of antelopes. . . . The next day they killed thirteen deer and antelopes, [and] jerked the meat" (Bidwell 1948: 49). Burcham (1957: 108) concluded from the 1840s expedition of Bryant (1848) and Charles Wilkes (1844, 1849), and the 1850 expedition of Lieutenant George H. Derby (Farquhar 1937), that the "antelope occurred in large herds from the San Joaquin delta north to Klamath Lake but were most abundant in the San Joaquin Valley where they formed herds up to two or three thousand animals. Antelope scattered in small groups in spring when fawns were being reared; with the approach of autumn they gathered into larger bands to pass the winter." In the San Gabriel Valley of southern California, Whipple (1856c: 8) of the Pacific Railroad survey stated that the party was "continually within reach of deer, antelope and hare." In the Central Valley east of Pacheco Pass, Muir (1974: 22) watched the "smooth-bounding antelopes."

Elk ranged through California from as far south as Los Angeles but were concentrated in riparian or swampy areas. The first European to record elk was Sebastián Vizcaíno at Monterey in 1603 (Bolton 1916). During the Portolá expedition, Crespí encountered elk for the first time in the San Gabriel Valley, possibly in the swamps of the San Gabriel River, where "tracks of very large animals are seen and droppings are found like those of cattle" (July 31, 1769; see Appendix 1 for locations of observations). Soldiers saw far off something like a mule, which they said might have been an elk (18). To the north, the expedition observed elk, appropriately, at Elkhorn Slough near Pajaro-Watsonville, where "a great many deer or large elk have also been seen . . . [with] droppings like those of a mule-like beast" (October 9, 1769). Crespí saw "as many as twenty-one beasts together, of all hues, with calves at their feet like cows. . . . Soldiers say they have seen some bands of these animals, long-eared like mules and with a short broad tail" (October 11). On his return from San Francisco, Crespí again recorded elk tracks at Elkhorn Slough (November 23). In a letter to the warden of San Fernando, Crespí stated that "at a throng of lakes 6 leagues north of Point of Pines [Elkhorn Slough] the scouts saw a great many large beasts in several bands. They must have seen over fifty of them in all . . . and they were as large as cows, with their calves at their feet, colored like deer, with a head and face like mule-

beasts, hoofs like cattle, and droppings like mules" (Brown 2001: 791).
The second Anza party in 1776 hunted elk near Berkeley.

At the Stanislaus River, Bidwell (1937: 28–29) saw elk tracks by the
thousands and later found "large herds of elk . . . grazing upon the plain."
Frémont (1845: 250) passed by "bands of elk and wild horses" at the
Rio Merced. Wilkes (1844) believed that elk was the most abundant game
animal in California. During his visit to the state in 1841, he estimated
that "an average of ca. 3000 elk and deer skins were shipped from San
Francisco each year." Bryant (1848) saw numerous herds of elk in the
Sacramento Valley east of Sutter's Fort, where they apparently preferred
marsh habitat. Burcham (1957) concluded that elk were gregarious and
were frequently found in herds of 1,000 to 2,000 animals.

William Davis (1929: 222) estimated that "from 1840–1843 there
were 2,000–3,000 elk on Elk Island in San Francisco Bay, the animals
swimming back and forth from the mainland." Elk were killed for their
hides and tallow by the rancheros, at the same time they slaughtered cattle. Likewise he learned from Captain Sutter that "the tallow derived from
the elk was an article of commerce" (222). Davis also recalled that "Don
José Ramon Estudillo . . . was fond of the sport of lassoing elk. Estudillo
remembered that elk once were seen grazing with cattle in the vicinity of
modern day Oakland [Rancho Pinole]" (34). Wilkes (1844: 183) saw an
increasing abundance of elk northward along the Sacramento River. He
observed that "at this season of the year, the rutting, they are generally
seen in pairs; but at other times the females are in large herds . . . thirty
to forty in number." When Vizetelly (1848: 10) passed by the valley of
San Juan Bautista, he saw "low hills on either hand . . . covered with dark
ridges of lofty pine trees [ghost pine, *Pinus sabiniana*] upon which herds
of elk and deer are seen now and then."

Deer was a widespread species in pre-European California, like today,
but did not attract the same attention as elk and antelope among the Franciscans, apparently because the animals were so common along the route
of the expeditions. Vizcaíno recorded deer at San Diego and Monterey
in 1602–03 (Bolton 1916). The southernmost account was at Arroyo de
los Álamos in the Sierra San Pedro Mártir of Baja California, "a delightful
spot for hunting because of the occurrence of rabbits and goats of handsome form . . . and deer" (Brown 2001: 781). Deer was again recorded
by Crespí to the west at Valle San Telmo (207). Anza saw deer in the San
Jacinto Valley on his first expedition in 1774. The most explicit accounts
were made at San Francisco, where Crespí observed "a great many
deer . . . upon the knolls." In another entry from his journal, he wrote
that "a great many deer have been seen together, while the scouts aver

that when they explored here, they saw whole bands of deer, and counted as many as fifty deer together in one" (293, 595, 599). The Spanish expeditions had deer meat at El Buchón, west of Lompoc and at Monterey. Deer skins were used in women's dress throughout California.

Accounts from the late-Mexican period reveal an abundance of large mammals in California. When the Walker party descended to the base of the Sierra Nevada to a point east of Sacramento in 1833–34, after near starvation in the mountains, they were elated with the abundance of game they found, including "deer, elk, grizzly bear and antelopes" (Leonard 1959: 83–84). They shot only wildlife, as cattle were not seen there. At Doctor Marsh's ranch in the northwest corner of the San Joaquin Valley, six miles east of Mount Diablo, Bidwell (1948: 51) remarked on the profusion of wildlife and the pristine quality of the land: "There were no other settlements in the valley; it was, apparently, still just as new as when Columbus discovered America, and roaming over it were countless thousands of wild horses, of elk, and of antelope." Near Lassen's ranch, Frémont (1848: 23) observed that "the game is very fat and abundant; upward of 80 deer, elk, and bear were killed in one morning." In a conversation with Captain Sutter, Davis (1929: 55) learned of the abundance of game in the Sacramento Valley in relation to a flood in 1839: "The winter of 1839–40 was severe, an immense quantity of rain falling, for forty days and nights." According to Sutter, "the whole country was flooded. . . . Among the stories he mentioned was one of seeing the deer, elk and other animals crowded together in large numbers on every little prominence which appeared above the waters, to protect themselves from being carried away by the flood" (55).

Perhaps the most striking account of wildlife was made by Davis (1929: 20) when he departed the newly established Sutter's Fort:

Captain Sutter gave us a parting salute of nine guns—the first ever fired at that place—which produced a most remarkable effect. As the heavy report of the guns and the echoes died away, the camp of the little party was surrounded by hundreds of Indians, who were excited and astonished at the unusual sound. A large number of deer, elk, and other animals on the plains were startled, running to and fro, stopping to listen, their heads raised, full of curiosity and wonder, seemingly attracted and fascinated to the spot, while in the interior of the adjacent wood the howls of wolves and coyotes filled the air, and immense flocks of water fowl flew wildly about over the camp. Standing on the deck of the "Isabel," I witnessed this remarkable sight, which filled me with astonishment and admiration, and made an indelible impression on my mind. The salute was the first echo of civilization in the primitive wilderness so soon to become populated, and developed into a great agricultural and commercial centre.

BUILDUP OF CATTLE

The Spanish mission system, pueblos, and presidios developed a large livestock base from relatively few animals driven to California in the first expeditions. Writings from the early Spanish explorations and by non-Spanish settlers beginning in the 1830s suggest that livestock numbers reached carrying capacities several decades after their introduction, possibly by circa 1800–1810. The relationship between grazing and the invasion of introduced species must consider a rapid increase in livestock numbers almost half a century after initial Spanish colonization.

The first Franciscan expedition of 1769 brought less than 200 cattle, and the total count of the entire mission system was 207 head by 1772 (Bolton 1927). The second Anza expedition of 1776 brought 1,000 additional cattle from Sonora, but many animals perished crossing the arid wastes of the Salton Sea trough (see Chapter 2). There were also contributions of 270 head from missions in Baja California (Bancroft 1888). Missions had many ranchos that gradually expanded out from one another until all of coastal California south of Sonoma to San Diego was devoted to stock raising (Burcham 1957: 131–33). The late eighteenth century saw a slow exponential increase in cattle numbers due to limited breeding stock. Numbers are estimated to have reached 1,700 by 1790 and greater than 6,000 by 1797 (Carmen et al. 1892). In 1800, the pueblos of San Jose, Los Angeles, and Branciforte "had a combined total of 16,500 head of cattle and horses, and about 1000 sheep . . . but consisted [of] 79,000 estimated for the entire mission system" (Bancroft 1888: 255). These values are surely below carrying capacities. Davis (1929: 408) estimated that it was "about 1800 when the missions were fairly started in the raising of live stock."

Cattle had doubtless reached large numbers by the early nineteenth century. In 1827, Jedediah Smith wrote that the herds of cattle had accumulated until they were nearly as numerous as buffalo on the plains of Missouri (Evarts 1958). Cronise (1868: 35) concluded the obvious, that cattle built up to large numbers largely because there was no demand: "From 1825 to 1834, the whole of this trade was in the hands of a few Boston merchants. A voyage to [the California] coast and back, during that time, was an enterprise of . . . two or three year. . . . there was no newspapers, telegraphs, nor stages . . . to inform customers of the ship's arrival. The [boat] crew had to travel all over the country to convey the news, which occupied considerable time." This provided incentive to keep cattle near the coast. Cronise also wrote that "between 1822 and 1832, the exports from California increased to an estimated

30,000 hides, [and] 7000 quintals of tallow" (61). Duhaut-Cilly (1929) stated that the San Diego Mission has 12,000 cattle, 19,000 sheep, 2,000 pigs, and a proportionate number of horses and mules. He found "Mission San Luis Rey to be a verdant valley, already enlivened by great herds . . . seen only as white and red spots, . . . as far as the eye could reach . . . 30,000 cattle, [and] 20,000 sheep" (226). The San Gabriel Mission had "immense herds" (248).

The English vessel *Blossom* ported at San Francisco in 1828, and Beechey (1831: 15), a member of the expedition, assessed the grazing lands on that part of the coast from a distinctly English perspective: "Almost all these [mission] establishments cultivate large portions of land, and rear cattle, the hides and tallow of which alone form a small trade." The area from Mission Dolores (south San Francisco) to Palo Alto was "wide country of meadow land, with clusters of fine free from underwood. It strongly resembled a nobleman's park: herds of cattle and horses were grazing upon the rich pasture" (44). The plain from Palo Alto to Santa Clara "continued animated with herds of cattle, horses, and sheep grazing; but the noble clusters of oak were now varied with shrubberies, which afforded a retreat to the numerous coveys of Californian partridges [quail]" (45). Duflot de Mofras (1937) estimated that the total livestock in the mission system in 1834, before the slaughter, was 424,000 cattle and 321,500 sheep and goats. The total number of mission cattle in 1834 was estimated to be 80,000 at San Luis Rey, 70,000 at San Juan Capistrano, and 50,000 at San Gabriel, with even greater numbers of sheep.

The first detailed surveys of mission livestock took place in the early 1830s in anticipation of the livestock slaughter due to secularization and were compiled by Forbes (1839) from mission records, Davis (1929) based on his experience as a San Francisco merchant, and by Cronise (1868) and Bancroft (1888) (see Table 3.1). Cronise's data was obtained from the Reverend Walter Colton, chaplain of the U.S. *Congress* and the first Protestant clergyman that resided in California in 1825. The total cattle in Forbes's account is 216,000 head. The other assessments give a consistent number of about one-half million head in the 1830s. Thornton (1848), one of many authors of gold rush books for an eastern U.S. readership, was apparently aware of mission records, and wrote that "in 1831 [just before the mission slaughter] the estimated horned cattle is 500,000, the number of sheep and goats, 321,000, and the number of horses, mules, asses, etc, 64,000." Forbes and Bancroft estimated that the highest number of cattle were in southern California, where the bulk of land grants were ceded, while data from Davis and Cronise suggest a more uniform distribution among the missions. The number of horses is

TABLE 3.1 ESTIMATES OF MISSION LIVESTOCK
IN EARLY HISTORIES
A

Mission/Land Grant	Cattle	Horses	Sheep
San Juan Bautista	7,070	401	7,017
San Carlos	2,050	470	4,400
Soledad	6,599	1,070	6,358
San Antonio	5,000	1,060	10,000
San Miguel	3,762	950	8,999
San Luis Obispo	2,000	800	1,200
Presidio Santa Barbara	7,900	300	—
La Purísima	10,500	1,000	7,000
Santa Inés	7,300	320	2,200
Santa Barbara	2,600	511	3,300
San Buenaventura	4,000	300	3,100
San Fernando	6,000	300	3,000
Pueblo Los Angeles	38,624	5,280	—
Presidio San Diego	608	625	—
San Gabriel	20,500	1,700	13,554
San Juan Capistrano	10,900	290	4,800
San Luis Rey	26,000	2,100	25,500
San Diego	6,220	1,196	17,624

SOURCE: Forbes (1839: 266).

roughly an order of magnitude less than cattle. The number of sheep is
roughly equivalent to cattle, but the grazing pressure of these smaller animals is far less.

Bancroft's data shows a spectacular decline in all livestock by 1842
as a result of the mission slaughter, as well as from extreme drought from
1840 to 1845 (documented based on tree rings) (Haston and Michaelson 1994). In 1842, the total head of cattle had declined to 31,000 and
sheep to 29,000. Davis (1929) estimated that nearly 100,000 head were
slaughtered in 1834 alone, the surviving cattle ending up being distributed among private ranchos. Bidwell (1937: 30) recalled that "Capt. Sutter has bought out the whole Russian settlement [Fort Ross] consisting
of about 2000 head of cattle, 600 horses, and 1000 sheep."

The hide and tallow trade can be estimated by the trading activity in
the California ports, as compiled by Davis's annual count of trading vessels (1929). Table 3.2 shows virtually no commerce from the beginning
of Spanish colonization, normally greater than ten trading vessels per year.
Before 1812, it was common for no vessels to appear at any California
port for an entire year. Trade was entirely controlled by the Spanish up
to the Mexican revolution of 1822. Hides had not begun to be of any

B

Mission/Rancho	Cattle	Horses	Sheep
San Luis Obispo	60,000	thousands	thousands
Sonoma	30,000	1,000	—
Santa Clara	65,000	4,000	30,000
San Juan Bautista	60,000	2,000	20,000
San Antonio	10,000	500	10,000
San Miguel	35,000	1,000	20,000
Soledad	25,000	1,000	10,000
La Purisima Conception	20,000	1,000	15,000
Santa Ynez	20,000	1,500	10,000
San Fernando	50,000	1,500	20,000
San Gabriel	80,000	3,000	30,000
San Luis Rey	60,000	1,000	20,000
San Juan Capistrano	20,000	1,000	10,000
San Diego	15,000	1,000	20,000
Santa Barbara	20,000	1,000	20,000
San Buena Ventura	25,000	1,500	10,000

SOURCE: Davis (1929: 389–95); based on his merchant and trading business.

C

Mission	Cattle	Horses	Sheep
San Francisco	76,000	2950	79,000
Santa Clara	74,280	6,100	82,540
San José	62,000	2,340	62,000
San Juan Bautista (1820)	43,870	6,230	69,500
San Carlos	87,600	1,800	7,500
Soledad	36,000	+	70,000
San Antonio (1822)	52,800	4,800	48,000
San Miguel (1821)	91,000	4,100	47,000
San Luis Obispo	87,000	5,500	72,000

SOURCE: Cronise (1868). Cronise states that "all the other missions [not on his list] were equally rich in livestock."
+ = greater than all the other missions

value, as the American traders did not commence to buy them until about 1820" (Cronise 1868: 38–39). In 1822, an English firm in Lima, Peru, established a branch of their house at Monterey, which was the first mercantile house opened on the coast. The annual exports for several years had averaged "30,000 hides, [and] 7000 quintals of tallow" (40). In the Mexican period, foreign trade increased to twenty to thirty vessels per year beginning in 1825 and continuing to the gold rush and California statehood in 1848–49. The annual hide and tallow trade was up to 30,000

D

	1834			1842		
Mission	Horned Cattle	Horses	Sheep, Goats, Pigs	Horned cattle	Horses	Sheep, Goats, Pigs
San Diego	12,000	1,800	17,000	20	100	20
San Luis Rey	80,000	10,000	100,000	2,800	400	4,000
San Juan Capistrano	70,000	1,900	10,000	500	150	200
San Gabriel	105,000	20,000	40,000	700	500	3,500
San Fernando	14,000	5,000	7,000	1,500	400	2,000
San Buenaventura	4,000	1,000	6,000	200	40	400
Santa Barbara	5,000	1,200	5,000	1,800	180	400
San Ines	14,000	1,200	12,000	10,000	500	4,000
Purísima	15,000	2,000	4,000	800	300	3,500
San Luis Obispo	9,000	4,000	7,000	300	200	800
San Miguel	4,000	2,500	10,000	40	50	500
San Antonio	12,000	2,000	14,000	800	500	2,000
Soledad	6,000	1,200	7,000	—	—	—
Carmelo	3,000	700	7,000	—	—	—
San Juan Bautista	9,000	1,200	9,000	—	—	—
Santa Cruz	8,000	800	10,000	—	—	—
Santa Clara	13,000	1,200	5,000	1,500	250	3,000
San José	2,400	1,100	19,000	8,000	200	7,000
San Francisco	5,000	1,600	4,000	60	50	200
San Rafael	3,000	500	4,500	—	—	—
Solano	3,000	700	400	—	—	—

SOURCE: Bancroft (1888: 339); includes data before and after the mission slaughter of the early 1930s.

hides and 700,000 pounds by the 1820s (Forbes 1839). Duflot de Mofras (1937) estimated an average export of 150,000 hides and 5 million pounds of tallow per year before the mission slaughter.

The time when cattle numbers reached carrying capacity can be estimated based on an unwritten principle in the operation of the ranchos. According to Davis (1929: 138), there was a rule among the rancheros that "an increase of one to every five head on the hacienda was the basis of the yearly estimate among the haciendados." Hence, an estimate of cattle growth can be obtained by assuming there were 1,000 cattle in all the missions by 1776. Annual growth can be estimated by multiplying this number by 1.2 in an annual time series, assuming no culling or mortality from catastrophic drought. (In the early years high mortality is unlikely, since grazing pressure from relatively few animals would preclude overgrazing, even in years with abnormally low precipitation and pasture productivity.) At this rate, the number of cattle would reach 2,074

TABLE 3.2 RECORD OF SHIPS IN CALIFORNIA PORTS,
AND DROUGHTS

Year	Arrivals	Drought	Year	Arrivals	Drought
1774	2		1812	0	
1775	0		1813	1	
1776	3		1814	2	
1777	0		1815	1	
1778	1		1816	8	
1779	3		1817	3	
1780	0		1818	2	
1781	0		1819	4	
1782	0		1820	0	
1783	2		1821	3	X
1784	1		1822	3	
1785	0		1823	7	
1786	4		1824	7	
1787	5		1825	33	
1788	2		1826	25	
1789	1		1827	33	
1790	0		1828	33	X
1791	5		1829	28	X
1792	0		1830	21	X
1793	0		1831	21	
1794	17		1832	24	
1795	7		1833	29	
1796	4		1834	31	
1797	5		1835	30	
1798	5		1836	16	
1799	3		1837	27	
1800	3		1838	21	
1801	1		1839	16	
1802	0		1840	19	
1803	6		1841	30	X
1804	4		1842	28	
1805	0		1843	4	X
1806	4		1844	10	X
1807	2		1845	16	
1808	0		1846	9	
1809	0	X	1847	22	
1810	0	X	1848	15	
1811	0				

SOURCE: Davis (1929: 397).

by 1780, 12,842 by 1790, and 79,512 by 1800, which is consistent with
estimates for the entire mission by Bancroft (1888: 255). At this rate, the
reported near half-million population by the 1830s would be attained
by around 1810. Hence, the low number of trading vessels before 1825
was in part due to a limited livestock resource. The total livestock in the

first decades of Spanish settlement may be somewhat overestimated because cattle were surely used for local subsistence.

Davis (1929: 408–9) made a second estimate of total cattle used for
export from California. Using a list of arrivals of vessels from Forbes
(1839), Davis estimated that the exports of hides and tallow in 1828–47
of thirty-three vessels was 1,068,000 hides. He made another estimate
from 1800, when "the missions were fairly started in the raising of live
stock, down to their impoverishment, but the enterprise was continued
by the *haciendados*" into the rancho period and "the vessels were [more]
numerous" than in the "primitive days." From 1800 to 1847, Davis estimated that there were 600 vessels that ported in California, of which
only 200 exported hides at a rate of 1,000 hides per vessel, giving a total of 9,400,000 hides for the forty-seven-year period. He believed that
it was "perfectly accurate to estimate the exportation of hides for forty-
seven years at 5,000,000 . . . , a deduction of nearly one-half from the
first calculation." He concluded that "cattle which were killed yearly as
an income to the department, kept those animals from over-running the
immense territory of over seven hundred miles of coast." He also concluded that "his first figures are the more correct of the two estimates."

Although the postsecularization growth of cattle was perhaps delayed
by long-term drought in the early 1840s (Haston and Michaelson 1994),
populations had built up to the levels of the premission collapse by the
gold rush. According to Davis (1929: 395), "After their downfall, the Missions became destitute and the lands were granted by the authorities of
the department [of California] to the citizens of the young country. Those
men became stock-raisers, and through the experience gained by their
observations of management by the fathers, they succeeded in reinstat-
ing the lost riches of California, which were taken from the missionaries; and they even accumulated more than twofold the former wealth
of the primitive land." Davis (1929: 31–33) estimated that in 1838–39 the
ranchos between Santa Rosa and San Jose had 150,000 head of cattle,
horses, and other livestock. Bryant (1848: 379), who accompanied Frémont in the conquest of California in 1846, saw abundant livestock across
the state. For example he saw the Santa Barbara plain from the Santa
Ynez Mountains with a "spyglass" with "herds of cattle quietly grazing
upon the green herbage that carpets its gentle undulations." Mission San
Fernando had "large herds of sheep and cattle were grazing upon the
plain in sight of the mission." The cattle trade was still in the tradition
of the Mexican period. Bryant (1848: 282) wrote that "the most lucra-
tive business in California is large cattle; their hides and tallow affording an active commerce with foreign vessels on this coast."

The largest herd was owned by General Vallejo. According to Davis (1929: 138), "The Nacional Rancho at Soscol [Suscol] had about 14,000 head of cattle, and a large number of horses. These cattle used to stray to a long distance along the margin of Suisun Bay. This rancho was under the control of General Vallejo from the time he found the military headquarters at Sonoma. Including Petaluma, Temblec, and another rancho, the total of cattle on all these estates reached the enormous number of fifty thousand head. This made the General the largest cattle owner in early California."

SELF-SUFFICIENT GRAZING ECONOMY

The open-range livestock grazing economy of the nineteenth century has been detailed in Cleland (1964). He represents the livestock era as one of "prosperity" in the mission and rancho periods, "a carefree life" and "a chronic dearth of money" (30–31). Leonard's account of California in 1833–34 reveals an economy almost entirely devoted to grazing: "The country was thinly populated with Spaniards and Indians with the poor Spanish population devoted to ranching and wealthier Spaniards conducting local farming, the labor being undertaken by Indians from the nearby mission. The primary crops were wheat, corn, and beans, plus vineyards for the manufacture of wine. Food was locally consumed" (1959: 101–2). Bidwell (1937: 38) noted that crops were seldom surrounded by fences and had to be guarded from the cattle and horses by the Indians, who were stationed in their huts near the fields.

After secularization, each ranchero was self-sufficient except for a few goods obtained from rare visits by trading vessels. According to Cleland (1964), each rancho was operated like a large, manorial estate that supported a population of several hundred people. The estate maintained household manufactures, produced its own grain, vegetables, and other foodstuffs, and grazed thousands of head of cattle, sheep, horses, and donkeys. Cattle was grazed in open range. Most ranchos were located near rivers and streams with moist ground in the summer drought.

The distribution of land grants (Figure 3.3) is consistent with the accounts of the best pasture during Spanish explorations (California Ranchos by County 2006; Cronise 1868). Virtually all ranchos were found along the coast from San Francisco to San Diego, avoiding the barrens of the Central Valley. Inland ranchos were granted mostly along Sacramento River in Butte, Colusa, Shasta, Tehama, Solano, and Yolo counties. The largest rancho in the San Joaquin Valley was at El Tejon, where early accounts document better pasture at higher elevations, and one

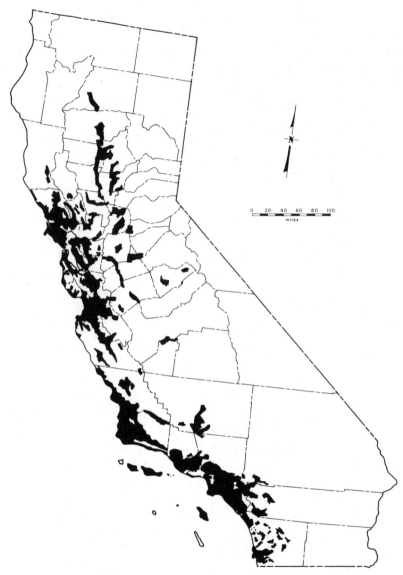

Figure 3.3. Mexican land grants (from Beck and Haase 1974).

along the San Joaquin River owned by Frémont. Coast Range ranchos extended as far east as Las Banos and Pecheco Pass (Cronise 1868).

The European presence in California was trivial in the Spanish period and remained so into the Mexican period, even after the Law of Colonization. Cronise (1868: 38) wrote, "In 1820 perhaps half a dozen foreign settlers in the whole country, . . . and only 8 ranchos belong[ed] to Mexican settlers, between San Francisco and Los Angeles. Monterey contained but six houses, besides the presidio; San Jose contained about 20. . . . No one, except the Governor and the missionaries, had wooden floors or doors to their houses, nor chairs nor tables, it did not require much lumber to supply the demand. . . . A few cabbages and other vegetables were cultivated on some of the missions as luxuries." There were few non-Spanish Europeans. According to Cronise (1868: 38–39), the first was John Gilroy, who landed in Monterey in 1814. The next permanent non-Spanish settler "was a pirate from Buienos Ayres, which arrived in 1819, captured the Fort, destroyed the guns, plundered the inhabitants and burnt the town." Cronise (1868: 46) estimated that perhaps five hundred foreigners lived west of the Sierra Nevada in 1830. Joseph Revere (1849a: 96) wrote that the pueblos comprised 4,500 *"gente de razon,* and 50,000 Indians" and that "22,000 were more or less Christianized, the majority of the Indian population being composed of gentiles."

Only small towns developed around the harbors. Describing San Francisco in 1840, Bidwell (1948: 72–73) wrote that "with the exception of the Precidio and the Aduana [custom house], all the buildings could be counted on the fingers and thumbs of one's hands." He estimated that Monterey had "possibly 200 people, [and] Santa Barbara 100 persons." In March of 1845, "Los Angeles had two hundred and fifty people and San Diego about 100." Bidwell (1948: 71–73) also listed the principle harbors: "San Francisco was the principal harbor; the next was Monterey. There was an anchorage off San Luis Obispo; the next was Santa Barbara, the next San Buenaventura, then San Pedro, and lastly San Diego" (cf. Gillis and Magliari 2003: 83). The hides were collected and brought to San Diego and there salted, staked out to dry, and folded so that they would lie compactly in the ship, and then they were shipped to Boston. The lack of good harbors near Point Conception may explain the low numbers of cattle in high-quality pastureland in that region.

Lumber was obtained by ax by the Indians, but there was virtually no demand, as "nobody except the Governor or missionaries had wooden floors or doors to their houses, nor chairs, nor tables. . . . The missionaries owned the whole country, and controlled all its inhabitants. The Indians did all the work required, as blacksmiths, carpenters, and weavers" (Cronise 1868: 40).

DISTRIBUTION OF CATTLE

The writings of Leonard, Bidwell, and Frémont establish where livestock was grazed and what lands were still dominated by native fauna during the Mexican period. Cattle were seen primarily in coastal mission, pueblo and presidio lands where livestock could be grazed on secure year-round pasture. Cattle were not seen in the Central Valley except at Sutter's Fort and other ranchos near the Sacramento River (compare Figures 3.1 and 3.3).

When the Walker party traversed the Sierra Nevada and the Central Valley east of present-day Sacramento to the California coast in 1833–34, Leonard (1959: 96) saw cattle at Mission San Juan Bautista and San Francisco, but none from the Sierra Nevada to the coast range. When the party overwintered along the coast north of Monterey, they had to provide themselves with new shoes, and "hunters were dispatched to scour these hills for the purpose of getting cattle hides to make moccasins. . . . They brought the tongues in order to show the number each man had killed" (95). Bidwell (1948: 54) stated that the Livermore Ranch "was the frontier ranch [at the time and] defined the interior limit of cattle." Bidwell (1948: 57) observed that until Sutter established his ranch in what is now Sacramento, "no other civilized settlements had been attempted anywhere east of the Coast Range; before Sutter . . . the Indians had reigned supreme." Indeed, when Bidwell first arrived into the Central Valley near Sacramento, he saw cattle from Sutter's Fort to the Sacramento River delta, where "beef . . . was abundant and of the best quality" (1937: 28).

Frémont (1848) gave a similar picture of livestock a decade later. When he entered the Central Valley east of Sacramento, he saw "fresh green grass for eight or ten miles into the valley, [the] cattle feeding upon it" (30). At Butte Creek, he saw "cattle with a rancheria nearby," and at Deer Creek on the Sacramento River near Mount Lassen, he visited "the rancho of 'Mr. Lassen, a native of Germany' who established a rancho . . . , which he has stocked"(23). He saw "3000 head of cattle on a farm" at Yuba City (17). He wrote before the onset of the rainy season that " the country looked parched and dry, the grass eaten down by the cattle, which were quite fat and fine beef" (30). Approaching the northern shores of San Francisco Bay from Sonoma, he saw scattered "herds of cattle and bands of horses" (30). Thereafter, he saw cattle along coast between San Francisco and Monterey, where he saw the grass was eaten down by large herds. In both of his expeditions he never saw cattle in the Central Valley south of Sutter's Fort (Frémont 1845, 1848). In southern California, cattle were encountered near the coast and inland to the foothills of the mountains. Land grants suggest cattle were grazed in mountain meadows at Lake Henshaw and at Cuyamaca. When he arrived in Los Angeles, Frémont (1848) saw abundant cattle across the plains.

"WILDSTOCK" AND *MATANZA*

Grazing was done by a vast population of semiwild cows and other livestock that were left to fend in the wilderness. Coastal California was treated like a vast barnyard, except there were no barns (Cronise 1868). Bidwell's first impression of the coastal ranchos was that "Spanish cattle were semi-wild and skittish, practically indistinguishable in tameness from the native wildlife, and often hiding in the forests of the Coast Ranges to escape the depredations by the rancheros and native Americans" (1937: 38). At Livermore Ranch, Bidwell (1948: 54) stated that it "was the frontier ranch, and more exposed than any other to the ravages of the Horse-thief Indians of the Sierra Nevada." He added that the wild cattle at Livermore "were more dangerous . . . than grizzly bears" (54). The Walker party also saw abundant semiwild cattle at their camp near San Juan: "The wild cattle are very timorous keeping hid pretty much all day and feed at night. They are much wilder than deer, elk" (Leonard 1959: 96). Leonard speculated that "these cattle incline much to rough and hilly parts of the country, owing, it is supposed, to the Spaniards and Indians hunting them when found in the plains" (96). In his summary of early nineteenth-century grazing, Bancroft (1888: 261) wrote that "cattle . . . roamed in a half-wild state upon the plains, and wiry-limbed, swift horses of large size, of large size and longer neck than the Mexican prototype, were subordinated at times by nomadic ranchers."

The rancho lands were vaguely distinct from the pristine wildlands of California. The most detailed account of a rancho from the period was made by Leonard (1959: 103):

> These people have no fences round their cleared or cultivated land, although they raise an immense amount of stock, such as horses, mules and horned cattle—all of which range at large over these extensive pastures all seasons of the year, many being in a manner totally wild, so much so that, when they wish to milk a cow, they mount one of their horses, noose her, fasten the cord to a tree, and then tie her feet, when she is forced to be quiet. During our whole stay in this country I have never seen anything like a stable or a barn, as the shelter for the dumb brutes—nor did I ever see anyone feeding an animal, unless it was a favorite cow or horse that was sick. This, however, is not at all singular as any number of animals could subsist, and be in good order all seasons of the year, on these plains, as in many spots the grass is green the whole year round. The months of August, September, and October are the least enticing to animals, as it is the warmest and driest season of the year. As soon as August sets in, the beasts inhabiting the dry prairies and hills repair to the low wet ground, where they can get enough to subsist upon until the dry season passes away. . . . During the [winter]

wet weather the animals grow fat, and the inhabitants employ the principal part of this time in catching and domesticating them.

Just before the gold rush, Vizetelly (1848: 9) was astonished that "cattle are suffered to run loose over the neighboring pastures" between Monterey and San Juan Bautista. Likewise, Revere (1849c: 77) also saw "extensive herds of wild cattle, [that] . . . betake themselves to the woods or ravines among the hills at the approach of a stranger." Cronise (1868: 353) explained the lack of infrastructure on the ranchos: "The climate of California is so mild in winter, which is in fact is the season of verdure, that very little feed and shelter is provided. Barns are almost unknown." Thornton (1848: 16) concluded that "the only trouble the people of California have in raising their cattle is to brand them annually with the peculiar mark of each owner." Only a few cows were kept near the house for milking. Butter and cheese were rarely made (Bidwell 1937: 39; Bancroft 1888: 347; Davis 1929: 35). At Jolon Ranch in the interior Santa Lucia Mountains, Brewer (1966: 95) noted that "the cattle here over the hills are very wild; they will run if they see a man at a distance of forty or fifty rods off. The *rancheros* consider it desirable that their cattle be wild—they are less liable to be stolen or be caught by wild animals." Cronise (1868) had the advantage of observing livestock in the transition between the Mexican period and the domestic livestock grazing that developed after the gold rush. He admired the positive qualities of Spanish cattle: "The wild cattle of the Mexicans are poor, long-horned and lank—but they cross well with imported stock, carrying the fine points of the latter and the endurance of the former. Their flesh is tough, and their milk scant" (370).

According to Charles Wilkes (1844), the rearing of cattle required minimum labor, mostly *vaqueros* for herding and branding. Cattle were allowed to roam except for slaughters for hides and tallow. John Bidwell (1937: 39) had a similar perspective: "Of all places in the world, it appears to me, that none can be better adapted to the raising of cattle than California. The cattle here are very large, and a person who has not a thousand is scarcely noticed as regards stock." Bidwell also observed that "it is a proverb here (and a pretty true one) that a Spaniard will not do any thing which he cannot do on horseback—he does not work perhaps on an average, one month in the year—he labors about a week, when he sows his wheat, and another week, when he harvests it, the rest of the time is spent riding about" (38). Leonard (1959) provided considerable detail on rodeos, use of ropes, and horse riding by the Mexicans along the coast.

Davis (1929: 241), in his frequent travels across the state in the 1840s, observed the exchange of cattle between ranchos through informal agreements:

> When cattle, old or young, were transferred from one rancho to another . . . it was generally done in the spring of the year, the new feed being then plentiful, and they were easier aquerenciado, or domesticated, in their new pasture than at any other season. A band of cattle taken to another rancho, would be placed under the charge of *vaqueros*, and watched and herded at first very carefully. Becoming accustomed to the new place, and less restless and uneasy, they were allowed more liberty of range, and at night were corralled. After some weeks, they were habituated to their new surrounding, and turned in with the other cattle, becoming a part of the general band belonging to the rancho.

Davis also pointed out a rule that "all *orejanos* (calves without ear-mark or brand) not following the cow, were considered as belong to the rancho on which they were found" (40).

Since cattle lived essentially as wild megafauna along the coast, it is perhaps remarkable that the herds never expanded into the Central Valley. The rancheros doubtless chased down strays, but more likely the absence of cattle may be due to predations by Native Californians, who developed a taste for the flesh of domestic livestock. Cattle would also have had difficulty swimming across or trodding the mudflats and rivers.

The harvest of cattle was an "emergency process." According to Leonard (1959: 104), whenever a trading vessel anchored on some portion of the California coast "the news is spread over the whole country like wildfire. The owners of cattle, who are of the wealthier class, collect together all the poorer Spaniards and Indians for the purpose of catching and butchering the cattle, in order to get their hides. This is the commencement of their sporting season. . . . After they strip off the hides and take out the tallow, and sometimes the choicest part of the meat, the remainder of the carcass is left on the ground to be devoured by the wolves, or for the wild beasts of the forest to feed on." Davis (1929: 256) learned that "in the Matanza, the killing season is summer when cattle are fattest, each bullock producing an average of ca. 25 lbs. In winter, when cattle were killed for home consumption and for use of the vessels, the tallow would average about 6 pounds to the bullock." The spring was the dullest season of the year, as the cattle then became quite poor, and not many were killed. Cattle were killed for the use of the ranchos in winter (184).

Bancroft (1888: 340) provided detail on the *matanza* (*nuquear*), or slaughter:

When a hacendado wished to *nuquear* or slaughter his cattle, he sent six men on horseback, who rode at full speed over the fields, armed with knives. Passing near an animal, one gave it a blow with the knife in the nerve of the nape of the neck, and it fell dead. These *nuqueadores* passed on, and were followed by a flock of hungry vultures, by dozens of *peladores,* who took of the hides. Next came the *tasajeros,* who cut up the meat into *tasajo* and *pulpa;* and the funeral procession was closed by a swarm of Indian women, who rapidly gathered the tallow in leather hampers. The fat was afterward tried out in large iron or copper kettles, and after cooling somewhat was put in skin botas, containing on an average 20 *arrobas,* or 500 pounds. . . . a field after the *nuqueo* looked like Waterloo after the charge of the old guard.

There were still rules in the *matanza,* which were also the basis of estimating the total head of livestock for each rancho (Table 3.1). Cleland (1964: 54) stated that "cattle were not slaughtered until they were about 4 years or older and the average calf crop was ca. 60%. Hence multiplying the number of branded calves by four would give an estimate of the total herd." The total slaughter also depended on the season. The size of the herd was roughly estimated at three to four times the number of branded calves (Cleland 1964: 54).

CATTLE EXPANSION TO THE INTERIOR
WITH THE GOLD RUSH

When the United States took possession of California in 1848, the Treaty of Guadalupe Hidalgo recognized the rights of Californians to their lands under the Mexican titles obtained from the Spanish and Mexican governments at Monterey (Davis 1929: 55; Cronise 1868; Cleland 1964). However, subsequent legislation of Congress required proof of their titles before a board of land commissioners created by the Land Act of 1851. The litigation and expense made for a new and perplexing experience for rancheros. In an environment of unscrupulous lawyers, rancheros sunk into debt and lost possession of their land. Of the 813 grants, the board of commissioners, which had the mandate to review records, approved only 553 (California State Archives). Other lands were lost to squatters, and children of wealthy families grew up in poverty.

From the gold rush through the 1850s, there was a cattle boom for local beef consumption, especially in the placers east of Sacramento, rather than for hides and tallow that could be exported from local ports (Cleland 1964). The number of stock increased rapidly and led to the expansion of livestock grazing into the Central Valley, nearly a century after the introduction of livestock grazing along the coast. For the first time, livestock ranchers faced the prospect of poor pasture in the interior barrens and soon discovered that they had to drive livestock into the mountains to find forage. Cattle drives from southern California to gold rush country followed superior forage near the coast, passing through San Buenaventura, Santa Barbara, San Luis Obispo, Santa Clara, San Francisco, and Sacramento. According to Cleland (1964: 104), "The cattle, which necessarily lived off the country, usually started north when the grass had reached maturity after the early winter rains. Since they seldom traveled more than ten or fifteen miles a day, a herd was usually a full month on the trail. Some southern ranchers leased grazing rights in the vicinity of San Jose, Sacramento, or San Francisco Bay, where the stock fattened after the long drive." Only in the spring growing season did southern California cattlemen drive livestock through Gorman and the Central Valley to the mines. By 1860, cattle demand generated by the gold rush had collapsed, and the trade in cattle between the Middle West and California had also declined. Cattle then perished in great numbers from drought from 1862 to 1864 (109). As stated by Cronise (1868: 370), "The cattle market was gutted by oversupply. A convention of stock raisers in 1860 said there were three million cattle in the state, far beyond the wants of consumption." Cronise estimated that cattle in California had declined to 600,000 by the end of the drought.

Cronise accurately predicted the future of livestock grazing: "As the Spanish grants became subdivided, the wild ranges grew smaller. As farms became more numerous they will be able to obtain legislation compelling the herdsmen to keep their stock for trespassing" (370). Stock raising became more and more restricted to the Coast Range and the Sierra Nevada foothills. The Americans put in about as much labor into cattle raising as their Mexican predecessors. According to Cronise, "The Tulare County tule swamps are excellent country for cattle feeding on the roots of these plants and on fresh water mussels. Owing to the heat of the climate in the summer, remoteness from market, etc., dairying is not extensively carried on—the most of the cattle raised being intended for the shambles [meat]. . . . Cattle thrive in this region the year round without housing or fodder, being rarely ever pinched by hunger or suffering from cold" (329).

Indeed, livestock grazing in the 1860s in interior southern California, distant from the gold rush, was a throwback to the rancho days decades before. Cronise wrote that the rancheros of the Temecula Valley lived "in the same style they did before the country became a State. One of these native Ranchers, living near Temecula, who owns several leagues of these plains, and has nearly five thousand head of cattle . . . never saves a drop of milk, or makes a point of butter—these being luxuries in little use here" (98–99). The main products were still hides and tallow at this late date. Much land was owned by Mexicans.

WILD HORSES

During the rancho period, horses were the primary mode of transportation and were vital in the various seasonal activities of livestock grazing, mainly for tending livestock (Davis 1929). Most rancheros maintained a ratio of about one horse to ten cattle. However, horses on the ranchos gradually dispersed and became feral in the Central Valley, as a result of horse stealing both by Spaniards and Indians, with the encouragement of Spanish law. The naturalization of horses became another source of grazing pressure in California pastures of interior California.

The theft of horses was not recognized as a crime because of the cheapness of the animals. If a better quality horse than your own was found, then it was appropriate to take it (Davis 1929). The dispersal of horses reflected a cultural conflict. What was owned by ranchers was wildlife to the Indians. Horse stealing was the basis for the place name "Horsethief Canyon" on topographic maps throughout the state.

It is unclear when horses first entered the Central Valley because their numbers in the ranchos may not have reached carrying capacities until circa 1810. As a result, horses may not have been an Indian food resource as late as four decades into the mission period. Nonmissionized Indians in the interior may have continued a traditional hunting and gathering economy exclusively, beyond the reach of Mexican control to the end of the eighteenth century. The earliest account of horses in the interior was during the 1805 Zalvidea expedition, when he recorded in his journal that he spotted a wild horse near the Cuyama River (Phillips 1993). Horses may have begun naturalizing in the Central Valley beginning at this time, but this is not documented in Spanish records. In 1815, Sergeant Juan Ortega with thirty men encountered two mounted Indians driving four horses (Phillips 1993). Beginning in the 1820s, the Spanish conducted assaults on Indian villages of the interior and frequently captured horses (Phillips 1993).

It is also unclear when Indians learned to ride horses. Ortega saw mounted Indians in 1815. In the 1830s, members of a Mexican expedition saw thirty Indians chasing deer on horseback in the northern San Joaquin Valley (Phillips 1993). In the early 1840s, John Bidwell (1948: 50) observed an Indian on horseback without a saddle near Mount Diablo. In the 1830s and 1840s, non-Hispanic explorers saw "droves of wild horses" in the interior beyond Mexican control. Leonard (1959: 86) was the first American to note the presence of wild horses, when the Walker party crossed the Central Valley prairies near Sacramento in 1833–34, noting "many places swarming with wild horses, some of which are quite docile, particularly the males, upon seeing our horses." He also recalled that Indians brought horses to their camp for trade. Captain Sutter saw vast droves of wild horses in the San Joaquin and Tulare valleys, bred from those stolen by Indians from the missions (Bancroft 1888). Likewise, Bidwell (1966: 15) recollected "countless thousands of wild horses, of elk, and of antelope roaming over the northern San Joaquin Valley." He estimated herds were in the thousands and sometimes "twenty miles long" on the west side of the valley (Bidwell 1904; Gillis and Magliari 2003). In his first expedition in 1844, Frémont (1845: 250–51) saw "bands of elk and wild horses" along the Rio Merced and "droves of wild horses" along the San Joaquin River. In his second expedition in 1846 Frémont (1848: 17) recorded "band of wild horses at the To-wal-um-ne [sic] River."

Davis (1929: 35), based on his visits to the ranchos, surmised that horses gradually filtered into the San Joaquin Valley, "which at the time was entirely unsettled." He wrote: "At times, a few mares, and perhaps a young stallion, would stray away from a rancho and get out of reach until in the course of time there were collected in that valley immense herds, thousands and tens of thousands of horses, entirely wild and untamed, living and breeding by themselves, finding there plenty of good feed to sustain them" (36). Wild horses were also reported in much of the Central Valley by Forbes (1839), Wilkes (1844), and Lieutenant George Derby of the topographic survey (Farquhar 1937).

Depending on one's point of view, the expansion of horses into the Central Valley was facilitated by horse stealing or, alternatively, by resource acquisition from the ranchos by the Indians. In the foothills of the Sierras east of Sacramento, "Indian hosts arrived to a camp of the Walker expedition with horses for the purpose of trading which were marked with a Spanish brand" (Leonard 1959: 87). The party concluded that the Indians' diet, in addition to the staple of acorns, was dominated by horsemeat. During his first expedition at the Kern River, Frémont (1845: 251) "cautioned Indians at camp with them for the night, not to steal the horses."

The hunting of horses also occurred in southern California. Davis (1929: 172) related that at Rancho Jurupa near present-day Riverside "they all noticed from the house a body of Indians in the distance, who were collecting horses they had stolen from the Mission San Gabriel and the Rancho Santa Ana. . . . As Bandini had but few men with him at the time, and the Indians were in large numbers, he did not deem it prudent to attack, and attempt the rescue of the animals. He therefore permitted them to move off to their retreats without any pursuit."

Bidwell (1948: 48) also found that "Horse Thief Indians [lived] chiefly on horse flesh; they had been in the habit of raiding the ranches even to the very coast, driving away horses by the hundreds into the mountains to eat." He recalled that Livermore Ranch, in the interior Coast Range, "was more exposed than any other to the ravages of the Horse-thief Indians of the Sierra Nevada" (54). Davis (1929) suggested that horse stealing increased late in the Mexican period. He recalled in 1839 that "at the time of the collapse of mission system the Sacramento Valley had no inhabitants but Indians, many of whom were Mission Indians who had left the impoverished Missions, returning to their former uncivilized life, making occasional visits to the different ranchos to steal horses" (19). He suggested that broken-down horses were also given to the Indians: "When the horses became disabled, or too poor for use, they were generally given away to the poorer people of the country, or to Indians who could make them useful" (40).

The facilitation of horse expansion in the interior is explained in the following anecdote by Davis:

> In the morning, a ranchero would discover that he was without horses for the use of the ranch. He would then borrow some horses from the neighbor, and ten or twelve men would collect together and go in pursuit of the raiders. They were nearly always successful in overtaking the thieves and recovering their horses, though oftentimes not without a fierce fight with the Indians, . . . who were armed with bows and arrows, and the Californians with horse carbines. At these combats the Indians frequently lost some of their number, and often as many as eight or ten were killed. Once in a while, but very seldom, the Indians were successful in eluding pursuit and got safely away with the horses. (63)

Bidwell (1948: 48) first encountered evidence of horse butchering and consumption in his first descent of the Sierra Nevada in 1839:

> We came to a place where there was a great quantity of horse bones. And we did not know what it meant; we thought that an army must have perished there. They were of course horses that the Indians had driven in there and slaughtered. A few nights later, fearing depredations, we concluded to stand guard—all but one man, who would not. So we let his two horses roam where they pleased. In the morning they could not be found. A few miles away we came to a village; the Indians had fled, but we found the horses killed and some of the meat roasting on a fire. (cf. Bidwell 1937: 27)

At another location he saw "the Indians . . . cutting the meat" of their abandoned horses that could not negotiate a steep slope (Bidwell 1948: 47).

Native American horse butchering even occurred in the Coast Ranges within eyeshot of the ranches. For example, Leonard (1959: 116) found that his six stolen horses were butchered by Native Americans for meat near the Mission San Juan: "They went into a thicket, where they found a large portion of their horses well butchered . . . the Indians having killed some of the horses, were engaged in drying the meat."

The rancheros frequently retrieved the horses. Davis (1929: 35) recorded in his narrative that "frequently during the summer, young men, the sons of ranchers, would go in companies of eight or ten or twelve to the valley on their best and fleetest steeds, to capture a number of these wild horses and bring them to the ranchos. When fifty or sixty of the wild horses were thus captured, they were taken to the ranchos, corralled at night and herded in the daytime, until they became sufficiently subdued to the introduced among the horses of the ranch."

The rancheros also retrived horses with other motivations. According to Davis,

> Some of the wild horses were rounded up as part of a conspiracy to avoid the military draft. The California parents had dread of their sons being drafted into the army, and the young men themselves had no liking for it. Some of the more wealthy rancheros had pre-arranged and reliable communications with their relations or friends living in Sonoma, who have them information whenever a squad of soldiers was about to be sent out to gather up recruits, and of the direction the squad would take. At such times young men would be sent off for a month of two from the rancho, either hunting game, or to the great San Joaquin valley to lasso some of the numerous wild horses there. (161–62)

When it was perceived that horses were degrading pasture or reproducing in excessive numbers, at the expense of cattle, the rancheros would conduct horse slaughters by running them over cliffs or steep slopes, like lemmings, even on the ranchos. Bancroft (1888: 346–47) found that by 1821–24 wild horses had become so "numerous that they would eat up the grass and spoil the pasture for the tame horses near the towns. The government resolved to hold a general slaughter in which the rancheros had large bands of breeding mares. In seasons of drought they would destroy large numbers of mares, and perhaps some of the horses, driving them over a precipice to get rid of them, and thus save feed for the cattle." Mission Santa Clara had another method. Davis (1929: 56) related that in an "exceedingly dry season" the head priest "ordered the destruction of several thousand head of horses and mares belonging to the mission, which was accomplished by drowning them in the Guadalupe river." Vizetelly (1848: 105) described another method that took place after California statehood: "As lately as ten years ago, it was customary to corral large numbers of wild and half wild mares and slaughter them with a lance, merely to check the natural increase in the equine race." Revere (1849c: 102) found that "in the plains of the Tulares corrals exist, formed by glens in the mountains . . . they would surround a large *caballada* of wild horses, . . . the Indians were always on hand, not to get horses to ride, but to eat."

It is uncertain whether horse was preferred meat, or whether horses were simply easier to raid from the ranchos than cattle. Several accounts from the 1830s and 1840s make clear that Indians would also eat beef. In the Mojave Desert, Frémont's party "killed three cattle, and the Indians had a great feast" (Frémont 1845: 259). During Frémont's second expedition near Butte Creek, in sight of Sutter's cattle and an Indian *ranchería*, Frémont (1848: 23) wrote that "some of the Indians gladly run races for the meat and offers of a fat cow which had been presented to us." However, semiwild cattle were dangerous and not easily herded to the interior. It would have also been difficult to guide cattle across the rivers and tule swamps of the Central Valley. Wilkes (1844: 174) wrote that "Indians stole only horses because they would not be able to avoid pursuit by the rancheros . . . if they took cattle." Even as late as the Pacific Railroad survey of 1853–54, Whipple (1856a:134) stated that at San Bernardino "the settlers deem it necessary to watch their herds with as much vigilance as if they were upon the prairies [Great Plains], in the midst of Indians."

Phillips (1993) argues that horse raiding of the rancheros was more than simple resource extraction. Instead, he views harvest of horses as an evolved response that lead to the expansion of economic freedom of Indians by providing them with a trade commodity that undermined the Spanish-Mexican system. After secularization, the coastal Indians went to the interior with Spanish culture, language, trading practices, and agriculture. The fundamental question is why Indians would raid domestic horses of the coastal ranchos instead of taking nearby wild horses in the relative safety of the Central Valley plains. Bancroft (1888: 336) argues that it was easier for Indians to steal horses than to tame them. However, Phillips (1993) asserts that Indians kept the fleetest horses for hunting elk. They also sought domesticated horses for trade to New Mexico. He concludes that the loss of horses from Indian attacks deprived ranchers of animals critical to the management of stock. This is supported by

Lymau (1850): "The swarms of Indians [the missions] once contained, have long since been scattered, and many of them, mingling with the wild tribes of the mountains, have become from their superior knowledge the most daring horse thieves."

Alternatively, a critical question is whether Indian raids were part of a war, or were they hunts for an easy kill. Why should they keep returning to ranchos to obtain tame horses, given the risk of Spanish violence imposed on them? If Native Americans were ineffective in harvesting wild horses, it follows that the dispersion of horses into the Central Valley resulted from continual escape of horses that became feral and bred. Phillips (1993) argues that raids as part of a war required an improved capacity to obtain horses, for example, by conducting raids on horseback. However, the deprivation of horses for ranchero activities would also take into account the high reproductive rate of horses.

Phillips's argument of Indian evolution may have merit. Indians may have been "neophytes" in horse acquisition in the Spanish and early Mexican periods, i.e., raids spontaneously led to the build up of feral horses. Raids after circa 1830 involved more sophisticated trade of domestic stock.

The feral horse population disappeared rapidly beginning with the gold rush. Bancroft (1888: 336) stated that during this period "Americans and Californians . . . lassoed them, catching all they wanted." Vizetelly (1848: 105) wrote that the "vast numbers of wild horses in California . . . have greatly diminished within a few years." Two decades later, Cronise (1868), in his exhaustive survey of land use in California, did not mention wild horses anywhere in the state, including the Central Valley. Apparently they were slaughtered or tamed by the invading U.S. settlers after the gold rush, helped along by mortality caused by the great drought of 1862–64.

SHEEP GRAZING AND OTHER LIVESTOCK

The increase of sheep in Hispanic California very likely paralleled that of cattle. About 100 sheep were introduced to Alta California by 1773 from Franciscan expeditions and stock provided from Baja California (Bancroft 1888). The sheep population was well below carrying capacities into the early nineteenth century because of limited breeding stock. The number of sheep was only 6,000 in 1797 (Carman et al. 1892), but increased rapidly to about 300,000 by the mission slaughter of the early 1830s (Duflot de Mofras 1937). By end of Mexican period in 1848, the distribution of sheep was similar to cattle within the rancho system from

San Diego to Sonoma. Sheep were never observed in the California interior by any non-Hispanic explorers of the Mexican period.

The importance of sheep varied among the missions and ranchos, although sources give conflicting data. Bancroft's data (1888) indicates that sheep populations were greatest at Missions San Gabriel, San Luis Rey, and San Juan Capistrano in southern California (Table 3.4). Forbes (1839) stated that 150,000 sheep pastured at Mission San Luis Rey alone. Similarly Davis's rancho estimates give the highest number in San Diego and Los Angeles counties, as well as for ranchos at Tomales Bay, Sutter's Fort, and at the Government Nacional (upper Salinas Valley). Nearly all the missions and ranchos had at least one band of sheep. However, sheep were less popular than cattle because they required reliable herders nearly all the time. Sheep were also vulnerable to predators, especially mountain lions, coyotes, and wolves (Bancroft 1888). California sheepmen "followed the time-honored practices that came from Spain into the arid provinces of the Southwest: owners drove their flocks into bands of 1000 to 2000 animals, and put a herder in charge of each band" (Cleland 1964: 140).

Sheep became feral only on Santa Cruz Island in the absence of natural predators (Brumbaugh 1980; Minnich 1980). In the 1860s, Santa Cruz Island had 30,000 sheep and Santa Catalina had wild goats (Cronise 1868: 89–90). While the number of sheep in California was large, the possible impact of sheep grazing compared to cattle was less because of sheep's small size. According to Davis (1929), there was a rule among the rancheros that one cow was equal to five sheep. Bidwell (1937: 39) observed that sheep "are small and the wool rather coarse."

Sheep numbers collapsed with secularization and the mission slaughter, as well as from drought in the early 1840s, possibly to as low as 31,600 by 1842 (Duflot de Mofras 1937). However, Bidwell (1937: 39) saw "a great many" in some places at that time, including 6,000 on at the Livermore Ranch. He saw 1,000 at Sutter's Fort. According to the 1850 U.S. census, there were less than 18,000 sheep in California (Cleland 1964: 138–40).

From the gold rush through the 1850s, a great influx of sheep, possibly one-half million head, was driven into California from New Mexico, first as a meat source and later for wool (Cronise 1868). Cleland (1964: 109) estimates that about 100,000 sheep crossed the Colorado River en route to California in 1858, and 35,000 sheep came to Los Angeles in February 1859. In the fall of 1859 it was said that 46,000 head from the Rio Grande Valley were passing through Arizona bound for California. In the winter of 1860 an estimated 80,000 head arrived at the ferry on the Colorado River. Cleland (1964: 138–40) writes that by 1860 the number of sheep in California had grown to about 1 million. Wool exports climbed from 175,000 pounds in 1854 to 2 million pounds in 1860 and 11.4 million pounds in 1870 (138–40).

Grazers quickly learned they had to employ transhumance—using mountain summer pastures in seasonally to assure year-round grazing for their herds—because forage was sparse in summer. Transhumance of sheep in mountain meadows began in earnest during the drought of 1862–64 (Burcham 1957). According to Twisselmann (1967), some flocks would spend almost an entire year away from their base in the Los Banos region in the central San Joaquin Valley, feeding south in late winter and spring, then north through the Sierra Nevada in summer and fall, returning home the following winter. In the 1860s, ownership of lands was of little concern to migratory herdsman (Twisselmann 1967). By the 1880s most of the valley grazing lands in Kern County were owned or controlled by resident sheepmen. The warm San Joaquin Valley, with its early growth in good years, was ideal for winter sheep range and for lambing the ewes. The "long trail" sheep grazed in the San Joaquin Valley in winter and spring, in the Mojave Desert in late spring, and in the Sierra Nevada in summer and fall, along a route from Bakersfield through the Tehachapi Mountains, north to Mono Lake and the high eastern Sierra Nevada. In southern California, livestock was driven to the San Bernardino and San Jacinto mountains for summer pasture from the Mojave Desert, the Riverside-Perris plain, and from the present-day Joshua Tree National Park (Minnich 1988).

Sheep grew to possibly 4 million in the 1870s, peaking at 6.7 million in 1876, partly in compensation for losses of cattle in the 1860s (Cleland 1964; Burcham 1957). Extreme drought in 1876–77 and increase cropping of winter lowlands began a decline of sheep pasturing that continued into the early twentieth century (Lockmann 1981). According to Cleland (1964: 209), pasture was so poor that "nothing green could be seen in wide stretches of the country" and sheep perished in great numbers. The subdivision of large landholdings and the expansion of agriculture diminished winter pasture (Cleland 1964: 209). The newly formed forest reserves in the early 1890s closed off summer pasture, as interests regarding public lands shifted toward recreation (Lockmann 1981). Sheep drives in southern California shifted to mountain pastures of northern Baja California (Minnich and Franco-Vizcaíno 1998).

Davis (1929) stated that mules were never in large numbers. Leonard (1959: 104) recalled that during the dry season mules were gathered into "large droves" and driven off to market at Santa Fe, New Mexico. Likewise, Cronise (1868: 370) stated that "mules are not numerous—being chiefly used for freighting goods into the mines and over the mountains. They are also employed for packing into districts where wagon roads are impracticable."

Hogs were doubtless found in small numbers on the ranches throughout California and appear to have gone feral in sheltered localities. Cronise (1868: 329) wrote that the Tulare County tule swamps were excellent for swine, which "were little cared for, and rarely seeing human beings, soon become quite wild, making it necessary for the owner to shoot them when he wishes to secure a carcass." Hogs also ran feral in the hills of Catalina Island (89–90).

EARLY EXPANSION OF FRANCISCAN ANNUALS

While the Franciscan missionaries introduced European annual grasses and forbs, they wrote little about the expansion of these species. An important early source of botanical data were mission bricks of precise age that contained plant remains from the beginning of the mission period (Hendry and Kelley 1925; Hendry 1931). Bricks were placed in water and they dissolved, leaving residual plant materials. Most bricks contained organic matter for binding and prevention of shrinkage, especially wheat or barley straw. Weeds of all kinds were extensively used, particularly those with fibrous stems, such as wild rye, sedges, tule, filaree, tarweeds, and various grasses, but the presence of other miscellaneous materials suggests that much of the general refuse from a mission was also used. As a consequence, the bricks bore a random sample of the local flora.

Perhaps the best evidence that Spanish missionaries introduced wild species either deliberately or as foul seed that escaped is the diversity of crop plants brought to California. In the second Anza expedition, Font saw that Palóu's garden at Mission Carmel included lettuce, cauliflower and artichoke (March 12, 1776; see Appendix 1). The mission bricks recorded numerous cereal crops, including Propo wheat (*Triticum vulgare graecum* Kcke., *Triticum vulgare albidum* Al.), Little Club wheat (*Triticum compactum humboldtii* Kcke.), California Club wheat (*Triticum compactum erinaceum* Kcke.), Sonora wheat (*Triticum vulgare albidum*), Coast barley (*Hordeum vulgare pallidum* Ser.), red oat (*Avena byzantina* G. Koch), and European oat (*Avena sativa* L.). The Spanish Friars also brought corn (*Zea mays* L.), red mexican or pink bean (*Phaseolus vulgaris* L.), garden pea (*Pisum sativum* L.), watermelon (*Citrullus vulgaris* Schrad), carrot (*Daucus carota* L.), fig (*Ficus carica* L.), olive

(*Olea europaea* L.), and the European grape (*Vitis vinifera* L.) (Hendry 1931).

Mission bricks document exotic wild species, including *Erodium cicutarium, Brassica nigra, Malva parviflora, Poa annua, Hordeum murinum, Medicago hispida, M. indica,* and *Trifolium wormskioldii* (Table 3.3). The Spanish missionaries were aware of these species' spontaneous expansion in California. *B. nigra* and *Malva parviflora* received a lot of commentary from the very beginning, partly because of their negative impacts around the missions. The fathers stated that they actually introduced species. For example, in 1792 Longinos-Martínez said that in California "the four seasons of the year are very like those of Spain. Thus it has been seen that all the trees and seeds that have been introduced from that country multiply with the same abundance and quality as in that climate" (Simpson 1961: 31). A concept that plants were deliberately introduced at multiple locations was ingrained among the Mexican population in California. Parish (1920) stated that "there is a persisting tradition among the Spanish-speaking population that the mission fathers were accustomed to carry the seed with them, and to sow by the way side."

Many of these invasives had expanded through large portions of California by the gold rush and the first government surveys in the mid–nineteenth century, but the frequency of records and age of bricks at individual missions suggests divergent rates of dispersal and expansion of the species in the mission period. Frenkel (1970) notes that many species are not commonly associated with *Erodium cicutarium* in the oldest (mission) walls, but frequently began to appear with them and with native species in later walls of the same buildings. This suggests a later arrival and a gradual expansion of these species compared to *E. cicutarium,* which expanded across California within a few years.

What follows is a summary of the early expansion of the most widespread and abundant Franciscan invasives. Then comes an assessment of regional patterns of European invasions and the persistence of indigenous floras in the mid–nineteenth century.

Brassica nigra *(black mustard)*

A deliberate introduction by the Franciscan missionaries (Hendry 1931; Frenkel 1970), this coarse annual was recorded in mission bricks in Baja California (Santo Domingo, San Vicente) and at Mission Soledad and Rancho El Sansal in central California (Table 3.3; Hendry and Kelly 1925). It had already invaded mission grounds by the 1770s. At Mission

Mission[a]	SFV	SAN	SDM	SVC	SDD	SFR	SJS	SJB	SFS	RES	RVJ	RNV
Date of founding	1769	1771	1775	1780	1791	1797	1797	1797	1824	1834	1834–45	1837
Date of adobe brick	1769	1787	1775	1780	1793–97	1797	1811	1805–13	1824	1834	1834–45	1837
EXOTIC ANNUAL SPECIES												
Erodium cicutarium	X	X	X	X	X	X	X	X		X	X	X
Malva parviflora	X		X		X	X	X					
Brassica nigra			X		X							
Avena fatua								?				
Poa annua								X				
Hordeum murinum							X					
Medicago hispida	X											
Melilotus indica	X					X						
Trifolium involucratum												
NATIVE SPECIES												
Elymus triticoides			X	X	X							X
Eleocharis spp.			X	X			X					
Scirpus spp.	X								X			
Carex spp.	X					X						
Stipa spp.												
Other bunch grasses												

SOURCE: Hendry (1931), Hendry and Bellue (1925), Frenkel (1970).
[a] California missions:

SFV, San Fernando Velicatá (Baja California)
SAN, San Antonio
SDM, Santo Domingo (Baja California)
SVC, San Vicente (Baja California)
SDD, Soledad
SFR, San Fernando
SJS, San Jose
SJB, San Juan Bautista
SFS, San Francisco Solano
RES, Rancho El Sansal (Salinas)
RVJ, Rancho Vallejo (Petaluma)

San Gabriel, Font referred to it as a "turnip," another member of the mustard family that "which from a little seed which was scattered took possession of the land" (January 4, 1776; see Appendix 1). In his summary of Alta California in 1792, Longinos-Martínez stated that "Mustard [is] a very common field plant" (Simpson 1961: 34). Parish (1920) speculated that "the fathers no doubt grew the plant in their garden" based on his view that it was grown as "a pot herb because the young leaves are relished by the Mexicans." He also acknowledged that "the seeds would be scattered by small birds, who freely eat them."

In 1827, August Bernard Duhaut-Cilly (1929: 219) first saw black mustard at the San Diego mission, 2 leagues north of the presidio, where the road to the mission "follows the edge of a long field of mustard whose flowers, of a beautiful yellow," were then in full bloom and dazzled the eye. In his first visit to San Pedro Bay, Duhaut-Cilly was deprived of a game hunt by the wild mustard. "What, at a certain distance, had appeared to us like a beautiful carpet of grass, . . . was found to be a thick vegetation of mustard, already reaching above a man's head . . . we plunged into this veritable forest, and soon repented for having done so. At every step we heard among the dried stalks of the past year, covering those of the present, the noise of rattlesnakes crawling about in so great numbers that a novice, accompanying us, killed two in a quarter of an hour" (165). In his second visit, in 1828 on his trip northward to Los Angeles, Duhaut-Cilly made the following observation, much like a modern restoration ecologist:

> For four leagues the way goes north, across a rather barren plain, but after passing the rancho I have already mentioned, it enters a great pasture ground stocked with large herds belonging to the inhabitants of the Pueblo of Los Angeles. Leaving these fields we encountered more than one forest of mustard, whose tall stalks were above the rider's heads, and made, as it were, two thick walls on the two sides of the way. This plant is [has] become, for some years, a terrible scourge for part of California. It invades the finest pasture lands, and threatens to spread over the entire country. The people could have fought this enemy in the beginning, by totally extirpating the first plants of this species, . . . but this neglect has permitted the evil to increase to an extent almost unremediable with so small a population [probably covered Dominguez Hills and other hills on the way to Los Angeles]. Fire even is an insufficient means which has been employed unsuccessfully. . . . When the stalk is dry enough to burn, it has already sown a large part of its seed, the fire serves but to make the ground more suitable for the reproduction of the plant one wished to destroy. (246)

Mustard was also seen in 1834 by Richard Henry Dana (1911: 116) on a trip from Santa Barbara to San Diego: "We coasted along down, the country appearing level or moderately uneven, and for the most part, sandy and treeless. . . . The land was of a clayey quality, and, as far as the eye could reach, entirely bare of trees and even shrubs. . . . The soil was, as it appeared at first, loose and clayey, and, except the stalks of the mustard plant, there was no vegetation." A decade later, Bryant (1848: 393) described mustard in the plains of Los Angeles. About the same time, Emory (1857–59: 18) of the U.S.-Mexican boundary survey observed at San Diego that "where the soil is rich the surface is covered by a rank growth of wild mustard (*Sinapsis nigra*) [*Brassica nigra*]." William Blake (1856: 75) of the Pacific Railroad survey wrote of the San Gabriel Valley, "The surface in some places was covered with a dense thicket of dead stalks of the wild mustard, which grows there to great height."

Black mustard was viewed as a pest by the rancheros of southern California. According to Cleland (1964: 57), "In southern California, the growth of wild mustard was even more remarkable than that mentioned in Christ's striking parable. During the late spring, a sea of yellow bloom flowed over the valleys, plains and foothills; and the thickset stalks [*Brassica nigra*], higher than a man's head, made an ideal hiding place for cattle. Even when the bloom and the leaves died, a forest of dry, rustling stalks furnished ample cover for livestock. In badly infested districts, neighboring ranchos and their *vaqueros* consequently united for a few days to carry on what was colloquially known as a 'run through the mustard.'" Cleland also found a letter to Abel Stearns, the manager of the Rancho Los Alamitos, who spoke of "a run of two or three days through the mustard at the Nietos and said that the lugos and four or five other ranchers had been invited to participate. 'The Temples, Manual Domínguez and the Coyotes will be there sure,' he added, 'and most likely a sufficient number of people will be brought together to effect some good'" (57).

Cleland further noted that W. H. Spurgeon laid out the town of Santa Ana on part of the 62,000-acre Rancho Santiago de Santa Ana, a Spanish grant dating to 1810. It is said that the wild mustard grew so thick and tall on the fertile land that Spurgeon was forced to climb a sycamore tree to get a bird's-eye view of the proposed townsite.

Wild mustard also was an important dominant over large areas of coastal central California (Bryant 1848; Cronise 1868). Most references suggest that *B. nigra* had a coastal distribution. Frémont (1848: 16) wrote generically of coastal California that the "fertile valleys are overgrown with wild mustard." In the state survey, Brewer and Watson (1876–1880) stated that "black mustard [is], a most troublesome weed and difficult to eradicate, covering large areas, particularly in the more fertile valleys

of the southern half of the state, sometimes forming a dense growth."
Nearly all late nineteenth-century botanists agreed that black mustard
expanded across California beginning in the mission days (e.g., Parish
1920; Smiley 1922).

Erodium cicutarium *(filaree)*

This member of geranium family, which has been given many common
names including filarie, alfileria, pin grass, pepper grass, pin clover, and
storksbill, expanded rapidly across the coastal plains, interior valleys, and
deserts of California from the beginning of the mission period. Filaree
was found in bricks at virtually all the missions in Alta California between 1775 and the 1830s, as well as at Santo Domingo and San Vicente
in Baja California (Hendry and Kelly 1925; Frenkel 1970; Table 3.3).
Erodium cicutarium was even found in packrat middens in Greenwater
Valley in the Mojave Desert in the past 250 years (Cole and Webb 1985).
The state survey's sources indicate it was widespread back to the Mexican period: "according to much testimony it was as common throughout California early in the present century as now" (Brewer and Watson
1876–80: 94). Parish (1920) and Brewer and Watson (1876–80) wrote that
the botanist Thomas Nuttall had already seen filaree far into the interior
western United States by 1836. *E. cicutarium* was also collected by the
Russian botanical collections of E. Voznesenski (Howell 1937), as well
as recorded in Hooker and Arnott's report (1841) on the botany of the
Beechey voyage (described in Burcham 1957). Hooker's locality is unknown, but must have been in the vicinity of Beechey's explorations between San Francisco and Monterey.

Hendry (1931) proposed that *E. cicutarium* may have naturalized
before the mission period based on its extensive distribution and its occurrence in bricks from many California and Baja California missions,
pointing out that "it has been found in the oldest walls of several mission buildings in widely separated localities, where they have occurred
frequently in the absence of other alien species, and in the total absence
of cereal remains. . . . This is reinforced by the knowledge that this species
is aggressive and capable of . . . becoming established in remote localites,
independent of man." For example, it is possible that *E. cicutarium* could
have been introduced to central Mexico in the sixteenth century with
Hernán Cortés's Spanish settlement, and then expanded northward
across the deserts, to which it is well adapted, into the southwestern
United States. However, the prospect of pre-Franciscan introduction of
E. cicutarium was refuted by Mensing and Byrne (2000), who found that

filaree pollen in varved sediments in the Santa Barbara Channel date to
1769. *E. cicutarium* seed apparently moves like dust.

   Filaree attracted the interest of many because it was an excellent fodder plant in the mid–nineteenth century. Frémont (1845, 1848) repeatedly described filiaree in his journals as seen across the Central Valley and in southern California. At Los Angeles he wrote, "The face of
the country was beautiful with a luxuriant growth of geranium, (erodium cicutarium), so esteemed as food for cattle and horses" (1848:
42). He also observed that filare was a dominant species in the interior:
"By the middle of March, the whole valley of the San Joaquin was in
the full glory of spring; . . . *geranium [Erodium] cicutarium* was generally in bloom, occupying the place of the grass, and making the upland a close sward" (1848: 20). Bryant (1848: 393) noted "an immense
quantity of the common pepper-grass" over California. In the State survey, Brewer and Watson (1876–80: 94) concluded that *E. cicutarium*
was "very common throughout the state," but was most dominant inland: "it is more decidedly and widely at home throughout the interior
than an other introduced plant." Brewer (1883) suggested that California pastures were dominated by plants other than grasses in more
arid portions of the state, such as with alfileria and burr clover, due to
the "rainless summers."

   According to Cronise (1868: 523), "Among the second species of nutritious herbs indigenous to California, and valuable to our herdsmen,
is the alfilerilla, *Erodium cicutarium*. . . . It stands second to none of the
grasses in its endurance of drought, and flourishes on hillsides where alfalfa grasses fail for want of moisture. To the eye, alfilerilla is a flattened
tuft, hugging the ground. It appears to give scarcely a fair hold to the
bite of cattle, but, if lifted up, it shows a great mouthful."

Erodium moschatum

Erodium moschatum was first recorded in the 1840–41 Russian botanical expedition of E. Voznesenski (Howell 1937). Because it was no doubt
confused with *E. cicutarium*, its early spread is unclear (Frenkel 1970).
Brewer and Watson (1876–80: 95), apparently unaware of the Voznesenski collections, indicated a widespread distribution of *E. moschatum*
that included "Los Angeles, Santa Inez Valley and northward as well as
southern in Mexico [Baja California]." Also unaware of the Russian collections, Parish (1890b) stated that *E. moschatum* "has not established
itself as widely as its companion [*E. cicutarium*]." He noted that "few

references to *E. moschatum* are found in botanical reports. It does not appear to have been found by the botanists of the Mexican boundary, but first appears in the collection of the Pacific Railroad Survey. The *Botany of California* returns it only from Los Angeles and Santa Inez Valley."

Malva parvifolia *(cheese mallow)*

A deliberate Franciscan introduction, this member of the mallow family (Malvaceae) was a pest in mission grounds by the 1790s. This ruderal species expanded rapidly through California by the mid–nineteenth century but was restricted to chronically disturbed ground. Hence, it was never a significant component of exotic annual grassland. Cheese mallow was recorded in the bricks of several missions, including at Soledad (east of Monterey), San Jose, and San Fernando in California, as well as at Santo Domingo and San Vicente in Baja California (Hendry and Kelly 1925; Hendry 1931). It was recognized as an escape early in the mission period. In 1792, Longinos-Martínez wrote, "The common apothecary's mallow [*Malva parviflora*], which is not known from these counties, has been propagated for some seeds, which was sent mixed with others. It grows with such vigor that because of it one cannot walk in the immediate vicinity of the missions" (Simpson 1961: 35). In a botanical collection, Torrey (1856) noted that plant (named *Malva borealis*) as a common weed in California. At San Diego, Emory (1857–59) later affirmed that mallows, *Malva obtusa* (*M. parviflora*), "does best on freshly disturbed ground, plowed fields, roadsides, etc." In the state survey, Brewer and Watson (1876–80) refer to it as a common weed on the West Coast by the 1860s.

Avena fatua *(wild oat)*

The wild oat (*Avena fatua*) was the most significant introduction of the Franciscan period. More than all other invasive species, it transformed the coastal prairies and it was the primary pasture for livestock before the gold rush. It is unclear whether the Spanish missionaries brought *A. fatua* in the first decades of the mission system, as is generally assumed. While it would have been preferred vegetable material in adobe bricks, it is only recorded only at Mission San Juan Bautista in 1810, where there was "a solitary wild oat kernel of doubtful authenticity believed to be *Avena fatua*" (Hendry and Kelly 1925; Table 3.3). Wild oats were widespread along the coast and along floodplains in the interior by the gold rush, effectively its modern distribution.

Hendry (1931) and Hendry and Kelley (1925) deduced that wild oats were introduced in the mission period based on circumstantial evidence: "it is difficult to account for the entire absence of wild oats—with this doubtful exception—except upon the grounds that its wide distribution in California has occurred at a comparatively recent time" (Hendry 1931). Likewise, Parish (1920) speculated that wild oats must have been among the earliest introductions of the mission era and, being well suited to the conditions, spread with rapidity. However, without direct evidence of the rate of dispersal, one cannot know whether wild oats spread across California at the beginning of the mission period or decades later. It is perhaps striking that wild oats were not recognized by Longinos-Martínez in 1792, nor collected by the Malaspina expedition in 1791, nor was *avena*, the common Spanish name for oats, observed by Zalvidea and Muñoz (of the Moraga expedition) in the Central Valley in 1805–06. Neither were oats described by Duhaut-Cilly at Los Angeles in 1827, even though he placed mustards there. He also described the plain beyond the mustard fields as barren, an unlikely portrayal of an oat field.

Nonetheless, wild oats were widespread by the gold rush, although the taxonomic acuity of early writers is problematic. At San Diego, Dana wrote in 1835 that the country between the presidio and the mission was "rather sandy . . . but the grass grew green and rank" (Dana 1911: 134), possibly in reference to wild oats. The U.S.-Mexican boundary survey recorded oats in the same area only a decade later when A. W. Whipple's route across the foothills and valleys east of San Diego led "over steep hills, uncultivated and barren excepting a few fields of wild-oats" (Whipple 1961: 25). Wild oats had apparently reached the Central Valley by the 1830s. Zenas Leonard (1959:114) of the Walker party in 1833–34 recalled that their encampment on the Sulphur River (San Joaquin River) was "beautifully located on a rising piece of ground, with a handsome river gliding smoothly along immediately in front, an extensive oat plain stretching as far as the eye will reach to the rear." The party then continued south along the Sulphur River, "passing through a fine country, most of which is prairie, covered plentifully with wild oats and grass." Leonard also traversed "a large prairie covered with wild oats along the San Joaquin River near the Sierra Nevada foothills" (120). In the spring of 1834, he wrote that the area 40 miles east of San Juan (Bautista) in the interior Coast Range "produced heavy crops of wild oats and grass—

affording excellent pasture for horses, at this season of the year" (114). Cronise (1868: 521) and Hittell (1874: 104), based on unclear sources, both described the expansion of wild oats on the north side of San Francisco Bay: "The wild oat in the year 1835 was found south of the bay of San Francisco; but about that time, when the whites began to cross frequently from the southern to the northern side of the bay, this grain being sown in a natural way by horses and cattle, spread rapidly over the Sacramento Valley and the coast region, its range now being very extensive. . . . The wild oat grows both on the plains and the hills." Brewer and Watson (1876–80) also suggested that wild oats came later than *Erodium cicutarium.*

In the same region, Frémont (1848: 20) crossed an oat field on the northern shores of San Francisco Bay: "it assumes . . . a beautiful and cultivated appearance. Wild oats cover it in continuous fields." Likewise, Wilkes (1849: 177) reported in 1841 that "the hills are thickly covered with wild oats" in the region that included San Pablo Bay and the Carquenex (sic) Strait. Frémont (1848: 34) wrote, "The country around Suisin Bay presents smooth low ridges and rounded hills, clothed with wild oats."

Brewer (1883: 963) viewed the invasion of wild oats to be inextricably linked to livestock grazing: "Wild oats (*Avena fatua*) . . . came into the state from the south with the Franciscan fathers and their herds, and spread northward." Based on his interviews with older residents, he suggested that wild oats had variable abundance in the mid–nineteenth century. He wrote that oats "were most abundant between 1845 and 1855, when hundreds of thousands of acres were clothed with it as thick as a meadow," but that *Erodium cicutarium* increased "until 1865 or 1870" due to extreme drought and overgrazing in 1862–64 (963).

The broadscale dominance of oats in California was indicated generically by several writers. Bryant (1848: 449) asserted "that [in the] portion of California between the Sierra Nevada and the Pacific . . . , oats and mustard grow spontaneously, with such rankness as to be considered nuisances upon the soil. I have forced my way through thousands of acres of these, higher than my head when mounted on a horse. The oats grow to the summits of the hills, but they are not here so tall and rank as in the valleys." He also wrote that in "most of the . . . valleys between the Sierra Nevada and the coast . . . , the hills are covered with oats and grass" (351). Bryant, as well as other many writers from the period, used the phrase "oats and grass," an apparent redundancy because oats *are* a grass. Bryant appears to be confusing oats and *Erodium,* a species so abundant that it was viewed as a "grass," or possibly he is referring to grass as a Spaniard uses *pasto* or *zacate*. Indeed, *Erodium* was called "pin grass." J. S. Newberry (1857: 13) of the Pacific Railroad survey, in what appears to be an exaggeration, claimed that "throughout central and southern California, wherever the ground was not occupied by forests, wild oats covered surfaces of many hundreds of miles in extent as completely as the grasses cover the prairies of Illinois." He asserted that "at this early date they were even more abundant than at present, the increase of cultivation having curtailed their area. They are frequent in cultivated fields especially as 'volunteers' in grain fields, but they also occupy great tracts of hills and plains. They afford good pasturage, and in early years were extensively reaped for hay." The comments of Bryant and Newberry are problematic because many writers from this period did not find wild oats in the California interior.

Cronise (1868: 355) explained how oats could spread their seed rapidly:

> Though parched in the long summer, the grain held firmly in its capsule and supplied the most fattening pasture. It still prevails outside of cultivation, furnishing a large proportion of the hay in use in many localities. It differs from tame oats in being smaller, and in this peculiarity, that it has beared projections, with bended joints, likes the legs of the grasshopper. When the first rain comes [the seed] limbers out of the joints, which being dried by the sun, after the rain, shrink, causing the berry to hop about, and give it a wide distribution to the land.

He also foresaw the forage value of wild oats. Although the grass was parched in the long summer, "the grain held firmly in its capsule and supplied the most fattening pasture, . . . furnishing a large proportion of the hay in use in many localities" (355). Blake (1856) of the Pacific Railroad survey had a similar insight on the success of the wild oat: "But on close inspection of the bare surface, the grains of oats could be perceived scattered about, or collected in little depressions, sheltered from the wind. The fire, it seems, is not hot and lasting enough to accomplish more than the separation of the kernel from the husk; . . . in this condition it is fitted to fall into the cracks of the soil, and thus be entombed in readiness to spring up after the first shower of winter, or to afford sustenance to birds and bands of deer and elk."

Trifolium *and* Medicago *(clovers) and other grasses*

The mission bricks document several species of clover early in the Spanish period, including *Medicago hispida, Melilotis indica,* and *Trifolium involucratum,* which were valuable leguminous fodder plants in the nineteenth century. Historical accounts generally fail to distinguish species of clover. Native Americans adopted these plants as a food resource. For example, clovers (*Trifolium* spp.) were favorites of the Santa Barbara Chumash, the plants being eaten raw (Timbrook et al. 1982). Cronise (1868: 525) stated that *Trifolium repens* was often eaten by the Indians, who liked it both raw and boiled. Frémont's writings (1845, 1848) reveal that clovers were extensive in both coastal and interior California in the 1840s. The U.S.-Mexican boundary survey in mid–nineteenth century (Emory 1857–59) states that clovers were major components of pasture in southern California, along with wild oats and filaree. Brewer (1883) stated that twenty-five species of clover occured in California, of which some were important livestock forage. He also noted that, in addition to the clovers, many native legumes such as wild vetches and lupines were eaten by stock.

The most important introduced legume was burr clover (*Medicago hispida*), an early Spanish-period introduction that was recorded in mission bricks dated as early as 1769 (Hendry 1931; Parish 1920; Frenkel 1970). It was also recorded in Hooker and Arnott's report (1841) on the botany of the Beechey voyage, which visited California between San Francisco and Monterey (Burcham 1957). Bryant (1848: 449) claimed that he saw "seven different kinds of clover, several of them in a dry state, depositing a seed upon the ground so abundant as to cover it." Parish (1920), citing the U.S.-Mexican boundary survey, stated that burr clover was reported as "abundant throughout California" in 1859.

During the Pacific Railroad survey, Blake (1856: 75) observed that "large flocks of sheep were feeding on the burr of the California clover" in the San Gabriel Valley. In the nearby San Fernando Valley, he saw "herds of cattle . . . on parts of the dry plains, feeding on dried grass or the burrs of the California clover which covers the ground in the latter part of summer when all the grass has disappeared" (75). Without stating sources, Brewer (1883: 963) wrote that burr clover, *Medicago denticulata* (*M. hispida*), expanded into California "later and slower than *Erodium cicutarium,* spreading with sheep along the lines of their drives, and along the lines of wagon roads across the great central valley." Cronise (1868: 525) noted that cattle were also extremely fond of burr clover, which was found "in all the settled parts of the State," and cattle did not prefer it when it was green, but after the burrs fell to the ground after they had dried.

Mission bricks at San Fernando Velicatá in Baja California document that *Melilotus indica* (sour clover) was also introduced at the beginning of the Spanish period (Hendry 1931; cf. Frenkel 1970). Although there was little description of the species in the nineteenth century, it was widespread by the gold rush. Brewer and Watson (1876–80) stated that it was common in the state, and Parish (1920), citing the boundary survey, assumed that sour clover was abundantly naturalized in cold, damp soils. The earliest collection was taken at Los Angeles in 1853–54 (Torrey 1856). Other clovers dating possibly to the Mexican period are *Medicago lupulina* and *Melilotus albus* (Frenkel 1970).

Several *Trifolium* clovers, including *T. variegatum, T. involucratum* (*wormskiodii*), *T. olivaceum,* and *T. microcephalum,* were recorded in mission bricks dating between 1771 and 1797 (Hendry and Kelly 1925; Hendry 1931). *T. albopurpureum* was collected by David Douglas in Monterey County between 1830 and 1833. In the state survey, Brewer and Watson (1876–80) stated than *T. gracilentum* was common in California.

Other introduced species, such as *Festuca, Hordeum,* and *Poa annua,* elicited few comments by early explorers because the plants were not conspicuous nor did they form substantial biomass. Wall barley, also known in the nineteenth century as farmers foxtail (*Hordeum murinum*), was recorded in mission bricks as early as 1810 (cf. Frenkel 1970). Mid-nineteenth-century accounts consistently indicate that this grass was most common in degraded, overgrazed pasture. It was collected at San Francisco by the state survey. Cronise (1868: 523) wrote that "wall barley was found near roads and vacant lots." He added that "if allowed to insinuate itself into meadows it injures the hay and lessens the value of the crops. Its strong beards (arms) hurt the mouths of horses" (523). Brewer (1883: 961) was alarmed by the occurrence of this species in pastureland, concluding that it

> has become a great pest in California. . . . While green and growing it has some value as feed, but not much. It is annual, and comes in where the pastures are overstocked, particularly on the lower hills and valleys of the coast ranges. When the heads ripen, they break up, and the barbed seeds and awns work their way in to every crevice: they insert themselves into the wool, and bore their way sometimes into the flesh of sheep and lambs; they get into the eyes of domestic animals, often destroying the sight; they get into the throats of horses and mules and cause inflammation; taken all in all, this weed is probably the most troublesome of California.

Annual blue grass (*Poa annua*) was also recorded in mission bricks, possibly by 1805, but was seldom described in the nineteenth century. It was collected in the Russian expedition of E. Voznesenski in 1840–41 near Santa Rosa (Frenkel 1970). Brewer and Watson (1876–80) stated that it was widespread in the state by the 1860s. The common fescue grass (*Festuca myuros*) was first collected by Torrey of the Pacific Railroad survey in 1856, but is widely believed to have arrived to California in the Spanish period (Frenkel 1970), and the state survey found it to be common throughout the California by the 1860s (Brewer and Watson 1876–80).

COASTAL AND INTERIOR VALLEY PASTURES

A growing body of landscape descriptions from the time of Frémont to 1880 suggest that introduced wild oat, black mustard, and clovers dominated the coast. The indigenous wildflower flora was holding its own against the onslaught of invasive annuals in interior California, as well as in areas with poor soils along the coast, such as beach dunes and serpentines. Herbaceous vegetation was ever changing due to the progressive expansion of introduced species and year-to-year rainfall variability. However, the pattern of expansion was unique for each species because they arrived at different times and had different competitive relationships with the indigenous herbaceous flora. There were few reports of wildflower fields before Frémont. The Hispanic and Indian populations in missions, presidios, pueblos, and later living on land grants were mostly illiterate. The gold rush brought a population explosion to central California, far greater in magnitude than in the south, including many writers who gave detailed descriptions of the landscape.

Southern California

While the Franciscan record of the pre-Hispanic landscape reveals that forbs were a significant component of herbaceous vegetation in southern California, by the mid–nineteenth century wild oats and black mustard were prolific in rich soils of the coastal plains and coastal hills. Interior valleys and hills were dominated by filaree and clovers, the primary winter forage. Wild oats and burr clover were the primary forage species in summer, with black mustard and wall barley considered to be pests in cultivated and overgrazed lands. Wildflowers grew in the coastal plains before agriculture and remained extensive in the foothills, inland valleys, and Channel Islands.

Vague early accounts of coastal southern California by Dana (1911) in 1835 show that plant cover was almost entirely herbaceous, like that described sixty years before by the Spanish missionaries. From his vessel, he wrote, "The country appear[s] level or moderately uneven, and for the most part, sandy and treeless. . . . The land was of a clayey quality, and, as far as the eye could reach, entirely bare of trees and even shrubs" (116). He came ashore near San Juan Capistrano, where he scrambled up the coastal bluff, "walking over briers and prickly pears, until we came to the top. Here the country stretched out for miles, as far as the eye could reach, on a level, table surface, and the only habitation in sight was the small white mission of San Juan Capistrano" (172). Ultimately, he only recognized the mustard stocks. Duhaut-Cilly in 1827 saw vast fields of mustard along the road from port at San Pedro to Los Angeles (see p. 109). Duflot de Mofras (1937: 175) wrote that the San Diego coast northward to San Luis Rey de Francia had "vast, verdant prairies, dotted with a few scattered clumps of trees." Similarly, the *diseño* of Rancho la Cienega shows the Baldwin Hills with grass cover.

Botanical detail of the herbaceous cover, including the status of introduced species, began with a visit to the Los Angeles by Frémont in 1846, where he saw "a luxuriant growth of geranium, (erodium cicutarium)" (Frémont 1848: 42). Frémont was accompanied by Bryant (1848: 393), who wrote, "Crossing a ridge of hills we entered the magnificent undulating plain surrounding the city of Angels, now verdant with a carpet of fresh vegetation. Among other plants I noticed the mustard, and an immense quantity of the common pepper-grass (*Erodium cicutarium*) of our gardens."

At San Diego, mustard fields were seen by Duhaut-Cilly in 1827, and Whipple of the U.S.-Mexican boundary survey saw fields of wild oats in the nearby foothills two decades later. Emory (1857–59: 18) described the coast from San Diego to Los Angeles: "Let any one follow up the coast in the month of March, and pass over the verdant plains that stretch towards the sea; let him see every valley and hill clothed in the rich green of the wild oats." Likewise, in the foothills he wrote, "At a variable distance inland, . . . the rounded hills [e.g. the Puente hills] are covered with a deep rich loam, which in the spring produces a luxuriant crop of wild oats." Farther north toward Los Angeles, he gave a more complete picture:

> The valleys are clothed with a luxuriant growth of wild oats (*Avena fatua*), which is so extensively naturalized that it gives to every fertile tract the appearance of a cultivated field. The wide plains that border the sea in the neighborhood of Los Angeles are covered with the richest pasturage. The *Erodium cicutarium*, (called here "pin grass," and furnishing a highly esteemed fodder,) with several species of wild clover, (*Trifolium* and *Medicago*) are mingled with a variety of other herbage, and thus serve to give a meadow-like aspect to this teeming land. The wide plains that border the sea in the neighborhood of Los Angeles are covered with the richest pasturage. (18)

The Pacific Railroad survey of the 1850s had a similar appraisal of the southern California plains. Bigelow (1856: 16), referring to the Los Angeles valley (the entire southern California coastal plain), wrote, "Grass and wild oats are abundant from one end to the other. . . . Nature has peculiarly favored this region and adapted it to grazing by furnishing it with a succession of plants, . . . so that no trouble or expense is experienced in raising cattle and horses." During the course of a rainy season, he continued, "the first crop is called 'pin grass,' (*Erodium cicutarium*) . . . the next is a crop of leguminous plants, such as *Medicago* and several species of clover, (*Trifolium*); then follow wild oats (*Avena*), and other species of grass in greatest abundance." Whipple (1856a: 134) described a mixture of grass and flowers in the San Bernardino Valley. He saw a "great grassy plain . . . [that] contains numerous herds of sheep and cattle grazing on the immense sheet of tall and luxuriant grass . . . variegated with an abundance of bright flowers." In the San Gabriel Valley, (Blake 1856: 75) "large flocks of sheep were feeding on the burr of the California clover." At Point Conception, where flower fields were described by Crespí and Font eighty years before, Blake found a "growth of wild oats and weeds" (3). In the state survey, Brewer described grassland across the southern California plains. West of Los Angeles, the Triunfo Ranch was "covered with grass," the San Buenaventura area was a "fine grassy plain," and at Santa Barbara "we rode over grassy hills, with some timber, where many cattle and sheep were grazing" (Brewer 1966: 45–49).

In a rare first-hand account, Lougheed (1951) wrote that his father "marveled at the sight of mustard" when he moved to Los Angeles in the 1850s and gave remarkable detail of its distribution, largely in areas with heavy, clay-rich soils. His father found "great fields of mustard extended from the outskirts of the then Los Angeles city limits east and south through Alhambra, El Monte, and the Puente districts, Whittier, and Santa Fe Springs; . . . and [in] the San Pedro Redondo areas. . . . Mustard varied from 4 to 10 feet tall [1.3–3.0 m]." In 1874, the Reverend T. M. Dawson (1874) journeyed from San Pedro to Los Angeles, along Duhaut-Cilly's 1826–27 route, where he rode through "a plain decorated with great quantities of yellow wildflowers," doubtless the black mustard.

Cronise (1868: 111) assessed the grazing potential of Santa Barbara County: "The highest mountains being covered with grass or wild oats during the winter and spring, furnish nutritious pasturage for sheep and cattle during the entire year." Similarly, "Sitacoy and Santa Clara Val-

leys have a frontage on the coast of sixteen miles [Santa Ynez River val-
ley] . . . these valleys and plains produce immense quantities of wild mustard, which grows to the size of small trees in some localites" (113). In
Los Angeles County, he said "these plains are covered with wild grasses,
oats and clover" (105).

Archduke Ludwig Louis Salvator (1929: 81) described the seasonal
course of livestock forage at Los Angeles in 1876:

> During the winter and spring months are found the plants that afford [live-
> stock] considerable food; alfilerilla (*Erodium cicutarium*) and bunch grass.
> The form which is most abundant of all the native growing grasses [it is
> neither native nor a grass, but an exotic geranium] grows thickly in the hills
> and plains, affording with its light greenish-yellow color a soft tone on the
> landscape. This is one of the richest foods for cattle. In summer and autumn,
> wild oats and burr clover furnish excellent fodder. The former, which the
> cattle eat when green, retains its nourishing seeds for a long time; the latter
> contains seed in a little spherical burr. . . . Even after the meadows turn earth
> brown in color, cattle and sheep can subsist on these seeds on land which
> to the experienced appears to be a desert. Practically all the pasture lands
> are devoid of weeds.

John Hittell (1874: 104–5), who observed the Los Angeles region in
the 1860s, stated that *Avena fatua* "grows very luxuriantly and in some
places surpassed in the height, size and abundance of stalks, as a field of
cultivated oats which I have ever seen." The burr clover *Medicago hispida,* which he called the white Californian clover, "has a large yellowish-
white bloom. It grows 2–3 feet high [0.6–1.0 m] in moist places while in
dry places it will mature its seed without rising more than two or three
inches. . . . The *Melilotus officinales*. . . . likes a very moist soil and
crowds out nearly everything else."

The interior valleys of southern California also afforded excellent pasture. Brewer (1966: 34) described inland plains near Corona as "a grassy
carpet." Cronise (1868: 98) wrote that the Temecula area, which had the
largest population in the Riverside-Perris plain at that time, had "the finest
grazing lands in the southern portions of the state, being covered with
wild oats, clover, and other nutritious grasses, furnishing pasture for thousands of cattle, horses, and sheep." The *diseño* of Rancho Cucamonga
(1840s) showed pasture in the southern part of the land grant (Black
1975). In 1860, the land grant received a description of the seasonal
pasture cycle comparable to other parts of southern California: "The
country from Cucamonga to San Bernardino for 18. . . . In the winter
this desert is covered with herbage, called alfilaria [*Erodium*] grass, upon
which the stock feed and fatten . . . in spring, . . . these now sterile hills

are covered for leagues with heavy crops of wild oats and fragrant clover"
(Black 1975: 45–46).

There are few reports of wildflower fields from the Spanish and Mexican periods. California wildflowers were symbolized in playing cards.
Francis M. Fultz, who provided many articles on wildflowers to the *Los
Angeles Times*, recounted that the jack-of-spades playing card from the
early Spanish days had an image of *Baeria* (*Lasthenia gracilis*), or goldfields.[2] The California poppy (*Eschscholzia californica*) was a cultural
icon. The Spanish Californians gave poppies pretty names—*amapola,
torosa,* and *adormidera,* "the drowsy one," the latter on account of the
way the petals fold up and go to sleep in the evening, and *copa de oro*
(cup of gold).[3] The *Times* also reported the origin of the genus name
Eschscholzia.[4] Leonard H. Eschscholz of Charleston, South Carolina,
viewed for the first time the poppies named after one of his ancestors.
The poppy was named for Russian surgeon Johann Friederich Es-
chscholtz, who was part of the expedition that visited California in 1816.
One of his shipmates was Adelbert von Chamisso, who gave the California poppy its scientific name, *Eschscholzia californica.*

The earliest description of wildflowers in Los Angeles was made in
February 1847 by Revere (1849c: 281): "In the plain itself, the richest
and most brilliant wildflowers flourish which far transcends all art; All
colors, all shades of colors, all hues, all tints, all combinations are there
to be seen. And the endless variety bewilder the senses. Perennial incense
ascends to heaven from these fragrant plains; and the size which some
of the gorgeous flowers attain, would seem fabulous to an eastern florist.
Among them are the poppy and a tulip whose flouting and gaudy hues
attract the humblist daisy and the meekest violet." The following year
Bryant (1848: 417) recalled. "In the Couenga plain [sic, southern San
Fernando Valley] . . . a few days [after a rain] have made a great change
in the appearance of the country. The fresh grass is now several inches
in height, and many flowers are in bloom."

In the U.S.-Mexican boundary survey of the early 1850s, Emory (1857–
59: 18) referred to the coastal San Diego region where he was based:

It is in the latter part of winter and during the earlier spring months
that California puts on her richest floral garb. In February the moistened
ground becomes arrayed in an assemblage of varied tints. The pale blossom
of the elegant *Dodecatheon integrifolium* [*clevelandii*] nod on every hill side,
blue *Lupines* and rainbow colored *Gilias* deck the ground. . . . The *Ribes
speciosum* hangs its scarlet pendants, and the rich yellow flowers of *Viola*
pedunculata are abundant everywhere. . . . A large number of *Hydrophyl-
laceae,* including species of *Nemophila, Phacelia,* and *Eutoca* [*Phacelia*],

are among the early tokens of spring, while the orange colored flowers of *Escholtzia* (sic), the pale blooms of *Platystemon*, and the pink ones of *Meconopsis*, show that the poppy family contribute largely to make up the vernal flora.

Archduke Salvator (1929: 24) toured Los Angeles in 1876 and found "the fields in and about Los Angeles are particularly rich in flowers. Thus in March they appear to be cloaked in red; in April in blue; while in May they resemble masses of pure gold (*Eschscholzia*) which is aptly called *Copa de Oro*,—cup of gold." The *Times* reported in 1890 that "a well-known surveyor once told the writer that [he rode] through miles of these poppies in the old days."[5] Cronise (1868: 373) described wildflowers in the context of the honey industry: "California has a great variety and expanse of very gay flowers, like the escolchia [sic] . . . ; but, as a rule, the gayer the flower the less honey it has. . . . Honey has come to market from Los Angeles, and is so abundant and cheap (twenty five cents a pound) that the production does not seem remunerative at this time." At Dos Pueblos west of Santa Barbara, Brewer (1966: 74) saw "very green grassy hills, on which were in a profusion of wildflowers." Flowers also covered the inland valleys. Brewer (1966: 40) observed that "on hills east of Corona or Temescal, I found some exquisitely beautiful flowers of very small size, several species being less than an inch in height."

In 1879 the Los Angeles region was visited by a botanist J. F. James, who provided a rare early glimpse of southern California with a scientific eye (see also Anonymous 1880). He wrote that

> in the vicinity of Los Angeles . . . the plains surrounding the city, the hills, and the valleys are one mass of gorgeous, brilliant flowers. . . . Most conspicuous of all, both for its abundance and its color, is the Californian poppy, *Eschscholzia californica*. It covers acres of ground, and the bright golden-yellow or orange of its flowers is visible for miles. One patch on a bright clear day [was] too dazzling for the eye to gaze upon, . . . truly [a] Field of the Cloth of Gold. In places where the ground was plowed paths of it had been left, and they seemed like tongues of fire running over the ground.

Owls clover (*Orthocarpus purpurascens*), he said, "grows in dense masses, covering the ground for miles, and giving it a purplish hue." Wildflowers that were "so common as to cover acres of ground" were *Baeria* (*Lasthenia*) *gracilis* and *Amsinckia spectabilis*. Other common species were *Sidalcea malviflora, Playstemon californicus, Peonia* (*Paeonia*) *brownii, Penstemon* (*Keckiella*) *cordifolius, Scrophularia californica, Phacelia ramosissima, P. tanecetifolia, Nemophila insignis* (*menziesii*), *N. aurita*, three species of *Gilia, Mirabilis californica, Viola pedunculata, Zauschne-ria californica*, and members of the genera *Clarkia, Lupinus, Hosackia* (*Lotus*), and *Astragalus. Salvia carduacea* and *S. columbariae* grew in dry sandy soil. James also noted two "alfillerilla" (*Erodium cicutarium, E. moschatum*) as valuable forage plants. The pimpernel (*Anagallis arvensis*) was common in cultivated grounds. The most conspicuous exotic species was *Brassica nigra*, which James referred to as "one of the most pernicious weeds of the whole of Southern California, . . . [which] covered the ground in many places for acres, to the entire exclusion of other plants." James rode through fields of it early in spring when it was as high as the saddle on a horse. Another exotic was *Malva borealis* (*parviflora*), which grew "everywhere around houses and in waste ground; in old sheep and cattle corrals . . . so thick and strong that even a horse has difficulty in forcing his way through it."

Coastal Central California

The central California coast was generally covered by wild oats, black mustard, and clovers, similar to that in coastal southern California. Native wildflowers were infrequently described.

The Beechey account from 1828 recorded extensive herbaceous cover between San Francisco and Monterey, but the expedition left little information on the composition of California pastures. San Francisco Bay was "a country diversified with hill and dale, partly wooded, and partly disposed in pasture lands of the richest kind, abounding in herds of cattle" (Beechey 1831: 4). San Bruno Mountain "was covered with burnt-up grass" (42). The Llano del Rey (Monterey Bay) was "covered with a rank grass, and has very few shrubs," while on their return to San Francisco the Beechey party saw "immense plains of meadow land" (56, 60).

Wild oats and other exotics were widespread by the mid–nineteenth century. In 1841, Wilkes (1844: 153) wrote that the country about San Francisco "presented a rather singular appearance owing . . . to the withered vegetation and the ripened wild oats. . . . Instead of a lively green hue, it had generally a tint of a light straw-colour." At the Carquinez Strait "the hills were thickly covered with wild oats" (177). Newberry (1857: 13) of the Pacific Railroad survey had a similar first impression: "The hills and mountain sides bordering the bay of San Francisco and San Pablo are generally covered with the wild oat." Wilkes (1849: 41) found the Salinas valley to be "open country covered with grass."

Botanical information from Frémont's 1846 journey emphasizes the dominance of introduced European annuals along the central California coast. In the Sacramento delta and straits of San Francisco Bay, after he passed westward by "missions and large farms . . . established on the head of navigation [Stockton]," he found "the country around Suisin Bay presents smooth low ridges and rounded hills, clothed with wild oats" (Frémont 1848: 30). Approaching its northern shores from Sonoma, he said that the land "assumes . . . a beautiful and cultivated appearance. Wild oats cover it in continuous fields" (34). Referring to the Coast Range near San Francisco in the growing season, he wrote, "after the spring rains . . . covered in grass . . . covered in summer with four or five varieties of wild clover several feet high. In many places it is overgrown with wild mustard ten or twelve feet high [3–4 m]" (33). The country north of San Francisco Bay had "close growth of wild oats" (35). The Santa Cruz Mountains coast was covered with "luxuriant growth of grass, a foot high in many places" (36). He saw grasslands along the remainder of the central California coast. Near Monterey, "The grass, which had been eaten down by the large herds of cattle, was now everywhere sprung up; flowers began to show their bloom. In the valleys of the mountains bordering the Salinas Plains . . . wild oats were three feet [1 m]high and well headed" (37). Farther south the Santa Lucia Mountains were "covered thickly with wild oats" (38). Near Point Conception, "cold summers afford good green grass throughout the year" (38). He found the sheep and cattle to be fat in the driest part of the year. The Santa Ynez Mission was "well covered with grass of good quality—very different from the dry naked parched appearance of the country below Santa Barbara. Throughout this region the fertile valleys are overgrown with wild mustard" (37–38).

During Frémont's "conquest" of California in 1846, Bryant (1848) saw extensive fields of wild oats in coastal California. When he departed the Central Valley near Sacramento into the Coast Range, Bryant (1948: 305) wrote, "From this plain we entered a hilly country, covered to the summits . . . with wild oats." San Jose was "a highly fertile plain, producing a variety of indigenous grasses, among which I saw several species of clover, and mustard, large tracts of which we rode through, the stalks varying from six to ten feet [2.3 m] in height" (314). He arrived at Mission San Juan Bautista just after the after the first winter rains, and found "the hills and valleys are becoming verdant with the fresh grass and wild oats, the latter being, in places two or three inches high [5–8 cm]" (365). Between Morro Bay and San Luis Obispo, he wrote that "the hills and plains are verdant with a carpet of fresh grass" (377).

The Pacific Railroad Survey visited San Francisco Bay in July and August. At Benecia (the north side of San Francisco Bay near Vallejo), Blake

(1856: 4) wrote, "At this season, the wild oat, which covers the surface, is golden yellow; but here and there are long dry straw that has been set on fire, and broad acres are burned off, leaving a black charred waste." Near Mount Diablo, he noted that the hills in the spring and summer were covered by a luxuriant growth of wild oats (7–8). The botanist John Torrey (1856: 156), while recognizing the widespread range of wild oat, suggested that it had a coastal distribution: "It may have been introduced by the Spaniards; but it is now spread over the whole country, many miles from the coast." In the 1860s, Brewer (1966: 84) found the southern Santa Lucia Mountains to be "very steep but perfectly grassy slope, covered with wild oats," and the Santa Ynez Mission area was a "grassy plain." In his explorations of the Coast Range from Santa Barbara to San Francisco, Brewer (1966) described hillsides and plains covered with wild oats on the coastal slope of the Santa Lucia Mountains (84), along Carmel Valley (108, 109), in parts of the Salinas Valley (110), at San Juan (Bautista) (117), along the San Benito River (119), and in the Santa Clara Valley (p. 178). Clarence King (1915: 26) of the U.S. Geological Survey passed through Pacheco Pass, where he found "wind-bent oaks trailing almost horizontally over the wild-oat surface of the hills." Blake (1856: 8) even drew pictures of flagged oak trees. Likewise, Brewer (1966: 287) saw at this place "oak trees often lay[ing] along the ground in the direction of the wind."

Cronise (1868: 135) gave a similar view of the California coast. In Santa Clara County, he wrote that "the slopes around the edge of the valley are covered with wild oats and native grasses, and afford excellent pasturage for large herds of cows." San Jose proper had "thick crops of wild oats and burr clover," while the uncultivated parts of the Santa Cruz County coastal marine terraces "produce enormous crops of wild mustard" (149). The Pajaro region in northern Monterey Bay was "exceedingly fertile, and almost level. On either side of it for several miles, there is a range of low smoothly rounded hills, well watered by numerous creeks, and but little less fertile than the bottom-land, which produces find drops of wild oats, bunch grass, and a variety of clover and native grasses" (122).

Early writers gave some suggestion of the distribution of exotic species. Wild mustard was an important dominant, particularly on better lands in the coastal valleys (Bryant 1848; Cronise 1868). Muir (1904) suggested burr clover was more abundant in flats and creeks than on steep slopes. Bryant (1848: 49) suggested that wild oats were taller and more rank in the valleys than on the hills.

Accounts of wildflower fields along the central coast are remarkably sparing and suggest that native forbfields had already been displaced by wild oats, black mustard, and other introduced European annuals. Flower fields grew either in poor soils or appeared intermittently after ideal rain years. The *diseño* for Rancho Bolsa del Potrero y Moro Cojo at the mouth of Salinas River (Becker 1964) depicts *"lagunas del la Herba de los Mansos,"* or *"Yerba Mansa"* (*Anemopsis californica*), a white wild-flower resembling the anemone, now common to this area. In January 1846, Frémont (1846: 35) saw "a few . . . flowers were already in bloom on the sandy shore of Monterey Bay. Among these were the California poppy, and *Nemophila insignis* [*menziesii*, baby blue eyes]."

After the first rains near Salinas, Frémont (1848: 37) "noticed geranium (*Erodium cicutarium*) in flower in the valley; and from that time the vegetation generally began to bloom." Near Point Conception, the cold summers resulted in "flowers blooming in all the months of the year" (38). Newberry (1857: 13) of the Pacific Railroad survey wrote that "the lowlands bordering the [San Francisco and San Pablo] bays . . . [have] a great variety of flowering annuals and on the richer slopes of the hills [they] dispute possession with the wild oat." J. G. Parke (1857: 10), also of the survey, described in general the lands between Los Angeles and San Francisco: "The grasses are chiefly annuals, and are reproduced by nature. . . . There are also a great variety of flowers intermingled with these grasses, whose brilliant colors, contrasted with the rich verdure, give these plains an Eden like aspect in the early spring months, rendering the country then as beautiful and inviting as it is bare and forbidding just prior to the rainy season." When William Perkins—a merchant for three years at the town of Sonora in the gold rush district—first arrived in California in May 1849, he saw at San Francisco harbor that "many of the islands, and the highlands surrounding the bay, are covered with a handsome growth of heavy timber, and . . . by large patches of brilliant flowers" (Morgan and Scobie 1964: 85).

Recalling the Mexican period, Davis (1929: 74) described flowery landscapes of San Francisco in the context of courtship and marriage: "The day when we were heaving anchor Captain Russom handed me a spy glass and directed my attention to two young ladies who were seated on a natural carpet of flowers covering the brow of a commanding hill which overlooked the vessel. . . . One of these ladies was picking the wild flowers within her reach." He also recalled that "the Californians were fond of having a good time out of doors on their ranchos. At *meriendas,* or barbecues of tender *terneras,* or yearly heifers, many rancheros would gather together, the guests seated on the ground relishing the good food spread before them on snow white linen, [and] the ladies found time to gather and arrange miniature bouquets of the flowers within their reach" (139).

The proliferation of landscape descriptions during the gold rush suggests that wildflowers were widespread across California, even to the coast. Merrifeld (1851) distinguished the California flower fields from other prairies in the western United States: "No country in the world, perhaps,—certainly not the prairies of the west, can compare with California, in the number, variety and richness of its wild flowers. From February until June, the plains are literally covered with wild flowers of almost every hue and color, varying from pure white to the deepest crimson and golden yellow." Referring to the gold rush, Merrifeld wrote, "If the floral wealth of California could be transplanted to our Atlantic borders, there are some, I think, who would prize it as highly as its mineral treasures." Weary after four months in the "gold diggings and speculation," Bayard Taylor went home via Mexico. Crossing California, he wrote that "everywhere the beauty and fragrance of innumerable wild flowers charm the senses" (Lymau 1850). Hunt (1859) wrote that "wild flowers . . . abound everywhere, and are of every hue and form, for the modest violet to the gorgeous poppy." An anonymous letter from California (1857) stated that the "plains and valleys are covered with green grass and the wildflowers begin to appear."

Flower fields may have been more extensive in the inland valleys of the Coast Range. The flank of the Gavalon Mountains descending into San Joaquin Valley was "clothed in the floral beauty of advanced spring" (Frémont 1848: 38). L'Enfant (1848) wrote of the Santa Clara Valley, "the whole plain is covered with wild flowers . . . the yellow wild oats waving in the breeze." Revere (1849b) wrote that the lands from Monterey to the Salinas plain were like most of the region: "delicious plains, richly enameled with those exquisite wild-flowers varying from the palest blue to brightest flame-color which are produced spontaneously in all parts of California." Revere (1849a) also suggested that wild oats at that time were most extensive in the plains: "the heights on which we rode, although thousands of feet above the level of the sea, were lavishly strewed with wildflowers of a hundred hues, large and more beautiful than any I had ever seen. On the plain below was growing a crop of oats, sowed and cultivated by natures hand along, and the bright green stems shot up more thickly and luxuriantly than in any cultivated field I have ever beheld." Likewise the Reverend T. Dwight Hunt suggested that wild oats grew primarily in plains along the coast: "Vast tracts of wild oats grow up to the coast range, the dry tops and the fallen grain of which afford the best of pasturage when the grasses of the valleys are brown and withered."

In the late 1860s, John Muir often took "the slow trip" from San Francisco to Yosemite through the San Jose plain and Gilroy, where he saw a mixture of wildflowers and introduced grasses. (The alternative route was by steamer to Fresno.) Along his route, Muir (1974: 19) found that the Diablo Range and Santa Cruz Mountains "still wear natural flowers, which do not occur singly or in handfuls, scattered about in the grass, but they grow close together, in smooth, cloud shaped companies, acres and hillsides in size, white, purple, and yellow, separate, yet blending to each other like the hills upon which they grow. Besides the white, purple, and yellow clouds, we occasionally saw a thicket of scarlet castilleias and silvery-leafed lupines, also splendid fields of wild oats (*Avena fatua*)." Brewer (1966) saw extensive wildflower fields in the Coast Range in the spring of 1861. Approaching Santa Inez, he saw "very green grassy slopes, on which there were a profusion of wildflowers with brilliant colors" (74). At Nipomo Ranch, Brewer (1966: 76) reported that "large oaks scattered here and there, the green grass beneath, and the great profusion of flowers, made it look like a fine park." The hills near Lompoc "were covered with pasture, or grass, with a great profusion of flowers" (38). Farther north on the ranch, Brewer wrote that "the profusion of wildflowers, beautiful elsewhere, now tired us with their abundance and sameness" (78).

Central Valley

The great Central Valley of California was covered by vast sheets of wildflowers mixed with *Erodium* and clovers. Wild oats and black mustard appear to have been restricted to the floodplains of the Sacramento and San Joaquin rivers and their tributaries.

The earliest description of the Central Valley northeast of Sacramento, by Jedediah Smith in 1828 (Evarts 1958), is ambiguous: "The winter in this valley is the best season for grass and at the time of which I am now speaking the whole face of the country is a most beautiful green, resembling a flourishing wheat field." His use of "grass" may be akin to the Spanish *zacate,* i.e., green cover of any kind, and it is uncertain whether "wheat fields" represent wild oats, a plant name never used in his diary. In 1833, Zenas Leonard (1959: 90), who had no botanical background, crossed the San Joaquin Valley, apparently near the San Joaquin River during the dry season, recalling: "This day our course lay through a large prairie covered with wild oats . . . on the south side of the river." When he approached the San Joaquin River, he made a confusing description of wild oats, which may in fact refer to some native such as *Hemizonia* (tarweed): "This day our course lay through a large prairie covered with wild oats—which at this season of the year when nothing but [the] stock [of the plant] remains, has much the appearance of wild oats" (88).

More explicit descriptions reveal that clovers were abundant species by the mid–eighteenth century. Bidwell (1937: 33), writing in 1839, said, "Here on this side of the Bay [Sacramento Valley], is an abundance of red and white clover growing with the grass." He described his Rancho Chico as "one of the loveliest of places. The plains were covered with scattered groves of spreading oaks [valley oak, *Quercus lobata*]; there were wild grasses and clover, two, three and four feet high [0.6–1.3 m], and most luxuriant" (Gillis and Magliari 2003: 94–95). Bidwell recalled that in spring the bears chiefly lived on clover, which grew luxuriantly on the plains, especially in the little depressions (vernal pools) on the plains (Gillis and Magliari 2003: 93).

Frémont's two expeditions make clear that *Erodium* and clovers were important pasture in the Central Valley. In the Sierra Nevada foothills east of Sacramento, the "hills generally are covered with a species of geranium (*erodium cicutarium*) . . . with this was frequently interspersed good and green bunch grass and a plant commonly called burr clover" (Frémont 1848: 17). When the San Joaquin Valley was "in the full glory of spring," the "*geranium* [*Erodium*] *cicutarium* was generally in bloom, occupying the place of the grass, and making the upland a close sward" (20). The rangelands of Butte County, near the Sacramento River, " consisted of excellent grasses, wild oats in fields, red and other varieties of clover" (27). He wrote generically that *E. cicutarium* "covered the ground like a sward" in the Sacramento Valley, where "squaws were gathering the seeds for food" (Frémont 1845: 243, 253). *E. cicutarium* was abundant in the San Joaquin Valley. At Kern River, west of present-day Lake Isabella, he wrote that "instead of grass, the whole face of the country is closely covered with *erodium cicutarium*, here only two or three inches high [4–7 cm]" (253).

Wild oats and black mustard were concentrated in the floodplains. In 1834, Leonard (1959: 114, 120) saw extensive fields of wild oats along the San Joaquin River. At Colusa, on the Sacramento River, Bidwell once spooked a group of Indians who were eating wild oats, which grew in abundance there (Gillis and Magliari 2004: 88). Bidwell (1937: 33) stated that near his Rancho Chico mustard grew in abundance. He also said that wild oats overspread the Sacramento Valley before there was livestock grazing, stating, "Here are also innumerable quantities of wild oats, which I am told, grow nearly all over California, and grow as thick as they can stand, producing oats of an excellent quality; but as neither cattle nor horses are ever fed here, they are never harvested" (33).

Frémont's (1848: 24–27) explorations of the Central Valley record wild oats in the floodplains. At the Sacramento River near Sutter's Fort, Frémont found "wild oats . . . in the bottoms" (24). At another location on the river he reiterated that the river "bottoms were covered with oats" (26). At the Buttes north of Sacramento, "the range consisted of excellent grasses, wild oats in fields, red and other varieties of clover. . . . Oats were now drying in level places exposed to . . . the sun, remaining green in moister places and on the hill slopes." Bryant (1848: 300–1) wrote in late summer (September 15) from the Coscumne (sic) River that his "route has continued over a flat plain, generally covered with luxuriant grass, wild oats, and a variety of sparkling flowers," very likely a summer annual such as *Hemizonia*.

The occurrence of wild oats along the rivers is supported by land grant *diseños* from the 1840s (Figure 3.3; Becker 1964). In the Sacramento Valley, most land grant concessions were obtained along the Sacramento River where the best pasture grows along the floodplain. Legal records for the *diseño* for Rancho de las Flores (1844), 10 miles south of Red Bluff on the Sacramento River floodplain, state that "wild oats reached to the saddle skirt" (Becker 1964; and see Figure 3.4). The *diseño* shows the best pasture along the rivers on the eastern and southern borders of the land grant. Likewise, the *diseño* for Rancho Omochumnes, 10 miles southeast of Sacramento, presents an elongated property boundary along the best lands near the rivers (Figure 3.5). The Rancho de los Molinos *diseño* shows the ranch on land above the floodplain, land that also follows the river. Wild oats also invaded wheat fields. Bidwell stated that "wild oats and weeds . . . have rendered the wheat in some fields undesirable for grain, and so has to be cut for hay" (Gillis and Magliari 2003: 157).

Other expeditions of this period describe the abundance of herb cover along the rivers but do not identify oats. In the Pacific Railroad survey of the 1850s, Blake (1856) described riparian habitat at Ocoya (Posé) Creek in the San Joaquin Valley, where "low banks and bottom-land were timbered with a dense growth of cottonwoods and willows, and considerable grasses can be seen along its border," i.e., grasses did not occur beyond the floodplain. Williamson (1857: 26), also of the survey, wrote that "a luxuriant growth of wild oats covers a large portion of the valley, and gives it an appearance of high cultivation." In the same period, Lieutenant George Derby of the topographic survey wrote that the Tule River "had rich tracts of arable land, fertile with every description of grass" (Farquhar 1937). The surrounding lands were "a beautiful, smooth, level plain, covered with clover of different kinds and high grass." He wrote that farther south "the Kaweah River is a beautiful, smooth, level plain covered with clover of different kinds and high grass and

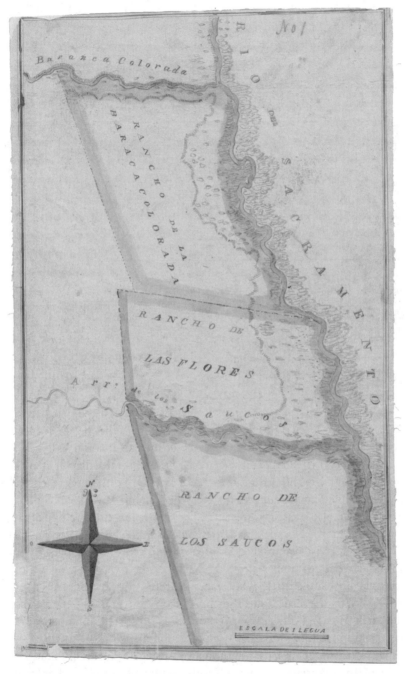

Figure 3.4. *Diseño* of Rancho las Flores. Courtesy of the Bancroft Library, University of California, Berkeley.

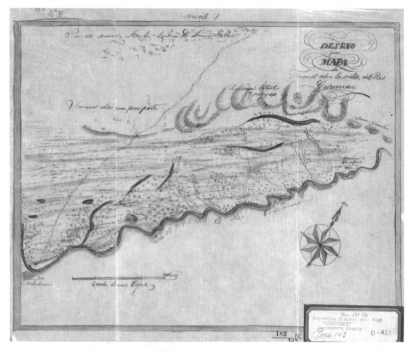

Figure 3.5. *Diseño* of Rancho Omochumnes. Courtesy of the Bancroft Library, University of California, Berkeley.

thickly shaded by one continuous groves of oaks." The oak trees, very likely *Quercus lobata,* clearly define a floodplain, as the species does not occur outside riparian zones in that part of the San Joaquin Valley.

Flower fields covered extensive areas of the Central Valley in association with *Erodium* and clovers. In 1834, Leonard (1959: 121) gave a broad picture of the San Joaquin Valley near the Sulphur River (San Joaquin River): "We also passed a great number of streams flowing out of the mountain [Sierra Nevada], and stretching afar toward the Pacific. The prairies were most beautifully decorated with flowers . . . interspersed with splendid groves of timber along the banks of the rivers—giving a most romantic appearance to the whole face of nature." In 1841, Bidwell (1937: 46) described the scene at his Rancho Chico : "Never did I expect to see the earth so beautifully arrayed in flowers as it is here." In another passage, Bidwell wrote, "The snow-capped mountains on each side of the valley seen through the clear atmosphere of spring, the plains brilliant with flowers, the luxuriant herbage, all truly combined to lend enchantment to the view. In fact this valley . . . was as new as when Columbus discovered America" (Gillis and Magliari 2003: 94–95). In the 1840s, T. John

Mayfield (1929: 9) wrote from the Coast Range near Los Banos, "The entire plain, as far as we could see, was covered with wild flowers. . . . Covered with great patches of rose, yellow, scarlet, orange and blue."

Flower fields caught the attention of John Bidwell, William G. Chard, A. G. Thoomes, R. H. Thomas, Job F. Dye and Major P. B. Reading in an expedition through the Sacramento Valley in search for lands appropriate for submission of a land grant, ultimately Rancho de las Flores, on the Sacramento River, 10 miles south of Red Bluff. They observed a mixture of exotic and native vegetation (Becker 1964): "All kinds of vegetation was most prolific, the wild oats reached to the saddle skirt. . . . They looked to the north, and saw that the coast range and the Sierra Nevada converge together . . . while to the south, a broad luxuriant plain stretched out before them laden with vegetation and sparkling with myriads of Nature's fairest gems." The abundance of flowers is suggested by the name Rancho de las Flores (Figure 3.4).

In his two expeditions, John C. Frémont made an extensive record of spring wildflowers along the length of the Central Valley. The lower Sacramento Valley was "gay with flowers, . . . some of the banks being absolutely golden with the California poppy (*eschscholzia crocea*)" (Frémont 1845: 245). A "showy Lupinus" adorned the banks of the "Mukelemnus River [sic]" (245). The Arroyo Calaveras had an "open grove of oaks, a grassy sward beneath; with many plants in bloom." South of Calaveras he "came again among innumerable flowers; and a few miles further, fields of beautiful blue flowering *Lupine*. . . . A lover of natural beauty can imagine with what pleasure we rode along these flowering groves. . . . The Californian poppy, of a rich orange color, was numerous today. . . . Under the shade of the oaks along the river I noticed *erodium cicutarium* in bloom, eight or ten inches high [20–25 cm]" (245). Frémont also "saw squaws gathering it on the Rio los Americanos" (248). On his second trip, Frémont (1848: 20) described the Sacramento Valley thus: "The higher prairies between the rivers presented unbroken fields of yellow and orange colored flowers, varieties of *Layias* and *Eschscholzia Californica*, and large bouquets of the blue flowering *nemophylla* nearer the streams." At another nearby locality, he saw "the California poppy, (*Eschscholzia Californica*) the characteristic plant of the California spring; *memophylla insignis* [sic], one of the earliest flowers, growing in beautiful fields of a delicate blue, and *erodium cicutarium*, were beginning to show scattered bloom" (19). At the Rio de los Americanos "the road led over oak timber . . . over ground slightly undulating, covered with grass intermingled with flowers" (21). The grass may have referred to *Erodium*, which at the time was called "pin grass." Near the Feather River, Frémont noticed "many plants were in flower, and among them the California poppy, unusually magnificent. It is the characteristic bloom of California at this season, and the Bear River bottoms, near the hills, were covered with it" (22). North of Sacramento, he made the remarkable comment, "Flowers were unusually abundant. The splendid California poppy characterized all of the route of the Valley" (25).

Flower fields also characterized Frémont's route in the San Joaquin Valley. Near the San Joaquin River, he wrote in his journal,

> On the third day out they approached the wide and deep San Joaquin river. They noticed that the soil was not the same as that of the Sacramento Valley. It was much more sandy, and there was not as much vegetation. This sparseness did not include the whole San Joaquin valley. They entered a region where flowering lupine gave the ground a bright color as thou some fantastic artist had smeared the earth with a giant brush. Beyond the thickets of lupine—some standing as much as twelve feet high [2–4 m]—past the sweet aroma of spiked flowers, there were great stands of live oaks. Then the color of the landscape changed to a rich orange as the men rode through acre after acre of California poppies. (Egan 1977: 233)

When he reached Tulare Lake at the terminus of the Kaweah River, Frémont (1848: 19) noted "the face of the country had been much improved by the rains. . . . Several humble plants, among them the golden flowered violet (*viola chrysantha*) and *erodium cicutarium,* the first valley flowers of the spring." His journal also provides detail on the distribution of individual wildflower species: "By the middle of March, the whole valley of the San Joaquin was in the full glory of spring; the evergreen oaks were in flower, *geranium [Erodium] cicutarium* was generally in bloom, occupying the place of the grass, and making the upland a close sward. The higher prairies between the rivers presented unbroken fields of yellow and orange colored flowers, varieties of *Layias* and *Eschscholzia Californica,* and large bouquets of the blue flowering *nemophylla* nearer the streams" (20). Between Lake Isabella and Walker Pass, the region was "enriched by a profusion of flowers . . . [and] our pathway and the mountain sides were covered with fields of yellow flowers [mostly likely goldfields], which was here the prevailing color. Our journey today was in the midst of an advanced spring, whose green and flora beauty offered a delightful contrast to the sandy valley we had left" (Frémont 1845: 21).

During the gold rush, William Perkins frequently made trips between San Francisco and Sonora though the Central Valley and Sierra Nevada foothills along the Stanislaus River, which he called the Estanislao. In his first California spring in 1850, he wrote, "The plains at this season [the middle of April] are literally covered with flowers. No garden can com-

pare in the beauty to the banks of the Estanislao. I think I never witnessed such a profusion of colors, and such brilliancy of hues; and this,
not confined to small and isolated spots, but the whole country is one
immense flower bed. The hills look like gigantic bouquets, and the *llanos*
[plains] like a huge Persian carpet" (Morgan and Scobie 1964: 148). In
another trip in 1852 Perkins wrote:

> Let any stranger to Californian scenery ride through the plains at this season
> of the year and he will find no language to express his astonishment and
> admiration. None of the most carefully cultivated gardens of the other lands
> can display the magnificent and gorgeous assemblage of colors with the
> plains, the hills, the river banks as far as the eye can reach, are pained. Out
> of the road track, the eye cannot rest upon a spot which is not studded with
> floral gems. In some places is a patch of an acre completely yellow; another
> is all purple, another is blue, others are all white, while alongside is another
> all red and scarlet; again there are patches literally crowed with flowers of
> all hues and all descriptions. (315)

Other explorers saw wildflowers in the Central Valley. When Thornton (1848: 89) left San Francisco for the Central Valley, he wrote, "In
the flower season, its blossoms make a beautiful display." Blake (1956)
of the Pacific Railroad survey recognized the spring wildflower season
in retrospect. When he crossed the Central Valley, he remarked that "the
season of green grass and flowers had passed" (9). Newberry (1857: 12–
13) of the survey wrote in 1855 that during the rainy season the southern portion of the Sacramento Valley was "perpetually in bloom. . . . With
the return of the autumnal rainy season vegetative life is again called into
vigorous action, and the country, which a few weeks before was a desert,
is now transformed into a flower garden." In 1862 near Mount Diablo,
Brewer (1966: 262) contrasted the vernal season with the "brown, dry,
dusty, and parched landscape" he saw the previous summer. He wrote
that this region "is now green and lovely, as only California can be in
the spring. Flowers in great profusion and richest colors adorn the hills
and valleys." In the 1860s, Clarence King and John Muir saw the San
Joaquin Valley in spring, using the road through Pacheco Pass from San
Juan Bautista and Gilroy. Their travels suggest that oat fields in the Coast
Range were replaced by wildflower fields in the Central Valley. C. King
(1915: 26) wrote that at Pacheco Pass he saw "the wild-oat surface of
the hills," similar to Bryant's (1848: 365) observation of "dense oat grass"
at Mission San Juan Bautista. Once over the pass, C. King (1915: 28)
described the Central Valley before him: "Brown foot-hills, purple over
their lower slopes with "fil-a-ree" blossoms, descend steeply to the plain
of California [Central Valley], a great, inland prairie sea, extending for

five hundred miles, mountain-locked, between the Sierras and the coast hills, and now a broad, arabesque surface of colors. Miles of orange-colored flowers, cloudings of green and white, reaches of violet which looked like the shadow of a passing cloud, wandering in natural patterns over and through each other, sunny and intense along near our range, facing in the distance into pale, bluish-pearl tones."

When Muir (1974: 21) traveled through Pacheco Pass into the Central Valley, he went "through the foothills, past the San Luis Gonzaga Ranch, and wading out in the grant level ocean of flowers. This plain, watered by the San Joaquin and Sacramento Rivers, formed one flower bed, nearly four hundred miles in length by thirty in width." Muir took a sample, "taken at random" near Hills Ferry, of all the flowers "counted one by one" to estimate the richness of this flower-field (Table 3.4). From his view, his sample reflected the sea of flowers across the valley as a "cupful of water from a lake." From his data, Muir appears to have crossed a sea of goldfields (*Lasthenia gracilis*), which he described, saying "the yellow of the Compositae is pure, deep, bossy solar gold." He also saw a purple stratum, very likely some species of lupine, and a greenish yellow of a moss stratum. Some legumes may have been species of clover. The Geraniaceae is very likely the introduced *Erodium cicutarium* and possibly *E. moschatum*. The thinly foliaged grass "with scarcely any leaves [that] did not interfere with the light of the other flowers" appears to be the fescue (*Festuca myuros*) introduced in the Spanish period. As Muir continued east from his sampling site, he described the Sierra Nevada as "that mighty wall up-rising from the brink of this lake of gold," in apparent reference to hillslopes dominated by goldfields.

Muir (1904: 338) summarized his trip across the Central Valley: "The Great Central Plain of California, during the months of March, April and May, was one smooth, continuous bed of honey-bloom, so marvelously rich that, in walking from one end of it to the other, a distance of more than 400 miles, your foot would press about a hundred flowers at every step. Mints, gilias, nemophilas, castileias, and innumerable compositae . . . one sheet of purple and gold." Facing west from the Central Valley, he saw that "the last of the coast range foothills were in near view all the way to Gilroy. Their union with the valley is one by curves and slopes of inimitable beauty. They were robed with the greenest grass and richest light I ever beheld, and were coloured and shaded with myriads of flowers of every hue, chiefly purple and golden yellow."

The following is Muir's famous description of the flower fields of the Central Valley in his essay on bee pastures (Muir 1904: 340–44), a panorama to be seen by many in the twentieth century:

TABLE 3.4 MUIR'S WILDFLOWER SAMPLE AT HILLS FERRY

Natural Order [family]	No. of Flowers		No. of Species
Gramineae	29,830	Panicles 1,000	3
Compositae	132,125	Heads 3,305	2
Leguminosae	2,620		2
Umbelliferae	620		1
Polemoniaceae	401		2
Scrophulariaceae	169		1
————?	85		1
Rubiaceae	40		1
Geraniaceae [*Erodium?*]	22		1
Musci	1,000,000 Funaria and Dicranum		

SOURCE: Muir (1974).
NOTE: Number of natural orders, 9 to 10; of species, 16; total number of open flowers, 165,912; mosses, 1,000,000.

Descending the eastern slopes of the Coast Range through beds of gilias and lupines, . . . I at length waded out into the midst of it. All the ground was covered, not with grass and green leaves, but with radiant corollas, about ankle-deep next the foot-hills, knee-deep or more five or six miles out [into the valley]. Here were bahia [baria], madia [tarweed possibly *Madia elegans*], madaria (now synonymy at *Madia*), burrielia [a member of the sunflower family], chrysopsis [*Chrysopsis sessiliflora*, golden aster, closely related to *Heterotheca*], corethrogyne [perennial herb *Lessingia filaginifolia*, cud-weed aster], grindelia, [gum plants, yellow sunflower], etc., growing in close social congregations of various shades of yellow, blending finely with the purples of clarkia [*Clarkia purpurea* or others], orthocarpus, and oenothera. . . . Because so long a period of extreme drought succeeds the rainy season, most of the vegetation is composed of annuals, which spring up simultaneously, and bloom together at about the same height above the ground, the general surface being but slightly ruffled by the taller phacelias, penstemons [sic], and groups of *Salvia carduacea*, the king of the mints. Sauntering in any direction, hundreds of these happy sun-plants brushed against my feet at every step, and closed over them as if I were wading in liquid gold. (Botanical interpretations by Andrew Sanders, pers. comm.)

Perkins made a more informal "sample" of the wildflowers than Muir's, remarkably comparable to the taxonomic sophistication of the Spanish journals, while having a smoke during a rest on a trip along the Stanislaus River:

The imagination may hardly conceive any thing so superb as is the sober reality of this country. Once we dismounted . . . and throwing ourselves down on a mattress of flowers, mingled the fragrant smoke of an *Havana* with their perfume. Without moving from where I lay, I plucked the flowers about me, and counted seventeen distinct hues: four of yellow, five of purple and blue, four of red, verging from brown to scarlet, one white, and the others perceptible shades without names. The Lubin [Lupine], and the marvel of Peru, the marigold, and amaranth, the Margarita, several varieties of Fuchia, the Larkspur of all covers, the golden Tulip [California poppy], the violet [*Viola* or *Nemophila*], the cowslip, the golden rod, the wild flowering current, and scores of others that I cannot name, all mingled in a most glorious and resplendent confusion, as if Nature meant to ridicule the comparative puny efforts of artificial cultivation (Morton and Scobie 1964: 316).

Boudet (1880) witnessed the same brilliant landscape as King and Muir before him, "a level expanse of pasture-land, lying between the Merced and San Joaquin Rivers, . . . which, for various and multitudinous color kaleidoscopic permutations, and general rainbow effects I have never seen equaled. Thousands of acres—not hundreds—dotted with the orange of the poppy, the purple of the violet, the yellow of the buttercup, the pink of the clover, the blue of the larkspur, and the deep scarlet of the silene, seem to weave a carpet harmonious in color and exquisite in its blending and contrasting tints." He also wrote that "the grass meadows of the San Joaquin and its tributaries . . . present a wilderness of bloom."

John Hittell (1874: 370–71) described the Central Valley wildflowers during a ride on the railroad:

Along the railroads on either hand runs continuously the rich radiant bloom. Your sight becomes pained, your very brain is bewildered, by watching the galloping rainbow. There are great fields in which flowers of many sorts are mingled in a perfect carnival of color; then come exclusive family gatherings where the blue, crimsons, or the purples, have it all their own way; and every now and then you come across great tracts, resplendent with the most gorgeous of all wild flowers, the yellow or orange poppy, which . . . a botanist [gave] the name *Eschscholzia*, but which long ago some poetic Spaniard . . . christened it *El Copo de Oro* (the golden cup). Every such tract where the sumptuous blossoms stand thick, reminds one of the "Field of Cloth of Gold." They are peculiarly joyous looking flowers, massed together, dancing, and hob-knobbing, and lifting their golden goblets to be filled by the morning sun.

Merrifield (1851) had a similar impression of floral mosaics: "One peculiarity noticed, which I have never seen in our western prairies, That is the arrangement of different kinds of flowers in masses almost entirely by themselves. Thus you may sometimes see patches of several acres, covered almost entirely with bright yellow flowers, as the Yellow Gerardia, for instance; and near to this another patch of a deep purple or scarlet hue. On the other side, perhaps may be seen a similar mass of white flowers."

California wildflowers also drew the attention of Cronise (1868: 232–33) in his environmental and land-use history of the state. North of Sacramento,

> the open spaces among the foot-hills, and more especially the prairies that skirt them, bloom in spring time with fields of wild flowers of every form and hue—all exceedingly brilliant and graceful, but generally deficient in odor. Sometimes a single variety will occupy several acres, to be followed by another patch equally extensive, covered by a different kind. It would be vain to seek in the most carefully cultivated gardens, where the choicest floral treasures of the world have been gathered, for anything more exquisitely shaped or tinted than can be found growing wild and uncared for in these immense parterres.

Cronise also remarked on the barrenness of the valleys in summer and fall, saying that after those seasons, "a few months later, . . . the whole country is not only verdant, but [also under] a luxuriant vegetation, mixed with millions of wild flowers, everywhere greeting the eye as the spring advances" (321).

Flowers also carpeted the Sierra foothills. According to Auber Forestier (1870), "Growing in among the *chapparal* [sic] on the hill sides in the forest glades, down ravines, in the fields, everywhere, we saw abundance of wildflowers. Oh these flowers! How shall I ever find words to convey the faintest idea of their profusion and loveliness." Referring to the poppy, "First we saw them in scattered patches, then extending over whole fields alone or intermingled with blue flowers, and white flowers and diverse other shades and varieties of flowers." Cooledge (1872) gave the same impression of the Sierra, saying that "never before in our lives had such wealth of flowers, such variety of kinds, and shapes, and tints come within our knowledge. Every ravine, every tangled thicket was hung and garlanded with blossoming things. Great belts of blooms sweep up the mountain walls; each mile of ascent revealed a new Flora; the very way-sides teemed with color." Cooledge's view of the Sierra Nevada was a microcosm of the state: "and yet this display, so unusual and beautiful in our eyes, was but the scanty gleanings of the magnificent harvest which each spring turns California in to one great flower-garden from Sierras to the shore." Coffin Jones (1881) wrote that "the valleys and foot-hills were clothed with freshest green, and wild flowers bloomed on a thousand slopes."

In his lay summary of the California flora, Cronise (1868: 525) wrote that wildflowers

> are widespread and numerous in California, the purely native all differing from the same species in other countries. . . . So numerous are these flowers in their season, as to form a marked feature, not only in the botany, but also

in the landscape scenery of California. In the spring of the year, the time for most of them to bloom, they cover not only the plains and foot-hills, but growth in many places to the very tops of the mountains. The forests are nearly everywhere filled with them, and even the arid prairies and deserts are often adorned by their presence. The different classes and genera do not usually intermix, but grow segregated in patches, some of which cover acres, and sometimes even square miles of space. Nothing can be more gorgeous than these vast fields of wild flowers, when arrived at full perfection. In the months of April and May, the whole country decked with its floral jewelry, set in the deep-hued verdure, presents a picture not easily found outside of California.

As the Spaniards had suggested a century before, Cronise speculated that Native American burning encouraged wildflowers at the expense of exotic grasses: "That the hills everywhere produced spontaneously from year to year a luxuriant crop of oats, and that the valleys, burnt up as they were in summer and autumn, were sure to be transformed into flower gardens in the spring" (346).

California wildflowers had a remarkably long blooming season that began soon after the first autumn rains. On his second journey, Frémont (1848: 19) noted that the California poppy and baby blue eyes become conspicuous in midwinter: "At the end of January, the California poppy, (*Eschscholzia Californica*,) the characteristic plant of the California spring; *memophylla insignis* [sic], one of the earliest flowers, growing in fields of delicate blue, and *erodium cicutarium,* were beginning to show scattered bloom" (cf. Wester 1981). The San Joaquin was in the full glory of spring by the middle of March. On April 11, Frémont (1848: 20) observed that "the California poppy was every where forming seed pods, and many plants were in flower and seed together." On May 25, "The plants were now generally in seed."

Muir (1974: 23–24) described California pasture with an inverted floral calendar: "February and March is the ripe spring-time of the plain, April the summer, and May the autumn." May to December is the "winter—a winter of drouth and heat." He also gives a more detailed phenology of wildflower fields:

By the end of January four species of plants were in flower . . . but the flowers were not sufficiently numerous as yet to affect greatly the general green of the young leaves. Violets made their appearance in the first week of February, and toward the end of this month the warmer portions of the plain were already golden with myriads of flowers of rayed compositae. This was the full spring time. . . . new plants bloomed every day. . . . In March, the vegetation was more than doubled in depth and color; claytonia [miners lettuce, *Claytonia*], calandrinia, a large white gilia, and two nemophilas were in bloom, together with a host of yellow compositae, tall enough now to bend in the wind. . . . In April, plant-life, as a whole, reach its greatest height, and the plain, over all its varied surface, was mantled with a close, furred plush of purple and golden corollas. By the end of the month, most of the species had ripened their seeds. (Muir 1904: 345)

Hittell's (1874: 370–71) calendar is similar to the others: "Of wild-flowers, there are a great variety and abundance in California, and they have different seasons of blooming. . . . In the spring time, the hills are frequently covered with it, and the red, blue, or yellow petals hide everything else. Each month has its flowers." Cronise (1868: 341–42) wrote that "the grass starts as soon as the soil is wet. . . . At Christmas, nature wears her green uniform almost throughout the entire state, and in February and March it is set with floral jewels. The blossoms increase in variety and profusion until April, when they are so abundant in many places as to show distinctly the yellow carpeting on hills five miles distant."

The phenology of the wildflower season was both earlier and later in the year near the coast. Thornton (1848: 89), who lived in the cool foggy summers at San Francisco, wrote that the greatest number of flowers occured during the months of May and June. Likewise, Brandegee (1892) stated, "The plants have a much longer growing season than in localities farther inland removed from the sea, and many perennials, especially herbaceous ones, are more or less in bloom during the whole year." In winter, according to Frémont (1848: 36), the vegetation along the coast between San Francisco and Monterey "appeared much more green and spring like, and further advanced than in the plains."

Flowers persisted later into early summer after wet winters. For example in 1880, the *New York Times* reported that "although springtime is now left far behind, . . . yet the fields and hillsides are in bloom through many interior valleys. . . . Especially this year, there are more flowers in June than in April."[6] Flowers persisted later into early summer after wet years: "Although the Springtime is now left far behind, and a popular prejudice very justly associates the Spring with wild flowers, yet the fields and hillsides are in bloom through many interior valleys."

Botanical inventory of California wildflowers

The first flora of California was compiled by Brewer and Watson (1876–80) based on collections obtained across California during the early 1860s. The flora provides only general, terse descriptions on the distribution and abundance of the most important wildflower species, e.g.,

TABLE 3.5 SELECTED WILDFLOWER SPECIES RANGES FROM THE STATE SURVEY, 1876–1880

Species	Range
Amsinckia intermedia	Dry open grounds. Common in the interior country.
Baeria (Lasthenia) gracilis	Open ground, San Francisco Bay southward.
Calandrinia menziesii (ciliata)	Abundant . . . in the valleys and on sunny hillsides. Vancouver I. to Baja Calif.
Castilleja foliolosa	Hillsides, Mendocino Co. to San Diego, most common southward
Chaenactis glabriuscula	Open grounds . . . along the foot-hills of the Sierra to Los Angeles
Chrysopsis (Heterotheca) sessiliflora	Santa Cruz to San Diego
Clarkia (elegans) purpurea	Valleys and hillsides, Mendocino. to Los Angeles and foothills of Sierra Nevada
Corethrogyne (Lessingia) filaginifolia	Open places, San Diego to Santa Cruz, and in the interior to Tejon and Yosemite
Cryptantha (Eritrichium) intermedia	Open grounds throughout the state
Eschscholzia californica	Sunny exposures, particularly valleys and low hills, throughout the state, . . . often in great abundance. Most conspicuous flower of the state flora and sometimes large areas are made painfully brilliant by its intense glow in the bright sunshine.
Eucrypta chrysanthemifolia	Shady grounds, not uncommon from Bay of San Francisco to San Diego
Filago californica	Open ground, through the western part of the state
Gilia achilleifolia	Hills and sandy ground. Common through the western part of the state.
Gnaphalium (decurrens) californicum	Cudweed. Common on hillsides, from San Diego through Oregon.
Heterotheca grandiflora	Near coast, on sandy plains, Monterey to San Diego
Layia platyglossa (elegens)	Common throughout the western part of the state
Lepidium nitidum	In winter and early spring, from above San Francisco to Los Angeles
Lupinus truncatus	From San Francisco to Los Angeles
Lupinus hirsutissimus	In dry places from Sacramento to southern California
Madia elegans	Hills and plains, throughout California
Mirabilis californica	On dry hillsides, southern California and eastward
Nemophila aurita (Pholistoma auritum)	Low shady grounds, Sacramento Valley to San Diego

Nemophyla insignis (menziesii)	Common in low damp grounds, . . . bright blue flowers from the earliest spring
Oenothera californica	Valley bottoms
Orthocarpus purpurascens	Common on the hills and mountains of the coast . . . so abundant as to give the ground a purple hue for miles in some places . . .
Pectocarya (penicillata) linearis	Common in sandy or gravelly soil along and near the coast
Penstemon spectabilis	Dry plains and hills . . . one of the handsomest species
Phacelia ramosissima	Dry ground, . . . San Francisco Bay to the southern limits of the State
Phacelia distans (douglasii)	Open grounds, from Monterey southward
Phacelia whitlavia (minor)	Los Angeles to San Bernardino. Prized in cultivation.
Plagiobothrys canescens (californicus)	Open grounds, common throughout the state
Platystemon californicus	Very common . . . on lower hills and in valleys Mendocino to S. California
Senecio californicus	Low grounds, common from Santa Barbara to San Diego
Stephanomeria (paniculata) virgata	Hills and plains, common throughout the state
Tropidocarpum gracile	Valleys and low hills in the coast ranges from Los Angeles to the Sacramento
Viola pedunculata	In coast ranges . . . southern California to San Francisco and probably northward
Zauschneria californica	In dry localities from Napa and Plumas Counties to S. California
Emmenanthe penduliflora	Open ground, not rare from Lake County to San Diego
Salvia columbariae	Common throughout the State
Camissionia	On slopes

SOURCE: Brewer and Watson (1876–80).

"common," "abundant," compared to the standards of even somewhat modern floras, e.g., the Jepson manual (1925). Hence, comparisons of the present flora with distributions in the Brewer and Watson's state survey are hazardous at best. All of the listed species are found extensively in California today (Table 3.5).

A few species inspired comments on their past abundance. For example, *Orthocarpus purpurascens* (owls clover) was so abundant as to give the ground a purple hue for miles in some places. The California poppy (*Eschscholzia californica*) elicited the remark, "Sunny exposures, particularly valleys and low hills, throughout the state, . . . often in great abundance. This is the most conspicuous flower of the state flora and sometimes large areas are made painfully brilliant by its intense glow in the bright sunshine" (23). In the Pacific Railroad survey, Torrey (1856: 64) stated that *Eschscholzia californica* was "common in most parts of California." The dominance of the poppy was captured as a metaphor for the gold rush by an anonymous writer in the *New York Times:* "Of the wildflowers along the road, the yellow were holding out best. By the way, the prevalence of this color in California landscapes is always noticeable—as it were the floral symbol of the aureate treasure hid under so much of the soil for many centuries."[7]

Brewer's work is also supported by lay articles, e.g., "Early Spring in California" (Cornhill Magazine 1883), which describe the same genera as the dominant ones across California.

PERENNIAL BUNCH GRASSLAND

Mid-nineteenth-century botanical observations and landscape descriptions reveal that bunch grasslands were found in the foothills of the Sierra Nevada and the Coast Range and in the Mojave Desert, where they are found presently (Beetle 1947), but were not described in the Central Valley. Moreover, *Nassella* (*Stipa*) and other bunch grasses were not found in mission bricks, even though the Spanish fathers often used the nearest refuse available (Hendry and Kelly 1925; Hendry 1931). While the absence of evidence may reflect differential preservation, the bricks record frail species such as *Erodium cicutarium*. Moreover, the importance of preservation is diminished by the "instantaneous burial" of plant remains by the hand of man, without natural selection forces of biological decomposition and erosion. Only the riparian *Elymus* was uncovered in brick samples. All the missions were located along the California coast plains in areas where bunch grasslands were not described in the nineteenth century.

In the Sierra Nevada east of Sacramento, Frémont (1848: 17) found "hills generally covered with a species of geranium (*erodium cicutarium*) . . . [and] with this was frequently interspersed good and green bunch grass and a plant commonly called burr clover." An "open prairie, partly covered with bunch grass" was seen near the Mokel-umne (sic) River (16). When Bryant (1848: 305) entered the Coast Range between Sacramento and San Francisco, he saw with the wild oats "tufts or bunches of a species of grass, which remains green through the whole season." Cronise (1868: 122) wrote that the moist Coast Range of the Pajaro region in northern Monterey Bay had "fine crops of wild oats, bunch grass, and a variety of clover and native grasses." In his general description of California vegetation, Cronise (1868: 356) stated that "bunch-grass is a peculiar herbage on many dry hill sides, and affords a perpetual pasture. It occurs always in detached bunches, sufficient in size to make a small mouthful and seems to be proof [adapted] against drought—but is not cultivated." He found "Festuca Scabrella (Torr.) [*Poa scabrella*]" growing on "north hillsides and lightly shaded woods," most abundantly on "shady hillsides of the Coast Range" (523).

In the Central Valley, coarse grasses grew in riparian tule swamps. In the Pacific Railroad survey, Blake (1859) found Ocoya (Posé) Creek to be a "swamp, without any vegetation except tufts of coarse grass and the common 'grease bush.'" At Tulare Lake was a dense growth of rank grass and tule, which in a footnote is identified as *Scirpus lacustris*.

Botanical distributions in the state survey (Brewer and Watson 1876–80) are consistent with the fragmentary observations of Frémont and the Pacific Railroad survey, and also suggest that bunch grasses were not pervasive in the state. Purple needle grass, *Stipa setigera* (*Nessella pulchra*) "is common on the coast ranges and on the foot-hills of the Sierra Nevada and. . . . , is the most common and valuable "bunch-grass" of the dry hills" (286). The state survey gave geographic ranges but did not assess the abundance of other bunch grasses. *Koeleria cristata* ranged from San Francisco and Santa Ynez northward to Oregon . . . to Pennsylvania." *Melica imperfecta* grew "in various localities from San Diego northward to Oregon" (303). *Atropis tenuifolia* (*Poa scabrella*) was "frequent throughout the State from San Diego to Oregon and northward," and was "one of the most valuable of the bunch grasses" (310). Other bunch grasses listed are riparian species of tule swamps. *Elymus (sibiricus) glaucus* was found in "San Francisco, Ukiah, and elsewhere," while *Elymus triticoides* "occurred at Fort Tejon, San Juan, Monterey, and San Francisco" (326). Hendry and Kelly (1925) record *Elymus* sp. in a brick from Mission San Jose.

In *Pasture and Forage Plants*, Brewer (1883: 961–62) summarized that eight grass species are known in California: "They are mostly bunch grasses. . . . M. imperfecta and M. stricta are the most abundant." *Stipa* (*Nassella*) *occidentalis* was a common bunch grass in the Sierra Nevada. The tule swamp plant wild rye (*Elymus*) was most common in the interior.

SUMMER BARRENS IN THE INTERIOR

The interior foothills and plains were barren in the hot summer climates of the interior foothills and plains of southern California and the Central Valley, similar to accounts from the Spanish period. Wildflowers,

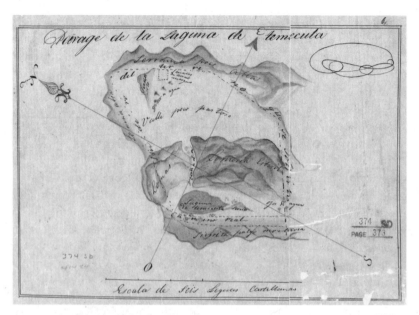

Figure 3.6. *Diseño* of Rancho la Laguna de Temecula. Courtesy of the Bancroft Library, University of California, Berkeley.

Erodium, and clovers dried out and disarticulated, leaving sparse, dry plant residue. Several writers contrasted the interior barrens with the luxuriant summer pastures along the coast exposed to the cool foggy summer climate. While the barrens could be attributed to overgrazing, most interior lands did not see livestock until the 1850s. Fires were rarely documented, apparently for lack of fuel.

The *diseño* for the Rancho Laguna de Temecula land grant extends from Elsinore to Wildomar. The map depicts *"Valle poco pastoso"* (valley with little pasture) north of Lake Elsinore. The nearby hills were *"lomería esteril,"* i.e., they lacked pasture (Figure 3.6). Brewer (1966: 40) contrasted verdant grazing lands near the coast with the interior valleys south of Riverside. Referring to "Temescal," a canyon north of Lake Elsinore, he "passed over the lovely plains of San Gabriel, El Monte, and Los Angeles, with their thousands of cattle, horses, and sheep feeding. Tens of thousands were seen, in pleasing contrast with the barren hills of Temescal."

Herbaceous cover was also scant in the San Fernando Valley. In the Pacific Railroad survey, Blake (1856: 74) wrote that "the plain was without trees or verdure." He also wrote that "herds of cattle were seen on parts of the broad plain, feeding on the dried grass, or the burrs of the California clover, which covers the ground in the latter part of the summer when all the grass had disappeared. This plain doubtless presents a beautifully green surface in the winter and early summer when watered by the rains" (75). Archduke Salvator in 1876 wrote, "Even after the meadows turn earth brown in color, cattle and sheep can subsist on these seeds (wild oats and burr clover) on land which to the experienced appears to be a desert" (Salvator 1929: 81). At Rancho Cucamonga, "The country from Cucamonga to San Bernardino for 18 miles . . . [has] scarcely any vegetation . . . [in the summer] the whole country presents a dry and parched appearance at this season, especially the mountains and ranges of rolling hills, having in some places the appearance of a fire passing over them" (Black 1975: 45–46). Theodore S. Van Dyke (1886: 27) described what appears to be disarticulation and desiccation of pasture: "The soils appear to be almost pure sand. . . . Vegetation dries up and disappears by slow pulverization."

The Central Valley elicited numerous observations of barren, desert like landscapes from the Mexican period and after the gold rush. The earliest non-Hispanic record of barrens was made in 1832 by John Work's party of trappers in the Sacramento Valley, who lamented the "very poor feeding here for the horses" (Burcham 1957: 98). In 1834, the Walker party suffered for lack of pasture in the arid southern San Joaquin Valley. At one point "they were in search for pasture for their cattle [and] forced to kill some of their beef " (Leonard 1959: 120). Kip recalled that "after leaving the grove by the [Kern] River we entered at once among the most desolate hills. Not a sign of herbage was see on them—not enough to attract a bee" (Wester 1981).

Bidwell (1937: 28), referring to the lands above the river of the Central Valley, said that "not a weed was to be seen, and the land was as mellow, and free from weeds, as land could be made by plowing it twenty times." He also recalled that "in many places the soil is black, and has every appearance of being as fertile as any land I ever saw, but I am informed that this is never sown, or planted in consequence of its drying too much in summer and cracking open" (31).

The *diseños* of the Central Valley contain few observations of vegetation except for riparian forests along the rivers (Ranchos New Helvetia, Llano Seco, Farwell, "Willy," and Pescadero). Sutter depicted on his *diseño* for his Rancho New Helvetia (Sutter's Fort) that the hills above were *"esteril,"* i.e., sterile or free of pasture (Figure 3.7). The *diseño* for Rancho Omochumnes on the Sacramento River is inscripted with the comment *"Lomeraís esteriles pedregosas"* (rugged sterile hills) and *"terenos* [sic] *altos con poco pasto"* (high terrain with no pasture) (Figure 3.5). The *diseño* at Rancho Tolenas in Solano County in the Sacramento Valley describes the slopes above the Sacramento River floodplain as *"lomas muertas,"* i.e., dead or barren hills. The *diseño* for Rancho de los Molinos shows the lands above the floodplain as *"tierra esteril,"* i.e., without plant cover or pasture good for cattle. Dust storms were apparently common in the Central Valley. On January 13, 1877, before cultivation was extensive in the valley, Bidwell's diary records "a dry north wind, bringing clouds of dust" (Gillis and Magliari 2003: 161). This remark recalls Father Ripoll's account of a dust storm in the southern San Joaquin valley in 1824 (see Chapter 2). In 1841, Charles Wilkes (1844: 193) described the Sacramento Valley as "barren and unproductive. . . . The high prairie [slopes above the floodplains] is in general barren, and as affording but little good pasture."

In 1841, Charles Wilkes (1849: 28–29), described the landform and vegetation in the Sacramento Valley in an interwoven logic, but he clearly distinguished the "upper prairie" and the fertile lands that were confined to the lower hillslopes and floodplains of the river:

> The upper prairie at the head of the Sacramento Valley is between two and three hundred feet [60–100 m] above the level of the river and inclines like the prairies to the south; its width does not exceed 5 miles . . . its undulating hills consist of a clayey or sandy loam, gravel and pebbles, while the soil of the lower prairie is rich alluvial . . . the southern part of the lower prairie on the west side of the river is covered with oaks, which likewise grow in the upper prairie, and as they approach the mountains, become more dense.

Wilkes characterized the upper prairie hills as consisting of rocky loamy soils overgrown with prairie, meaning that the land surface consisted of partially exposed soil, i.e., was free of herbaceous cover. His appraisal of the more arid San Joaquin Valley clarifies the status of pasture away from the rivers: "The lower prairie is almost entirely wanting. . . . The western side . . . is destitute . . . [and] suffer[s] all the effects of drought and excessive heat, being deprived even of the winds of the ocean . . . and is entirely barren and useless" (28–29).

During his first expedition, Frémont traversed the length of the San Joaquin Valley when the pasture was drying in April and May. His comments are strikingly comparable to those of Muñoz of the 1806 Moraga expedition. The San Joaquin River "prairie bordering the river had little grass" and at the Rio Merced "the country had lost its character of extreme fertility" (Frémont 1845: 250). While crossing the southern San Joaquin Valley, another entry of his diary states, "Today we made another long journey of about 40 miles, through a country uninteresting and flat, with very little grass" (253).

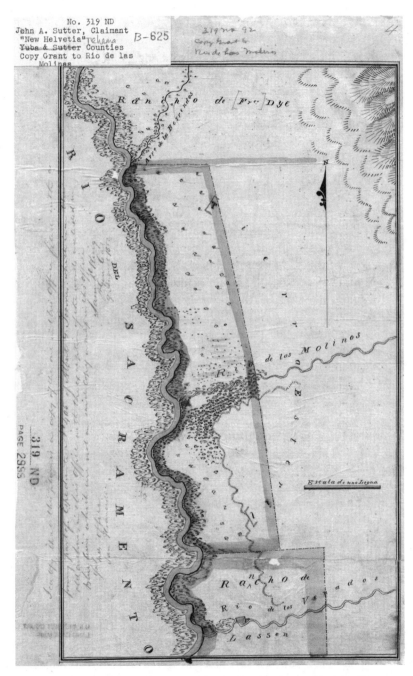

Figure 3.7. *Diseño* of Rancho New Helvetia (Rancho Sutter, Rancho Rio de los Molino). Courtesy of the Bancroft Library, University of California, Berkeley.

Lieutenant George Derby was also disparaging of the San Joaquin Valley in April and May of 1850. He called the southern San Joaquin valley a "desert" where "the vegetation was best along the margins of creeks" (Farquhar 1937: 253). In another passage of his journal, the area consisted of "barren, decomposed soil and no trace of vegetation but a few straggling Artemisias, except along the margins of the creeks" (256). The riparian belts offered no relief: "The country over which we passed between these sloughs [of the San Joaquin River] was miserable in the extreme, our animals suffered terribly for want of grass" (260). Derby described the entire region accordingly: "The Tularé Valley, from the mouth of the Mariposé to the Tejon pass at its head, is about one hundred and twenty miles in extent, and varies from eight to one hundred miles in width . . . is little better than a desert. The soil is generally dry, decomposed and incapable of cultivation, and the vegetation, consisting of artemisias and wild sage, is extremely sparse" (265). Perkins witnessed the drying of the flowers into barrens several times in his three years in California (Morgan and Scobie 1964: 152). In 1850 he wrote, "The flowering are disappearing from the *llanos*, and the grass is already turning brown and crisp. . . . But by the commencement of June nothing green is to be seen except the leaves on the trees, and these are blackened by the dust and the constant sun." The following year he wrote that "the flowers are covering the hills. . . . But, 'all that's bright must fade,' and soon the summer's sun will banish the winged zephyrs to the mountain fastnesses of the *Sierra Nevada*; dry up the gold and purple mists of the valleys; shrivel up the brilliant flowers on the plains and hill sides, and tan to an olive brown" (212). In 1852, Perkins wrote, "How soon all this beauty vanishes like a short dream! In six weeks, pass over these same plains, these fairy banks, and where now you cannot find a spot to stand on without crushing a score of flowers, all will be desolate, bleak, sun-scorched and brown" (316).

Blake (1856) of the Pacific Railroad survey also crossed the San Joaquin Valley in summer. At Livermore, he appreciated that the hot interior climate was unfavorable to wild oats: "The soil is loose and poor and does not appear to sustain the usual growth of the wild oat" (8). Blake contrasted the river vegetation and the surrounding plains. "The lower part of the San Joaquin River is bordered by numerous sloughs, and winds through low marshy ground covered with rushes and willows." The surface of the San Joaquin Valley "was level, and is almost a desert plain, consisting of gravel and pebbles brought down from the hills." He did see "a growth of sunflower, standing six to ten feet high, and the blossoms very small [a summer annual such as *Hemizonia*]" on the plains east of Mount Diablo (10). From the Chowchilla to the Fresno River,

Blake noted that "the surface is formed of sand and gravel . . . almost destitute of vegetation, and free from moisture" (17). At the Kings River, he wrote, "This beautiful plain, covered with luxuriant vegetation is a fine example of the effect of irrigation, for without these streams it would be a desert" (27). He also wrote that a "hot wind . . . sweeps over the dried weeds and gravel of the plain" (27). At Bear Creek near the Mariposa River, the "exposed strata contained a dwarfed bush or tree, here and there, the only vegetation" (15). Farther south Blake described the country almost identically as had Zalvidea and Muñoz a half century before. The area was "desolate and desert-like plains . . . skirt[ing] the foothills of the sierra" (27). Near Wasco, the Ocoya (Posé) Creek stood out as a "green strip of vegetation" in contrast to the "barren and parched surface of the surrounding hills" (33). The country west of the Tulare Lake had "no vegetation, other than a few dried weeds" (41). At the south end of the San Joaquin Valley near Grapevine Canyon, Blake wrote, "There is little or no vegetation, and the surface was dry and gravelly" (41). Blake perhaps best captured the contrast between the rivers and the plains at Four Creeks, a drainage south of the Kings River: "Here the aspect of the landscape is suddenly changed. Instead of the brown, parched surface of gravel to which the eye is accustomed on the surrounding plains, we find the ground hidden from view by a luxuriant growth of grass, and the air is fragrant with the perfume of flowers. The sound of flowing brooks, and the notes of the wild birds, greet the ear in strange contrast with the rattling produced by the hot wind as it sweeps over the dried weeds and gravel of the plain" (27). Generalizing for the entire San Joaquin Valley, Williamson (1856: 13) wrote that "the plains between the streams are destitute of foliage." The party found relief in the higher elevations of the Tejon Pass that borders the south end of the San Joaquin Valley, noting, "The Tejon is really a beautiful place. It . . . produces fine groves of oak, and an abundance of grass, and the green and fresh appearance of the spot presents a striking contrast to the parched and barren plain north of it" (Williamson 1856: 21). Likewise, Blake (1856: 38) observed that it was only at higher elevations at Tejon Pass that the expedition found "an abundance of excellent grass."

The botany of the Sacramento Valley was examined by Newberry (1857: 22) of the Pacific Railroad survey, who found the lands near Vacaville to be "a tableland . . . covered with a thin coating of the grasses and other plants . . . characteristic of other gravel surfaces of the valley." His description of the prairie in the dry season is similar to Charles Wilkes's report from 1841: "The center of the valley is occupied by a broad alluvial plain, . . . of which the soil is generally fine and fertile. . . .

The more fertile surface is covered with a growth of wild oats, or grasses, interspersed with a great variety of flowering annuals, while the gravelly and more unproductive portions support a thinner growth of coarser plants (*Eryngium, Hemizonia, Madaria* (*Madia*), etc.). Bordering the central plain is a second 'bench' or terrace [that] . . . forms a distinct prairie plain. . . . The scenery of the foot hills is frequently picturesque and beautiful, with its lawn-like slopes and clumps of spreading oaks" (Wilkes 1844). In *Rambles of a Botanist,* Muir (1974) noted that citizens visiting Yosemite from San Francisco invariably went by boat along the route of navigation through the bay and up the San Joaquin River to Fresno. The "slow route" by horse or wagon across the Central Valley was avoided not only because of lost time, but the valley was also undesirable. Muir wrote, "I have lived in Yosemite Valley three years, and have never met a single traveler who had seen the Great Central Plain of California in flowertime; it is almost universally remembered as a scorched and dust-clouded waste, treeless and dreary as the deserts of Utah" (1974: 17). Muir (1874) suggested that flower cover was thin during the growing season, especially on the hills, pointing out that with respect to "the floral profusions extend[ing] to the rolling foot-hills, . . . the redding tint of the soil shows through its vernal dressing." Muir (1904: 346) saw the barrenness of the Central Valley in relation to desiccation and decomposition of the flower fields: "The shrunken mass of leaves and stalks of the dead vegetation crinkle and turn to dust beneath the foot, as if it had been literally cast into the oven." In *Keweah's Run,* Clarence King (1915: 141) described the Central Valley as "a plain slightly browned with the traces of dried herbaceous plants." An extract from *Cornhill Magazine* (1883) describes the crossing of the Central Valley as a perilous adventure in which "the summer wanderer travels in choking, blinding dust clouds."

Cronise (1868: 323) distinguished the pasture on the western and eastern sides of the San Joaquin Valley. In Fresno and Merced Counties, "nearly a third of its territory comprising the western part is extremely dry; the most of it so arid as to produce but little grass, and being, at best, fit only for sheep pasturage." In Kern County, "near all of the western part of the county is valueless, for agricultural purposes" (118).

Brewer (1966) found that the plain from Visalia to the Tule River for "thirty miles" was very barren. The Tule River had "wide stretches of barren sand on either side" (38). Farther south near White River, "the road this day was through a desolate waste—I should call it a desert—the soil is barren and, this dry year [1863], almost destitute of vegetation" (38). Brewer (1966) examined the Central Valley by riverboat in early June 1864, after two consecutive winters of severe drought, landing at Firebaugh's Ferry near present-day Fresno on the San Joaquin River. He recorded in his journal that over a span of 25 miles "portions of this ride, for miles together, not a vestige of herbage of any kind covered the ground; in other places there was a limited growth of wire grass or alkali grass" (510). The following day at Fresno, "for the first ten miles the ground was entirely bare, but then we came on green plains, green with fine rushes called wire grass, and some alkali grass. The ground is wetter and cattle can live on the rushes and grass. We now came on thousands that have re-treated to this feed and have gnawed almost into the earth" (510).

Barrens were also described in the interior valleys of the Coast Range. The eastern slope of the Santa Lucia Mountains was described in the Rancho Guadalupe *diseño* as *"Sierra Árida"* (arid mountains) (Online Archive of California). The barrenness of the interior northern Coast Range was recorded in the Russian botanical collections of E. Voznesenski in 1840–41 (Howell 1937). The expedition left the north coast for the interior basin in summer, where they traversed hot and desiccated country. Eastward from Santa Rosa the botanical labels, faithful to Latin botanical nomenclature, have the word "desertum." According to Frémont (1848: 38), in the inland Coast Range "the plains become dry, parched and bare of grass. The cattle had climbed to the ridges of the Coast Range [where] good green grass [possibly perennial bunch grass-land] grows there throughout the year." Brewer (1966) had a similar impression of the Coast Range. At Salinas Valley "soil was dry and parched and the grass as dry as hay" (97). Scanty pasturage was seen at Guadalupe Ranch (97), New Idria (139), the Panoche plain (144), and the Diablo Range south of Mount Diablo (278, 279). Brewer observed or experienced clouds of dust and dust storms at many locations in the interior. The San Juan (Bautista) plain produced "a cloud of dust, raised by the high wind like a gauze veil two to three thousand feet height," and at Pacheco Pass "the wind shrieked through the canyons; it had annoyed us much on coming up the pass by the dust it raised" (150). At the south end of the San Joaquin Valley, "the storm . . . struck us; but instead of rain it was wind—fierce—and the air filled with dust and alkali. . . . Sometimes we could not see a hundred yards in any direction—all was shut out by clouds and dust" (382–83). Brewer recorded that he saw dust in the Cuyama plain (84; cf. account of Zalvidea in Cook 1960), Guadalupe Ranch (97), the Panoche plain (144), Mount Diablo (201), and the San Joaquin Valley near Coalinga.

Blake (1856) described the southern San Joaquin Valley as a labyrinth of dry ground, rivers, lakes, and swamps, reminiscent of Fages's account of a labyrinth in 1772 (see Chapter 2). From his vantage high up in the

San Emigdio Mountains that rim the southern margin of the San Joaquin
Valley, Blake wrote,

> Northwards the vision was unobstructed, and the broad, extensive valley
> of the Tulares was before me. On one side the heights of the Sierra Nevada,
> and on the other, the ranges of the Coast Mountains, stretched out toward
> the north until the remote peaks were lost in the smokey distance. They
> seemed like great arms holding a semi-desert plain and its shallow lakes
> between them. The two small lakes, Buena Vista and Posuncula [Kern],
> were distinctly visible below, and the lone line of timber on the sloughs
> of Posuncula River lay spread out before me as if on a map. . . . The whole
> region seemed peculiarly brown and barren. (44)

Burning rarely elicited comments from early American settlers, Muir,
Brewer, King, or from the various U.S. government surveys. Leonard
(1959) saw Native American campfires, i.e., "smoke rising from the
prairies in different places." He apparently once saw a broadcast burn,
but his description seems generic rather than a specific event: "At this
season of the year, when the grass in these plains is dry, if a fire should
be started it presents a spectacle truly grand—and if the flame is assisted
with a favorable wind, it will advance with such speed that the wild horses
and other animals are sometimes puzzled to get out of the road" (86).
Cronise (1868: 314–15) suggested that most burning occurred in the tule
swamps near the rivers, which he estimated covered 200,000 acres (60
km²), much of which was covered by a shallow water. In late summer,
he said, "large sections of these lands become dry on the surface—the
dense body of rushes, the growth of former years, having meantime wilted
and dried up, the latter often take fire, and burning with terrific fierce-
ness for days in succession, many thousand acres are burned over and
stripped of both the dead and living tules" (314). He also stated that fires
"break out in the grass and herbage, which late in the summer become
dry as tinder, and sweeping over the plains and mountains, leave millions of acres scorched and blackened, though the heat is not generally
sufficient to injure the forest trees or larger shrubbery" (315). It is unclear
how he came to estimates of such large fires in areas he also described
as barren in the dry season.

The verdancy of coastal pastures was frequently compared to sparse
pastures in the interior, first noted by Crespí, Font, and Fages in their
traverses of the Carquinez Strait, and the difference was attributed to
summer climate by several mid-nineteenth-century writers. At San Francisco Bay, Frémont (1848: 35) observed, "The country in July began to
present the dry appearance common to all California as the summer
advances except along the northern coast within the influence of the

fogs. . . . In some of these was an uncommonly luxuriant growth of oats, still partially green, while elsewhere they had dried up." Farther south in the Coast Range, he saw that "the plains become dry, parched and bare of grass," while "along the coast, the grass continues constantly and flowers bloom in all months of the year" (38).

Brewer (1966: 257–58) wrote, "The finest grazing district I have yet seen in the state in among these hills, lying near the sea, moistened by fogs in the summer when the rest of the state is so dry." The "grassy slope" of the coastal Santa Lucia Mountains (92) contrasted with the interior slope. Brewer wrote that "on passing the Santa Lucia the entire aspect of the country changed. It was as if we had passed into another land and other clime. The Salinas Valley [on the interior side] . . . is much less verdant than we anticipated. There are more trees but less grass" (92). Likewise, Hittell (1874) stated, "The natural pastures near the ocean keep green longer than those in the interior. . . . Fine pasture is found in some of the high parts of the Sierra Nevada, and many dairymen who have their homes in the valleys or foot-hills, drive their herds up into the mountains at the beginning of summer." Referring to the Santa Lucia Mountains, Cronise (1868: 116) wrote that "the grasses are green and fresh on the south side [San Luis Obispo] for more than a month after those on the north side [Salinas Valley] are dried and withered." In northern Monterey, near Pajaro, "The hills in the coast Range afford pasturage, in seasons when the plains and valley suffer from drought" (122).

Walker Pass presents another climatic gradient between the interior California valleys and the Mojave Desert, where Mediterranean ecosystems are replaced by desert scrub, similar to that described in the Anza expeditions across the San Jacinto Mountains (see Chapter 2). After seeing wildflower fields in the Lake Isabella region, Frémont (1848: 252) looked into the Mojave Desert from Walker Pass 30 miles east and saw that "the distant mountains are bold rocks again; and below if the land had any color but green." In the descent of this pass, he states that "the Erodium cicutarium finally disappeared." Frémont's Native American colleague, who was fluent in Spanish and who was on visiting relatives at the San Fernando Mission, said while "stretching out his hand" to the Mojave Desert before him: "No hay agua; no hay zacate—nada. There is neither water or grass, nothing."

WILDFLOWERS AND BUNCH GRASSES IN THE DESERT

Accounts of "palms" by Fages in 1772 (Priestly 1937) and "yuccas" by Frémont (1845: 257) identify the Joshua tree in the western Mojave

Desert. Frémont's writings make clear that a "zygophyllaceous shrub" (creosote bush) dominated the Mojave Desert, just as Font and Arrillaga saw decades before the *"hediondilla"* or *"gobernadora"* in the Sonoran Desert.

Surprizingly, the most frequent accounts of bunch grassland in California were made in the Antelope Valley in the western Mojave Desert, where *Nassella* (*Stipa*) *speciosa* is now abundant. Near Palmdale, Frémont (1845: 257) reached "the top of the spur, which was covered with fine bunch grass." Blake (1856: 63) found bunch grass to be common on the *bajada* extending northward from the San Gabriel Mountains, but it was "entirely brown or dry," possibly *Hilaria rigida*. At Antelope Valley, "the surface was gravelly and not covered by grass, but here and there an isolated dry tuft called 'bunch grass' could be found" (52). The Mojave Desert near the Spanish Trail near Barstow consisted of "a level prairie of great extent covered with green bunch grass" (Whipple 1856a: 125). Brewer (1883: 961) noted that *galleta* grass (*Hilaria rigida*) was "a hard grass occurring in . . . the semi-deserts of San Bernardino County. It is valuable for feed, is said in places to constitute three-fourths of the pasture."

Patches of wildflowers were seen in the Mojave and Sonoran deserts. Near the Liebre and San Gabriel mountains, Frémont's party came upon fields of rich orange-colored California poppy mingled with other flowers of brighter tints. Near the foot the Liebre Mountains, "red stripes of flowers" caught Frémont's attention, perhaps near the present-day Poppy Park (Frémont 1845: 257). On the north side of the San Gabriel Mountains, he "came to a most beautiful spot of flower fields: instead of green, the hills were purple and orange, with unbroken beds, into which each color was separately gathered. A pale straw color, with a bright yellow, the rich red of the California poppy mingled with fields of purple, covered the spot with floral beauty" (257).

Clarence King (1915) entered southern California from the Coachella Valley in late spring after the flowering season. He crossed barren lands like those described by Spanish priests between Yuma and Borrego Desert a century before. Leaving for San Gorgonio Pass from the Colorado River along the trace of present-day Interstate 10, he wrote, "Far ahead a white line traced across the barren plain marked our road" (15). Descending Coachella Valley, he saw that the basin was marked by "the shore-line of an ancient sea [prehistoric Lake Coahuilla]" (18). Near a spring just above the eastern floor of the valley, he saw "minute flowers of turquoise blue, pale gold, mauve, and rose" (18). In San Gorgonio Pass, he found that "scattered beds of flowers tinted the austere face of the desert" (18). Leaving the desert near Beaumont, he compared the new Mediterranean landscape with the desert, similar to accounts at the "pass of San Carlos" by Anza, Font, Garcés, and Díaz. C. King stated, "There are but few points in America where such extremes of physical conditions meet. Spread out below us lay the desert, stark and glaring. . . . Sinking to the *west* from our feed the gentle golden-green *glacis* sloped away flanked by rolling hills covered with a fresh vernal carpet of grass, and relieved by scattered groves of dark oak-trees [*Quercus agrifolia*]" (26).

Little pasture covered the desert. On the Mojave Desert side, Leonard (1959: 123) of the Walker party descended Walker Pass, named for the expedition, and remembered that "our horses and cattle were pretty fatigued. . . . The country on this side is much inferior to that on the opposite [San Joaquin Valley] side—the soil being thin and rather sandy, producing little grass, which was very discouraging to our stock." The Owens Valley was found to be almost entirely destitute of grass. Blake (1856: 62) described primarily shrub cover on the alluvial fan *bajada* on the north side of the San Gabriel Mountains: "The higher parts of the slopes are covered with a thin growth of the Yucca [*Yucca brevifolia*] . . . occupying a belt 3 or 4 miles in width. They are interspersed with cedar [*Juniperus californica*] which grow to be large shrubs. . . . Sage bushes, (Artemisias,) and many small thorny shrubs, grew thickly together [likely *Coleogyne ramosissima*] . . . between the cedars."

Grasslands were described along fault lines with perched water tables on the northern side of the Transverse Range. The Lake Elizabeth area, according to Blake (1856: 62) had "an enormous growth of grass," which was part of a "fertile strip" where "grass grows luxuriantly in most of the valleys." At Silverwood Lake on the north side of the San Bernardino Mountains, Frémont (1845: 257) found "tolerably good grass, the lower ground being overgrown with large bunches of the coarse bunch grass [*Carex aquatilis*, or possibly *Muhlenbergia*]," recalling Zalvidea's account forty years before. From there, Frémont left California, following the Spanish trail for Salt Lake City. He traveled through the hyperarid Mojave Desert between the Mojave River and Colorado River, which he described as "country . . . extremely poor in grass and scarce in water, . . . where the road was marked by the bones of animals," livestock left by previous travelers (259). From there to Baker, "the zygophyllaceous shrub [was] constantly characteristic of the plain along the [Mojave] river." He could not even camp on the Mojave River "because there was not a blade of grass in sight."[8]

CARRYING CAPACITIES AND PRODUCTIVITY
OF HISTORICAL CALIFORNIA PASTURES

Writings from the Spanish and American periods clearly show that coastal pasture was more productive than in the interior. An estimate of primary productivity can be obtained from historical data on cattle density and annual dry feed consumption (Table 3.6). The following estimates are based on the assumption that grazing is at carrying capacity; therefore, primary productivity of each rancho is proportionate to cattle density.

Estimates are made using the number of livestock on ranchos of a known size, determined from land surveys in Table 3.1. Data on cattle numbers for each rancho were compiled by Davis (1929: 389–95) using a "proven counting technique" (138) among the rancheros, in which it was a rule among the *hacendados* to slaughter as a yearly income about four-fifths of the yearly increase of the herds. Using Haciendado Vallejo as an example, Davis stated that the ranch "slaughtered eight thousand steers of three years of age and over in the *matanza* season. Ten thousand calves were castrated, earmarked, and branded about the first of March each year, or one-fifth of his great herds. . . . An increase of one to every five head on the hacienda was the basis of the yearly estimate among the hacendados. To verify the rule, they counted the cattle as they went out of the corral, before the number became too great on a hacienda." The areas of ranchos, measured in leagues, were obtained from the governor's office and are largely consistent with areas published by the U.S. government survey's general office (Table 3.6). After California statehood, most ranch holdings remained under original ownership under protection from the Treaty of Guadalupe Hidalgo. Davis's data was used for ranchos that were subdivided after California statehood. He listed only "the most productive ranchos," as the other ranchos approved by the Treaty of Guadalupe Hidalgo, largely from the California interior, "had little livestock." The areas of the ranchos were obtained from U.S. Surveyor General's office. If area data could not be determined, densities were estimated from Davis's estimate of land area (in leagues and acres), which is usually consistent with survey office data. Differences arise where the subdivision of land grants could not be determined.

Estimates of cattle density (animal units) (Figure 3.8) are consistent with the accounts of pasture by Spanish and American writers. The per capita density ranges from one head per 2–5 acres (1–2 ha) along the southern California coast, Monterey Bay, and San Francisco Bay. Per capita density decreases to one head per 5–10 acres (2–5 ha) in the southern California interior valleys and inland valleys of the central Coast

Ranchos (after secularization) [modern place name/location]	Cattle	Horses	Sheep	Land Area (Davis 1929)	Land Area (acres)	Acre/Head Cattle	Acre/Head Head
San Felipe/San Luis Gonzales [Tomales Bay]	14,000	500	20,000	—	48,821	3.5	1.4
Zanjones/Guadalupe Correos, Chualar [N Salinas Val.]	6,000	200	2,000	—	8,890	1.5	1.4
San Benito [San Benito Valley]	2,000	200	1,000	—	6,671	3.3	2.8
Laguna Seca [Carmel]	3,000	200	—	—	20,000	6.6	6.3
Ex-Mission Soledad/San Lorenzo [N Salinas Valley]	4,000	300	2,000	—	48,000	12.0	10.8
Toro [Monterey County]	3,000	150	2,000	—	5,660	1.8	1.6
Bolsa del Pajaro [Santa Cruz]	2,000	100	—	—	—	—	—
Rancho Los Corralitos [Santa Cruz]	5,000	300	2,000	—	15,440	3.1	2.7
Rancho Pajaro [N Monterey Bay]	4,000	200	—	—	—	—	—
Alisal [Salinas]	2,500	200	—	—	2,971	1.2	1.1
Las Verjeles [Monterey County]	2,200	100	—	—	8,759	4.0	3.8
San Miguelito de Trinidad [Monterey County]	4,500	200	—	—	22,135	4.9	4.7
Guilicos [Sonoma County]	3,000	500	—	—	18,833	6.3	5.4
Government Nacional [Monterey County]	15,000	200	—	6 leagues	26,334	1.7	1.7
Buenavista [Salinas]	2,000	200	—	3 leagues	7,725	3.9	3.5
Santa Margarita [N San Rafael]	4,000	300	2,000	6 leagues	17,734	4.4	3.8
Bolsa Nueva y Moro Cojo [Castroville]	6,000	500	500	8 leagues	28,827	4.8	4.4
San Bernardo [San Ardo, Salinas River]	3,000	100	—	3 leagues	13,345	4.4	4.3

TABLE 3.6 (continued)

Ranchos (after secularization) [modern place name/location]	Cattle	Horses	Sheep	Land Area (Davis 1929)	Land Area (acres)	Acre/Head Cattle	Acre/Head
Punta Pinos [Pacific Grove], Noche Buena, Salcito, San Francisquito	4,000	200	2,000	7 leagues	25,500	3.3	3.0
Los Aromitas y Agua Caliente [Santa Cruz]	4,000	200	4,000	3 leagues	13,233	3.3	2.6
Tularcito [Monterey]	1,000	50	—	—	4,394	4.4	4.2
Nipoma [Nipomo]	6,000	200	10,000	32,728 a	37,887	6.3	4.6
Lompoc	2,000	200	1,000	38,335 a	42,085	21.0	17.5
Monte Diablo	3,500	300	4,000	18,000 a	17,291	4.9	3.8
Sotoyomé [Santa Rosa]	14,000	1,000	10,000	11 leagues	48,836	3.5	2.9
New Helvita [Sutter's Fort]	4,000	800	10,000	11 leagues	48,839	12.3	7.2
Saucelito [Marin County]	2,800	30	—	19,571 a	19,571	7.0	6.3
Tomales/Baulinas [Tomales Bay]	5,000	150	5,000	2 leagues	—	2.7	2.2
Los Médanos [Pittsburg]	5,000	500	5,000	—	—	—	—
Potrero de los Cerritos [Union City]	4,000	200	2,000	3 leagues	10,161	2.5	2.2
Punta de la Concepcion	4,000	500	—	24,992 a	24,992	6.2	5.5
San Miguel	2,000	200	—	1 league	4,693	2.4	2.1
Tamalpias	2,000	200	1,000	2 leagues	—	4.4	3.7
Los Cerritos (N Long Beach)	14,000	1,000	5,000	5 leagues	27,054	1.9	1.7
San Jose (San Gabriel Valley)	8,000	800	—	22,720 a	—	2.8	2.6
Alamitos and other ranchos (Abel Sterns [Stearns]), Long Beach/Seal Beach	30,000	2,000	10,000	6 leagues	26,334	0.9	0.8
El Niguil [Laguna Niguel]	9,000	500	—	4 leagues	13,316	1.5	1.4
Los Coyotes [Long Beach]	10,000	1,500	5,000	56,980 a	48,816	4.8	4.5
Santa Margarita, Los Flores, San Mateo [San Diego County]	10,000	2,000	15,000	—	133,440	13.0	9.0

Sespi	5,000	1,000	5,000	6 leagues	26,465	5.3	3.8
Camulos [Santa Paula]	5,000	1,000	5,000	22 leagues	97,626	19.5	13.9
San Pedro	8,000	1,500	5,000	10 leagues	48,813	6.1	4.6
Santiago de Santa Ana [Corona/Riverside]	11,000	1,500	8,000	11 leagues	78,941	13.2	8.7
La Sierra				4 leagues	17,786		
El Rincon				1 league	4,394		
La Ballona [Culver City]	10,000	600	—	13,920 a	—	1.4	1.3
Los Verdugos [Eagle Rock/Glendale]	5,000	500	—	8 leagues	35,192	7.0	6.4
La Puente	5,000	500	5,000	48,790 a	48,790	9.8	7.5
San Joaquin [Irvine Ranch]	14,000	3,000	—	11 leagues	—	3.5	2.9
San Pedro	3,700	—	—	2 leagues	—	2.4	2.4
Los Palos Verdes	5,000	1,000	5,000	31,600 a	31,629	6.3	4.5
San Pedro	5,000	300	—	2 leagues	—	1.8	1.7
Ballona [W Los Angeles]	3,600	200	—	13,920 a	13,920	3.9	3.7
San Antonio/Chino	30,000	1,500	—	11 leagues	29,513	1.0	1.5
Santiago de Santa Ana [Orange County]	6,000	400	4,000	8 leagues	—	5.9	4.9
Santiago de Santa Ana [Orange County]	4,800	500	—	7 leagues	—	6.5	6.5
La Cienega [W Los Angeles]	2,000	1,000	15,000	—	4,439	2.2	0.8
Los Positos [Livermore]	2,000	200	2,000	2 leagues	8,880	4.4	3.4
Agua Caliente [Alameda County]	3,500	350	4,000	2 leagues	9,563	2.7	2.0
El Valle de San José [Pleasanton]	6,600	500	5,000	51,573 a	48,435	7.3	6.0
El Valle de San José [Pleasanton]	4,000	400	4,000	—	17,634	4.4	3.4
Corral de Tierra [N Monterey Bay]	2,100	200	—	1 league	7,266	3.5	3.2
San Pedro [San Mateo County]	2,000	200	—	2 leagues	—	4.3	4.0
Bolsa Chica [Huntington Beach]	2,400	500	—	2 leagues	—	3.4	2.8
Santiago de Santa Ana [Orange County]	3,200	300	—	2 leagues	—	2.8	2.5
Santiago de Santa Ana [Orange County]	2,500	400	—	2 leagues	—	3.5	3.1

TABLE 3.6 (*continued*)

Ranchos (after secularization) [modern place name/location]	Cattle	Horses	Sheep	Land Area (Davis 1929)	Land Area (acres)	Acre/Head Cattle	Acre/Head
La Ballona [W Los Angeles]	4,800	400	2,000	2 leagues	—	1.9	1.6
Los Palos Verdes	2,300	300	—	2 leagues	—	3.9	3.4
San Vicente [W Los Angeles]	5,000	500	—	38,000 a	30,259	6.1	5.5
San Bartolo [Los Angeles]	7,000	350	—	8 leagues	35,440	5.1	4.8
Molino [Santa Rosa]	6,000	200	—	3 leagues	17,892	3.0	2.9
Sur/Bolsao de Potrero y Moro Cojo [Monterey Co.]	3,000	220	—	4 leagues	17,898	6.0	5.6
Santa Ana/Quien Sabe [San Benito]	4,000	300	4,300	7 leagues	31,073	7.8	6.0
San José [Pomona]	3,000	500	—	2 leagues	8,880	3.0	2.5
Sauzal Redondo [Hermosa Beach]	4,500	500	—	5 leagues	22,485	5.0	4.5
Ex-Mission San Fernando [San Fernando Valley]	5,000	500	—	11 leagues	116,856	23.4	21.2

SOURCE: Ranchos and livestock numbers from Davis (1929); land area in leagues and acres from Davis (1929) (1 league = 4,439 acres) (1 league = 4,439 acres = 1,770 ha), in acres from California State Archives.

Figure 3.8. Estimated area per animal unit of cattle in 1830–31, based on data from Davis (1929).

Range, such as the upper Salinas Valley and Livermore. The number of horses and sheep, as well as total livestock for each rancho is generally proportional to cattle numbers.

Estimates assume that, on the rancheros, the forage requirement of one head of cattle was equal to one horse or five sheep (Davis 1929). The highest livestock densities (one head < 2 acres, 1.0 ha) were at Toro, Chualar, Aliso, and Government Nacional in the northern Salinas Valley and Monterey Bay, and at Los Alamitos, Las Cerritos, El Niguil, La Ballona, San Pedro, and San Antonio/Chino in coastal Los Angeles and Orange counties. In contrast, per capita head was greater than 10 acres (4 ha) at ex-Mission Soledad in the southern Salinas Valley, Sutter's Fort in the Sacramento Valley, Santa Margarita in San Diego County (mostly shrublands), Camulos in the interior Santa Clara River, Santiago de Santa Ana in the Riverside plains, and at the ex-Mission San Fernando, which covered most of the San Fernando Valley. Moreover, Davis (1929: 250) was not favorably disposed to the Cajon Rancho and San Jacinto Nuevo Ranchos, 60–80 miles inland from the southern California coast, which had "some cattle and horses" even though both land grants covered 11 leagues. Davis noted that "notwithstanding these large holdings of land he [Don Miguel Pedrorena, owner of both ranchos] was in rather straightened circumstances in his latter years, and much in need of money."

The carrying capacity of ranchos near the San Joaquin Valley is not known. In 1805, Zalvidea estimated that from the end of the Kern Lake to the river eight thousand head of cattle could be maintained. If it is assumed that the area from Kern Lake to the Kern River occupies an area of about 30 kilometers by 30 kilometers (900 km^2 or 90,000 ha), Zalvidea's estimates amount to one head per 10 hectares, i.e., one head per 25 acres, or roughly an order of magnitude lower than estimates along the coast.

The primary productivity of California pasture is estimated from modern studies of forage consumption of cattle, in which the forage-acre requirement is estimated from grazing a known number of livestock in a range unit of known area for a definite period of time (Burcham 1957: 121–22). Research at the San Joaquin Experimental Range for 1934–57 found that cattle weighing 1,000 pounds (0.5 metric ton) required 22.1 percent live weight ingestion or about 640 pounds of dry feed per year. Since hay production in wild oats ranges from 1.0 to 1.23 tons per acre (2.5–3.0 tons ha^{-1}), cattle require feed over an area of 7 acres per year. Acreage requirements of early work (Gordon 1883; Davy 1902) give lower estimates ranging from 2.5 to 4.8 acres per animal year. The lower estimates are more reasonable because it is unfair to impose modern cattle-raising standards on the Spanish-Mexican rancho system. The Spanish cow, which was leaner and meaner, weighed only 600–700 pounds (Burcham 1957: 123) and hence had lower forage requirements than those of a 1,000-pound animal. If one compares per capita forage requirements, a simple weight ratio of Spanish cows to modern livestock predicts a Spanish cow forage acreage requirement of 4–5 acres, which is consistent with Davis's data (1929) for cattle numbers and rancho sizes (Table 3.6). Hence, rancho cattle densities, held roughly in steady state on an interannual basis by *matanzas* and culling of four-fifths the natural increase (one-fifth was lost to predators), suggest that annual herbaceous biomass productivity in the 1830s ranged from 1.0 ton ha^{-1} along the coast, to 0.5 ton ha^{-1} in interior valleys and 0.2 ton ha^{-1} in the Central Valley. Quality pasturelands described by the Spanish before the invasion of exotic annuals may have had productivity comparable to the wild-oat grasslands.

DROUGHT, FLUCTUATIONS IN LIVESTOCK, AND OVERGRAZING

Cattle populations, left to fend for themselves, fluctuated with climatic variability as much as the wildlife, with changes in pastoral carrying capacities. The reduction in primary productivity from a dry year, combined with an antecedent legacy of high cattle numbers and carrying capacities, inevitably resulted in episodes of mass mortality to stock and corresponding heavy grazing pressure. While records from the Spanish period are sparse, evidence of livestock cattle crashes are lacking before 1810, when cattle and stock occurred in low numbers, even in drought (Table 3.2). Tree-ring records from the Sierra Nevada and near Santa Barbara show several episodes of drought and reduced ring growth between 1770 and 1810 (Haston and Michaelson 1994; Graumlich 1993). Once carrying capacities were reached, cattle became "rain gauges" for years of deficient rainfall, in the absence of instrumental records.

According to Bancroft (1888: 337), the first drought recorded in mission records took place in 1809–10 (Table 3.2), when "the missions and presidios suffered greatly for pasturage and crops." In 1820–21, "livestock, now numbering 400,000 had much difficulty in finding grass enough to keep them in condition fit for food" (337). Another drought spanned twenty-two months from 1828 to 1830, when "large quantities of livestock perished or were devoured by the wolves, coyotes and bears" (337). Bancroft's estimate (1888: 337) that cattle declined to less than twenty thousand by the 1830 drought is probably too low. Missions record one-half million cattle only five years later, and the landing of thirty trading vessels per year suggests a high demand for hides and tallow (Table 3.2).

Tree ring records reveal protracted drought from 1840 to 1844, possibly the worst in centuries in the San Luis Obispo region (Haston and Michaelson 1993). Bancroft (1888: 338) wrote that in 1840–41 "no rain worth mentioning fell for fourteen consecutive months . . . particularly south of Soledad." The drought and cattle die-off was also described by Wilkes (1844: 193). Bidwell recalled that "the summer of 1844 was an exceptionally dry one, because the previous winter had been almost rainless" (Gillis and Magliari 2003: 88). Cattle numbers decreased by an order of magnitude, but part of the decline was due to the mission slaughter (Table 3.1).

After the gold rush, a severe drought killed "hundreds of thousands of cattle" in California in 1856, estimated to be one-third of all the flock (Cleland 1964: 109; Salvator 1929). It was followed by a flood-drought

sequence of 1861–65 that led to the collapse of open-range cattle grazing in the state (Cleland 1964: 135). According to Brewer (1966: 242)
the 1861–62 flood turned the Central Valley into an "inland sea 250 to
300 miles long and 20–60 miles wide," and steamers on the Sacramento
River were moving cattle to higher ground. The flood was followed by
drought until 1865. In February 1864, Bidwell wrote, "Unless a very considerable amount of rain comes . . . I have seen one year (1844) dryer
(sic) than this" (Gillis and Magliari 2003: 157). At San Jose, Brewer
(1966: 506) wrote, "This drought is terrible. In this fertile valley there
will not be over a quarter crop, and during the past four day's ride we
have seen dead cattle by the hundreds." Cronise (1868: 371) stated that
half the cattle in Los Angeles region perished. Davis (1929: 255) wrote
that "nearly all stock perished for lack of feed. Vaqueros quickly took
off the hides in merchantable condition as cattle were dying en masse by
day." Cattle losses statewide were estimated at 200,000 and the number
of cattle decreased from 1,234,000 to 670,000 from 1860 to 1870 (Cleland 1964: 135).

During the drought of 1876–77, Bidwell stated, "But it has been a
strange winter—two months now and not a drop of rain! Unless it comes
soon much grain will be utterly ruined, and if there comes no rains at
all, or almost none as in 1843–44, there will be no harvest" (Gillis and
Magliari 2003: 161). Muir (1904: 368) wrote, "The year 1877 will long
be remembered as exceptionally rainless and distressing. Scarcely a flower
bloomed on the dry valleys."

The impact of drought was exacerbated by large increases in stock for
meat and by expansion of grazing into the unproductive pastures of the
Central Valley due to the gold rush. Cattle numbers doubled the maximum levels of the Mexican period. According to Cronise (1868), the collapse of the gold rush contributed to high stock levels by gutting the cattle market. Census data show that beef cattle dropped from one million
in 1860 to half a million in 1870 and remained at that level in 1880. Native herbivores did not contribute to overgrazing. According to Salvator
(1929: 27), "The principle cud-chewing animals are the antelope (*Antilocapra Americana*), which is, unfortunately, rapidly being exterminated, the mountain sheep (*Ovis montana*) and high up in the Sierras,
two kinds of elk, the American Elk (*Cervus canadensis*), and the black-
tailed Deer (*Cervus columbianus*)." Horses were slaughtered or rounded
up within years of the gold rush (Cronise 1868).

The failure of browse resulted both from the seasonal distribution and
total amount of winter rainfall. According to Cronise (1868: 349), "The
aridity of the dry season is a blessing in disguise. . . . The dried grass is

well preserved, after going to seed, and both stalk and seed afford nutritious food to sheep and cattle. . . . For this reason, our agriculturalists
desire no rain until late in the season." Drought, of course, also reduces
primary productivity of pasture. Cronise (1868: 371) stated it simply:
"When the winter rains fail, the summer pasture also fails; and when, in
the midst of winter rain, there comes frost to retard the growth of the
herbaceous [cover], the feed is cut off, and want of shelter, joined to the
want of food, kills of the cattle by thousands. The winter of 1862–63 is
an example of the latter, and the summer of 1864 of the former casu-
alty." Field burning may have also brought disaster to cattle. Cronise
(1868: 382–383) pointed out that

> in 1862–63 the pasture did not respond to the winter rains, by reason of
> the cool atmosphere—the stubble had been [frost] burned in many places
> and the straw, as usual, consumed by fire, to get rid of it. Had the latter
> been preserved it would have saved the stock from the terrible destruction
> that followed. Late rains are also undesirable. All grasses that dry standing
> cure like hay, and carry their usual nutriment which they retain on the field
> till the first rain [in fall]. The rain loosens the capsule, casts out the seed
> and rots the grass-hay beyond resuscitation—since it would not suffice
> to make new pasture from the seed, with one even several showers;—nor
> could it, even then, survive the arid sun and the newly backed surface soil.
> All cattle would inevitably perish; for the summer feed, prepared expressly
> for a long dry season, would be entirely destroyed. (cf. Salvator 1929)

Brewer's first impressions of California were during the great 1862–
64 drought, when cattle were perishing across the entire state from deficient pasture. But it is unclear whether he was unaware that barrenness
was also the natural condition of the newly grazed Central Valley in the
dry season, with or without drought. The Californios from the Mexican
period may have warned the Americans of the poor Central Valley pasture, but the Americans, fresh from the spoils of conquest, may not have
been listening. While there was little feed outside the tule swamps, the
absence of pasture does not mean that the Central Valley was overgrazed.
In *Pasture and Forage Plants*, Brewer (1883: 963) came to the view that
the barren lands he saw in 1862–64 were due to overgrazing, his views
premised on the assumption that annuals were easily grazed out and their
seed banks were extirpated (cf. Cronise 1868: 521; Bidwell's view in Gillis
and Magliari 2003: 170). Brewer (1883: 963) developed a conceptual
model to explain the change in California pasture in the 1860s. It is
quoted at length because this hypothesis became the basis for Fredrick
Clements's overgrazing hypothesis of the destruction of perennial bunch
grassland:

When new and natural pastures become occupied with cattle or sheep a condition new to the region is introduced, the old balance established by nature is disturbed, and immediately a change begins in the pasturage, as to both kind and quantity. . . . When a very considerable portion of the forage is of annual plants, as is notably the case in California, if the growing plant is eaten before the seed ripens, or if the seeds themselves are palatable and are eaten, then the natural seeding is prevented and the pasture rapidly diminishes, then new species come in, which are either less palatable to stock or have some natural provision by which the seed is protected and self-planted . . . in California, particularly in the valleys and the lower ranges west of the Sierra Nevada, . . . originally the winter and spring herbage of this region was especially rich in variety and abundant in quantity; . . . With settlement and herding the entire native herbage decreased and European species came in their place. Wild oats (*Avena fatua*) first came in. This species, originally from the Mediterranean area, naturalized on the Pacific coast of both North and South America, came into the state from the south with the Franciscan fathers and their herds, and spread northward. It was most abundant between 1845 and 1855, when hundreds of thousands of acres were clothes with it as thick as a meadow. Alfileria, or pin-clover (*Erodium cicutarium*), apparently came before it, some botanists believing it to be native, as Nuttall found it far in the interior in 1836, but did not increase so rapidly as wild oats, and never had such possession of the soil anywhere' but it increased until perhaps 1865 or 1870. This too, is a native of southern Europe, as well as bur-clover (*Medicago denticulata*) [*M. hispida*], which came in later and slower, spreading with sheep along the lines of their drives, and along the lines of wagon roads across the great central valley. All these are annual species, all are valuable for forage, but each has a provision [adaptation] for the protection or planting of the seed.

The dilemma in Brewer's account is whether cattle can cause displacement of wild oats by *Erodium* and clovers or whether this is a natural outcome of interannual precipitation variability. Plants inherently have high fecundity and produce far more seed than necessary to replace themselves, which likely persist in the soil surface or fall from the mouths of cattle. But it is unclear, based on the observations of Frémont, Bidwell, Muir, and others, whether wild oats were ever important in the Central Valley except on the floodplains. Oats have high water demand, and they thrive along the coast and along the floodplains of the interior. *Avena fatua* also comes from moist parts of Mediterranean Europe (Jackson 1985). But oats can perish in dry winters when germinating rains are insufficient to bring them to reproductive maturity. *Erodium* and burr clover survive in the interior valleys and deserts and thrive in drought, but are competitively displaced by taller wild oats in wet years. Frémont saw wildflowers in the 1840s, King, Muir, Hittell, and others saw California in wildflowers a few years later, and as will be seen, wildflowers were seen in the California interior well into twentieth century.

SPACE-FOR-TIME SUBSTITUTION
VERSUS HISTORICAL RECORDS

While there is broad agreement that lowland California has experienced catastrophic invasions of Mediterreanean annual grasses and forbs (Mooney and Drake 1986; Huenneke 1989; Bossard et al. 2000), the historical record since Spanish times has led to diverse views in the scientific community as to the makeup of the indigenous herbaceous flora. The argument fundamentally centers on the choice between two baselines: the aboriginal floral landscape descriptions of the Spanish versus the overgrazed lands seen after the gold rush, a classic example of the "shifting baseline syndrome" (Jackson et al. 2001). The choice is a trade-off between nonscientific observations before biological invasions versus the rigor of botanical science in already invasive-contaminated pastures, and a trade-off between descriptions of forbfields versus a hypothesized bunch grassland that had already "disappeared." Proponents of bunchgrass-grazing theory dismissed the Spanish record, because it failed to meet scientific protocol, relying instead exclusively on formal botany, largely beginning with the U.S.-Mexican boundary survey of 1849, the Pacific Railroad survey of the 1850s, and *Botany of the State Flora* of the 1860s.

It is more important that the Spanish described flowers across the state than that their taxonomy was incorrect. Spanish texts reflect the observations of intelligent explorers and should be taken in their historical context (Grove and Rackham 2001): they wrote in response to a mandate from the Spanish viceroy in Mexico City to evaluate the landscape for lumber, fuelwood, and pasture. Indeed, many misconceptions of past herbaceous vegetation among English speakers may be an outcome of mistranslations of Spanish by historians lacking the insights of ecology and biogeography.

The Spanish journals, the only texts of the California landscape before European species expanded across the region, provide sufficient descriptions of landscapes to distinguish bunch grasslands, pasture, and flowers. The journals of Crespí, Font, Anza, Palóu, and Fages document "pasture" (*pasto, zacate*) in the summer, i.e., forage good for livestock, along the length of the California coast. In the vernal growing season, they wrote of flower fields in numerous locations and retrospectively reported that flower fields were continually present along their routes. In interior California, forbs disarticulated into barrens in the dry season.

In principle, the fragmentary information left behind by the Spaniards should not be an issue. Paleoecologists reconstruct vegetation change from fragmentary data of macrofossils and pollen preserved by chance through geologic processes. The records of Spanish missionaries represent a chance route of their expeditions, guided by the texts of earlier Spanish voyages. The limited botany in the Spanish journals should not provoke mystery because botanists have since inventoried the same flora.

The translation of *pasto* and *zacate*—interchangeable words in Spanish—to "grassland" is a fundamental error in English versions of the Spanish journals (Minnich and Franco-Vizcaíno 1998). The best translation is "herbaceous forage good for livestock," or "pasture." Besides, why should there be unique mistrust of accounts of pre-Hispanic herbaceous cover? English speakers have adopted many Spanish names of plants in our deserts and chaparral, including palo verde, ocotillo, mesquite, manzanita, madrone, and chamise. We adopted Spanish place names for our streets and cities.

The Spaniards occupied the California coast as a remote outpost of the Spanish empire, with small settlements near harbors and with European and Native American population densities similar to those in pre-Hispanic times as late as the gold rush. The entire state was as a pristine wildland, only with semiwild cattle and horses living with native herbivores. The novelty of the Hispanic period was not the introduction of livestock, but the introduction of herbaceous annual species from Europe.

The hypothesis that bunch grasslands were converted to modern exotic annual grassland from overgrazing has little support from historical evidence. Annual forbs dominated pre-Hispanic California pastures and the California interior in the nineteenth century. The rapid expansion of introduced species into the interior valleys ahead of grazing suggests that colonization of Old World species took place spontaneously, independently of livestock. The transformation of California pastures in the mid–nineteenth century can be better explained using biological invasion theory (Mooney and Drake 1986; Huenneke 1989) and competitive exclusion of native wildflower floras by invasive species.

The bunchgrass-grazing hypothesis, using Clementsian climax theory, assumes perennial bunch grasslands as a pre-European vegetation baseline. Relict theory assumes that scattered stands of bunch grassland were remnants of former broader distributions. The problem is that time-dependent hypotheses are formulated from geographic rather than historical evidence—in effect, space-for-time substitutions. As pointed out by Blumler (1995), early twentieth-century ecologists, historians, and geographers, having an a priori model of vegetation, imposed their conclusions upon the historical data rather than letting the early accounts speak for themselves.

The bunchgrass baseline is defined by botanical collections that are scarce before the gold rush. Hence, the trend line begins with an ambiguous vegetation baseline of pastures already contaminated by introduced species. A more prudent approach would be to develop a model of vegetation change inductively from historical evidence. While early writers were not trained in scientific protocol, the "collective weight" of these historical writings lends little support to the view that pre-Hispanic vegetation was dominated by perennial bunch grasslands (Blumler 1995). Fields of wildflowers were documented time and time again across California. The writings of Frémont, King, Muir, and the less well-known Hittell, Salvator, Boudet, Cronise, Revere, Wilkes, and Cooledge confirm the spectacle, highlighted by the California poppy—the state flower—from the coast to the Sierra Nevada and the desert, through the nineteenth century. Wildflowers grew with *Erodium* and clovers in the interior, although coastal forblands were already displaced by wild oats and black mustard. Barrens of disarticulated forbs were seen in the interior (Wester 1981; Hamilton 1997) while grazing thrived through the Spanish and Mexican periods along the coast. Perhaps most compelling are descriptions of wildflowers from the same locations by different observers, such as the grand view of wildflowers across the Central Valley from Pacheco Pass by King, Muir, and Boudet. Jepson (1925) concluded that much of the Central Valley was dominated by native forbs rather than by perennial bunch grasses. Biswell (1956) took an intermediate position. He suggested that hundreds of annuals may have contributed to the composition of the aboriginal "grassland." He also suggested that perennial grasses may have decreased greatly along the coast, where conditions are most favorable for them, while native annuals may have formed most of the "lost portion" in such areas as the lower foothills of the western slope of the Sierra Nevada (cf. Beetle 1947: 311; Twisselmann 1967: 91–92). The interior barrens clearly represent an abundance of wildflowers. The accounts of Frémont, Muir, and others give every impression that the growth flush involved overlapping sequences of wildflower phenologies. In summer, desiccated forbfields left little organic residue. Lacking the high foliar silica content of grasses, native forbs produce limited cured biomass because dried foliage shriveled and disarticulated into fine parts. Fragments were utilized by microfauna or were blown away.

Mission adobe bricks dating from 1769 to 1820 (Hendry and Kelley 1925; Hendry 1931) and the accounts of José Longinos-Martínez, the only Spanish scientist to undertake land expeditions into California during the eighteenth century, do not record bunch grassland, even though it would have been a highly valued pasture resource. The Spanish missionaries never used explicit language for bunch grasses, preferring to use the generic *pasto* and *zacate* for dry pasture. Both words had the same meaning, "green cover," and were used interchangeably. Timbrook et al. (1982) found that grass seeds were barely discussed in the Harrington notes in Chumash territory of the Santa Barbara Channel. While Timbrook concluded that absence of evidence reflects the extirpation of bunch grasses, such evidence may directly account for the dominance of forbs in pre-Hispanic vegetation.

In the mid–nineteenth century, Frémont and others described bunch grass in the Coast Range and Sierra Nevada foothills, consistent with modern distributions (Beetle 1947; White 1967; Hickman 1993), in areas having at least 40 centimeters mean annual precipitation, in the same areas where bunch grass was described in historical accounts. Since no writers recorded bunch grassland in the semiarid Central Valley, a central dilemma of the bunchgrass-grazing model is that it must account for the vast disappearance of an assemblage of perennial grasses, which requires ad hoc explanations—grazing and drought (Clements 1934; Burcham 1957; Heady 1977). The simplest explanation is that bunch grasses are rare now because they have always been rare.

In *Flora of Kern County,* the rancher-botanist Ernest Twisselmann (1967: 91–92), who lived on the Twisselmann Ranch from birth, extracted himself from the dilemma of "lost" bunch grasslands from the standpoint of climate: "Various authorities . . . conclude that this region was once covered by a perennial grassland that has been destroyed by grazing. Impressive evidence can be marshaled to reject this assumption; I doubt that the scant rainfall could ever have supported a perennial grassland. . . . It is probably safe to assume that the primitive flora was largely one of native annuals that still occur but whose number has been greatly reduced by the dominance of immigrant annuals."

The grazing history of California provides a test of the hypothesis of whether livestock animals facilitated the expansion of European annual grasses and forbs. While the coast and Sacramento delta were extensively grazed by livestock, the California interior experienced little grazing from domestic herbivores. Indeed, the coastal expansion of the mission system was very likely based on fundamental observation that the best year-round pasture grew along the coast. The barrenness of interior pastures, as well as the hot climate and increasing conflict with Native Americans, discouraged Spanish and Mexican colonization of the interior. Hence, the question can be asked whether introduced species spread ahead of European settlement. In the mission and rancho surveys of the 1830s and 1840s, cattle and other livestock numbered almost a million animals and reached carrying capacities to a point that livestock perished in large numbers from drought. However, livestock numbers were well below carrying capacities before 1810 when droughts were not recorded. Native American horse raids of the ranchos led to a buildup of feral horses in the interior, with estimates as high as twenty thousand animals, but these numbers very likely represent densities far below carrying capacities in an area as large as the Central Valley, even with barren summer pastures.

There is abundant evidence that most introduced species spread ahead of Spanish and Mexican grazing, but at different rates and habitats because species have unique histories of selection, migration, and paleo-biogeography (Sauer 1988). Some species expanded "instantly," such as *Erodium cicutarium,* which was recorded abundantly in varved sediments in the Santa Barbara Channel within years of initial Spanish colonization (Mensing and Byrne 2000). Frémont also recorded *Erodium* and clovers almost daily in his journal of the Central Valley. The black mustard (*Brassica nigra*) was widespread by the time of Longinos-Martínez in the 1790s and was viewed as a biological invasion in Los Angeles by Duhaut-Cilly in the 1820s. Bidwell observed wild oats covering portions of the Sacramento Valley before it was grazed. Some introduced species have ruderal properties and proliferate with disturbance. Wall barley (*Hordeum murinum*) thrives in overgrazed pasture, and cheese mallow (*Malva parviflora*) grows along roads, trails, and vacant lots. The pace at which both expanded across California suggests that these invasives dispersed into new suitable habitats without disturbance.

Burcham's (1957: 138) claim that herds of wild cattle were in much of the Central Valley is unsupported by any expeditions before the gold rush. According to Twisselmann (1967: 3), "Although the [Central] valley was discovered in 1772 by the Spanish explorer, Don Pedro Fages, it wasn't settled until long after the coastal regions. As late as the mid-1860s, Colonel Baker's field at Kern River was the only fenced pasture between Firebaugh's Ferry near Fresno and Los Angeles where travelers could grass their horses untethered."

The gold rush drew a large human population to San Francisco, as well as to the Sacramento goldfields, but few settlers lived elsewhere in

California, where a Mexican-period livestock economy persisted until about 1880. The Americans expanded livestock grazing from the coast to the Central Valley, with transhumance into the Sierra Nevada and southern Coast Range, with disastrous results. Brewer did not appear to recognize the possibility that the interior barrens was a natural condition, deleterious to livestock grazing, perhaps because he witnessed California in the great drought of 1862–64. It seems likely that the Americans, frantically in pursuit of gold, were not paying attention to the livestock grazing experience of the defeated Californios. The livestock disaster of 1862–64 was related to overstocking well beyond numbers attained in the Mexican and Spanish periods. Grazing expanded into marginal pasture of the interior barrens and ultimately into the meadows of the Sierra Nevada and southern Coast Range.

The gold rush expansion of grazing into interior California provides a different test of the bunchgrass-grazing hypothesis, especially in the Central Valley. Many writers, including Wilkes, Derby, Muir, and King—who all saw the Central Valley after normal or wet winters, not drought—document that the Central Valley remained in wildflowers through the nineteenth century, just as Frémont had seen the area before grazing in the early 1840s. *Avena fatua* and *Brassica nigra* remained along the coast and along the rivers of the Central Valley. Apparently both species, native to moist regions of the Mediterranean basin (Jackson 1985), had reached their ecological limits, and encouragement of domestic herbivores failed to expand their ranges.

"Relict" theory was challenged because native bunch grasses thrive with disturbance (Biswell 1956; Wills 2000). Biswell (1956) pointed out that at the time of Clement's observations (1919–18), railroad rights-of-way were burned almost annually and that frequently *Nessella pulchra* was favored by burning. Biswell (1956), Naveh (1967), and Barbour et al. (1993) speculated that frequent burning by Native Americans may have favored purple needle grass and other bunch grasses, perhaps in areas where they are found presently in the foothills of the Sierra Nevada and the central Coast Range. Wills (2000) proposed that many areas of *Nessella pulchra* might be more common today with fire management. Burning experiments at the Santa Rosa Plateau in southern California show that frequent fires decrease the abundance of many native forbs and annuals while increasing *Nessella* dominance (Wills 2000; cf. Wells 1962). However, this finding may apply site-specifically, a constant dilemma for ecologists working in complex plant assemblages. Burning in coastal California by Native Americans appears to have contributed to the proliferation of wildflowers (see Chapter 2). This evidence casts doubt that *Nassella* stands are "relicts" and hence on the deduction from Clements's climax model that railway stands represent remnants of past extensive perennial bunch grasslands across California.

In an anecdote, Twisselmann (1967: 189) wrote that some of the best surviving colonies of bunch grass grew in the Pinole Hills of eastern San Luis Obispo County, where "an entire colony of German settlers abandoned their homesteads when their last horse starved to death in the dry year of 1898. . . . Thus, it is reasonable to assume that the present annual grassland of the upper San Joaquin Valley did not replace a perennial grassland in which *Stipa* (*Nassella*) *cernua* was dominant but a vegetation of native annuals; in this process, grazing played an important but far from fully understood role."

Proponents of the bunchgrass-grazing model cite present ecological studies to explain the past, i.e., the equivalent of uniformitarian thought in geology that present processes help explain the past. Heady (1977: 493) writes: "After 25-years of study, . . . I believe that plant succession tended toward perennial bunch grass dominants on nearly all well-drained upland sites, that numerous annual species were present, and that they dominated intermediate and low succession stages, just as they do in many other grasslands." The question is whether he is concluding inductively from data or fitting data deductively to a Clementsian model. Can one also extrapolate findings at the Hopland Research Station to all of California, and to the past from evidence in ecosystems already contaminated by exotic species?

An alternative hypothesis is that California pastures were displaced by invasive species independently of disturbance (Mooney and Drake 1986; Sauer 1988). Such species arrived in climate and habitat comparable to conditions in their indigenous ranges in the Europe and had life traits that permitted rapid expansion, including short life spans, self-pollination, heavy seed production, rapid seed dispersal, capacity to colonize mineral soils, vigorous germination with first rains, and high productivity and growth. The pace at which European annuals expanded across California is nothing less than extraordinary (Bartolome 1979; Wester 1981; Mooney and Drake 1986; Huenneke 1989; Blumler 1995; Carlsen et al. 2000). While domestic animals may have accelerated seed dispersal in the first years of grazing through disturbance, over long time scales these species would have expanded without human facilitation. Mooney et al. (1986) cast doubt on the bunchgrass-grazing theory, based on Wester's view (1981) that bunch grasses grew in favorable sites in the Coast Range and Sierra Nevada foothills, and that "the original grassland was composed mostly of annuals."

Native herbivores are an unknown grazing force. California pasture supported abundant wildlife that also included semiferal cattle and horses in the Spanish and American periods. Clearly, the native fauna coexisted with the native flora over geologic time scales. While the number of animals is unknown, one can conclude that the consumption of biomass and trampling of soils was not unprecedented. Burcham (1957: 108) concluded that "none of the native megafauna formed large migratory herds throughout the year after the manner of the bison." However, Leonard, Sutter, Bidwell, Frémont, Davis, and others all described an abundance of native herbivores such as antelope, elk, and deer in the early to mid–nineteenth century. Twisselmann (1967: 316) pointed out the obvious: "The land was always grazed." He commented that "the very large herds of tule elk and San Joaquin antelope that roamed the valley before the coming of white man must have had the same effect on the plant cover in dry years as the sheep had" (4). While the southern San Joaquin Valley was severely grazed in times of drought, "what permanent effect this had on the flora and vegetation is a matter of conjecture and debate" (316). He noted, "Certainly the enormous number of sheep that wintered in the San Joaquin Valley must have had an effect on the native flora. However, as their food preferences and habits are largely the same as those of the great herds of tule elk and San Joaquin antelope that roamed the valley before the white man came, it is doubtful the sheep actually eliminated many species or even greatly altered the character of the vegetation" (32). Another unexplored topic is the effect of trillions of rodents and birds (Schiffman 2000). Kimball and Schiffman (2003) assert that the native California prairie assembled in the absence of grazing herds; invasive European grasses were exposed to grazing for centuries. .

The role of grazing should also be viewed in geological time scales because the evolution of the California flora coincided with the presence of a diverse megafauna that exerted cattle-like disturbance until the end of the last glaciation, only 10,000 to 12,000 years ago, and that included bison, horses, camels, antelope, mammoths, and mastodons (Bell et al. 2004; Woodburne 2004; Paleobiology Database). Wildflowers had long been part of California's heritage at least to the last glacial maximum. The same genera of desert flowers, many closely related to species found along the coast, have been recorded in packrat middens since the late Pleistocene, including modern dominants in *Phacelia, Amsinckia, Eschscholzia, Salvia, Lepidium,* and *Cryptantha* (Van Devender 1990; Spaulding 1990; Table 3.7). California's wildflower heritage in evolutionary time scales very likely extends through the Quaternary and perhaps long before.

TABLE 3.7 FOSSIL WILDFLOWERS AND BUNCH GRASSES
IN PACKRAT MIDDENS IN THE MOJAVE AND SONORAN DESERTS

Source	Mead and Phillips 1981	Spaulding 1983	Cole 1986	King 1976	King and Van Devender 1977	Cole and Webb 1985
Location	Grand Canyon	Marble Mtns., Owl Canyon, of Rocks	Picacho Peak Point	Lucerne Valley River Valley	Lower Colorado	Greenwater Valley
Period	Pleistocene	Late Pleistocene	Late Pleistocene	Late Pleistocene-Holocene	Late Pleistocene- Holocene	Late Holocene
Amsinckia	X	X	X	X	X	X
Argemone	X	X				
Aristida			X	X		
Astragalus		X	X			
Castilleja			X	X	X	X
Chorizanthe		X	X			
Circium	X	X		X		
Cryptantha		X	X		X	
Dithyrea						X
Draba			X			
Eriogonum		X		X		
Eschscholzia				X	X	
Euphorbia			X			
Gilia		X			X	X
Hilaria			X			
Lepidium	X	X	X		X	X

TABLE 3.7 (continued)

Source	Mead and Phillips 1981	Spaulding 1983	Cole 1986	King 1976	King and Van Devender 1977	Cole and Webb 1985
Location	Grand Canyon	Marble Mtns., Owl Canyon, of Rocks	Picacho Peak Point	Lucerne Valley River Valley	Lower Colorado	Greenwater Valley
Period	Pleistocene	Late Pleistocene	Late Pleistocene	Late Pleistocene-Holocene	Late Pleistocene-Holocene	Late Holocene
Lupinus			X	X		
Malvastrum				X		
Mentzelia		X				
Mirabilis		X				X
Oryzopsis		X	X			
Penstemon					X	
Perityle			X			
Pectocarya		X				
Phacelia	X		X	X	X	
Plagiobothrys			X		X	
Plantago		X			X	
Solanum					X	
Stephanomeria						
Stipa (Nassella)		X	X			X
Vulpia (Festuca)			X			X
Other grasses					X	

Ornduff (2003) speculates that herbaceous plants respond to selective pressures of repeated ice age cycles, which resulted in extinction of many species but also in rapid speciation of survivors with less competition. The composition of desert wildflowers appears to have remained unchanged since the Pleistocene, even though there were large changes in woody cover (Van Devender 1990; Spaulding 1990). Because annual forb species persist on seed banks of long viability, selection may have been lacking for long-range seed dispersal, even with climate change as large as that in the Pleistocene-Holocene transition. Native forbs adapted to intense grazing pressure by prolific seed production and long seed life in the soil. An alternative view from the standpoint of ecology and evolution is that the California flora just completed a period of abnormally low grazing pressure until the introduction of cattle and horses.

A related view is that grazing and intensive land uses for millennia selected for the weediest, most aggressive annuals, many self-fertilized—annuals that benefited from millennia of grazing and other intensive land use in Mediterranean Europe and could rapidly invade new habitats to dominance (Naveh 1967). This hypothesis is problematic because Pleistocene megafauna were also abundant in Europe until the early Holocene. Did a weedy flora emerge from domestic animals or paleoherbivores?

Grazing alone is unlikely to extirpate species or change the composition of the California wildflower flora. Plants produce annual seed crops that replace the progenitors by several orders of magnitude. Hence, the seed bank will never become extinct because of grazing. Even if livestock reduced flowering, the growth of a new generation requires only a fraction of seed to replace the parent generation.

The bunchgrass-grazing model is a classic case of too little data and too many ideas. Hypothesis testing has become a random process rather than being constrained by empiricism. It is striking that the bunchgrass-grazing model, founded on deductive Clementsian space-for-time substitutions, persists as the primary paradigm in ecological research of California pastures (e.g., Sims and Risser 2000). Hamilton (1997) proposes that theories other than Clements were not given serious attention because of Clement's prestige. Clements also wrote during an era of Dust Bowl politics and soil conservation that encouraged government interference with science (Grove and Rackham 2001).

This account of historical vegetation reveals that introduced annual grasses and forbs invaded annual forblands, a transformation still in progress. By the mid–nineteenth century invasive species had won many battles, but introduced species also have weaknesses that make them vulnerable in California's version of the Mediterranean climate, resulting in complex outcomes. Invasives outcompeted and even displaced wild-flowers along the coast, but wildflowers persisted as dominants in the interior valleys and deserts.

Cronise (1868: 521) advocated the introduction of "hay species" to improve the pasture in the interior. He noted that "while the grasses of California are numerous in variety, and the most of them valuable for pasturage, few are well adapted for making hay; wherefore, it may yet become necessary to import foreign varieties for meeting this want, provided such can be found suited to our peculiar climate. . . . It may be easy to find grasses adapted to those portions of the state situated within the foggy regions along the coast, especially west of the redwoods. But to find those that will survive the long dry summers in the interior valleys, and on the foot-hills, will be difficult, if not impracticable" (521).

Perhaps Cronise got his wish because new invasive grasses expanded across California after his time, notably red brome (*Bromus madritensis*), ripgut brome (*B. diandrus*), and the slender wild oat (*Avena barbata*). None were deliberately introduced, and perhaps they arrived as foul seed in grain sacks, as ballast off sea vessels, or were transported by train from the eastern United States. It is vital to evaluate the history and geographic expansion of these new invaders because they displaced or extirpated the wildflower flora across much of California in the twentieth century.

CHAPTER 4

A Century of Bromes and the Fading of California Wildflowers

THE ALTAR CLOTH OF SAN PASQUAL: POPPY LANDS OF THE MESA

But once in the year is this cloth of gold
Unfolded for men to see
And spread at the feet of the mountains old
For the Easter jubilee;
I am kneeling in rapture before its shrine.
Have you seen it, and worshiped, O neighbor
Mine?

—Elizabeth Grinnell in 1891[1]

The diffusion of both bromes continued with increasing rapidity, and in a very few years large patches of either could be found in all parts of the valley and the surrounding hills. They are now among the most wide-spread, abundant and well established grasses of the region.

—Samuel Parish (1909)

The pinnacle for southern California wildflower lovers at the turn of the twentieth century was the famous poppy field of San Pasqual, the mesa that now hosts the suburbs of Pasadena and Altadena. Citizens of Los Angeles annually visited this landmark by rail every spring from the early 1880s to 1920. The mesa flowers may have been the inspiration of Charles Fredrick Holder's proposal to hold the now-famous Rose Parade in 1890. While oat and black mustard pastures grew along the coast, wildflower fields still covered interior California well into the twentieth century. Many Los Angeles suburbs celebrated annual flower festivals as late as the 1920s. Since that time, the cloth of San Pasqual has been cut up for urbanization, but residents and tourists of the Los Angeles megalopolis visited wildflower fields ever farther away from the expanding city, first traveling to the inland valleys, then to the San Joaquin Valley, where flower festivals were held at Arvin and Shafter from 1930 to 1960, and ultimately to the deserts.

The shrinking flower fields not only succumbed to land clearing, but also to the rapid expansion of a new suite of invasive European grasses beginning in the 1890s, termed "second wave exotics" (Heady 1977). The most virulent of the new invaders are two bromes, red brome (*Bromus madritensis* [*rubens*]) and ripgut brome (*B. diandrus*), which the Samuel Parish epigraph describes above, accompanied by the slender wild oat (*Avena barbata*), summer mustard (*Brassica geniculata*), and other new exotic species that raced ahead of grazing and agriculture, swallowing up most of California's wildflower fields by the mid- to late twentieth century. Now the California state government has dedicated flower reserves to protect the state flower, the California poppy. While this floral symbol is unlikely to experience the fate of its official state colleague, the extinct grizzly bear, the future of the poppy bodes poorly given modern land management and land development across California.

Wildflowers illuminated virtually all of California during the pastoral days of the Spanish and Mexican periods, whether wildlands or livestock pasture. As stated by Brewer (1883: 963), California lands "were the delight of botanists and the paradise of cattle." However, the dearth of capital prevented the development of California's natural resources (Cleland 1964). Citrus and stone fruit cultivation began locally in Los Angeles in the late 1860s, and corn was grown in bottomlands near rivers (Cleland 1964: 143). Wheat and cotton farming developed in the Central Valley in response to demands of the Civil War. Livestock grazing continued to dominate until circa 1880 due to the Spanish-Mexican land grant system; ranchero owners refused to subdivide their vast holdings for settlement by small farmers. Immense livestock landholdings persisted, and the Trespass Act favored grazing interests at the expense of settlers. Inefficient transportation before the railway restricted markets. The irrigation required to cultivate lands beyond the rivers had not expanded since the secularization of the missions in 1833. The federal census of 1850 shows only 2,648 acres (1,000 ha) in Los Angeles County under cultivation (Cleland 1964: 138–39). The county's rangeland was 1 million acres (400,000 ha) and its ranches grazed over 100,000 head of cattle. As long as California remained cattle country and the rancheros held their immense estates intact, a large increase in population was impossible, since, obviously, a region so restricted in economic interests and so limited in opportunity offered no inducement to immigration.

These trends were reversed with the completion of the national railroads, increases in land value, and land subdivision. After 1880 land conversion to agriculture developed in the Central Valley and southern California coastal plains. By 1900, most of the southern California plains were under cultivation in grain crops, vineyards, orchards of stone fruits, and citriculture (Cleland 1964: 183). Land use in the Central Valley made a successful transition to agriculture after the gold rush. Agricultural land clearing began in the fertile lands near the Sacramento delta, the Sacramento and San Joaquin rivers, and locally along the Kern River near Bakersfield (Kahrl 1979). Private companies built dams in the Sierra Nevada and southern Coast Range and sold water to farmers. Irrigation lands expanded from the rivers to the plains on east side of the Central Valley in the early twentieth century and to the west side in the late twentieth century, due to the California Water Project. Urban centers developed throughout the state. The central Coast Range and the foothills of the Sierra Nevada have remained in cattle grazing. The higher mountains of the central and southern Coast Range and of the Sierra Nevada came under the jurisdiction of the national forest system and the national park system for watershed, fire protection, recreation, and land preservation (Lockmann 1981).

A NEW EXOTIC ANNUAL GRASSLAND

In the late nineteenth century, coastal pastures were still dominated by introduced Franciscan annuals, including the wild oat (*Avena fatua*) and black mustard (*Brassica nigra*). Exotic species of interior California were largely forbs, dominated by *Erodium* spp. and clovers, which coexisted with wildflowers. Franciscan invaders were still the focus of attention in the scientific community. In a public lecture before the State Horticultural Society in August 1891, the noted professor Eugene W. Hilgard, soil scientist from the University of California, Berkeley, stated that "nearly every bad pest of the kind with which we have to contend in California has been imported. Among the worst," he mentioned "the black and yellow mustard [*Brassica nigra, B. alba*], the common radish, large fennel, the chess or rye grass, the plantain, meadowlark clover, tar weed [a native genus], poison chickweed, American licorice [sweet fennel, *Foeniculum vulgari*], [and] foxtail [*Hordeum murinum*]."[2]

In a paper read in 1905 before the State Federation of Women's Clubs,

Alice Merrill Davidson, the wife of botanist Anstruther Davidson, stated:

> Of course we are not going to bewail the conversion of pristine nature into homes and cultivated fields, but the zeal of the real estate agent is sure to outrun the demand for improved property, hence unsightly areas of rubbish and weeds in the place of our beloved wild gardens. Indeed I suspect the most potent factor today in the destruction of our native flora about towns is this invading host of foreign weeds that has come with the tilling of the soil. Nature has many resources, and left to herself would soon repair the havoc of the grader's plow, but she is powerless against these plant immigrants. . . . Our native annuals vanish like snow before them, and our sturdy perennials succumb sooner or later."[3]

The "farmers foxtail" (*Hordeum murinum*) was particularly noxious. A *Los Angeles Times* editorial in 1890 described swollen "sore eyed chickens," and the irritation caused by the "seed of the foxtail grass."[4] In 1896 it was reported that "haycutting looked very well . . . barring an occasional lot that shows up too much foxtail."[5] Even as late as 1922, the State Highway Commission stated that the wild mustard (*Brassica nigra*) is the "most common menace" due to its dry and inflammable condition.[6]

From around 1890 to 1920, two brome species—*Bromus madritensis* and *B. diandrus*—accompanied by *Avena barbata* (bastard oats, barbed oats, slender wild oats) and *Brassica geniculata* (summer mustard), began invading Franciscan pastures along the coast and native forbfields in the interior. Although the earliest records of bromes date to the 1860s, botanists make clear that these grasses increased rapidly only by the turn of the twentieth century and had begun dominating California grasslands by the 1920s.

In contrast with Franciscan exotics, the expansion of "second wave" exotics is well documented by botanical studies and floras (Davidson 1893a,b, 1895, 1907; Abrams 1904; Reed 1916; Parish 1890a,b, 1909, 1920; Davidson and Moxley 1923; Jepson 1925), by botanical collections of introduced species (Appendix 3), and in books written to attract immigrants to California (including Lindley and Widney 1888; Van Dyke 1886; Saunders 1914, 1931).

EARLY INVASION OF BROMES AND OTHER "SECOND WAVE" INVASIVES

It was perhaps Samuel Parish (1920), an outstanding botanist who resided in San Bernardino from the 1880s to his death in 1928, who best documented and foretold the new transformation of California pastures: "Here belong a number of Mediterranean plants, dating from the mission era, now widely spread over the state. . . . The early immigrants [Franciscan species] possessed qualities which made them a valuable addition to the plant population, but the later-comers are entirely worthless, but good and bad alike have greatly modified the native vegetation."

The initial expansion of *Bromus madritensis (B. rubens)* and *B. diandrus (B. maximus)*, occurred in the late nineteenth and early twentieth centuries. Parish (1890b) wrote that the bromes "are of very recent introduction." Only two collections of *B. madritensis* and *B. diandrus* in Plumas County and San Francisco had been published by the state survey (Brewer and Watson 1876–80), and four other collections were made (two by Parish himself) by that time (Appendix 3). By 1920 "a few pestiferous bromes recently introduced [were] disseminating themselves with great rapidity" (Parish 1920). In the San Bernardino Valley, *B. rubens* first appeared in the spring of 1888 in Reche Canyon, south of San Bernardino. *B. diandrus* was first observed the same year in a small stubble field of lower Waterman Canyon. A few plants had already established in San Bernardino by 1890 (Parish 1890b). Twenty years later, according to Parish (1909),

> The diffusion of both bromes continued with increasing rapidity, and in a very few years large patches of either could be found in all parts of the valley and the surrounding hills. They are now among the most wide-spread, abundant and well established grasses of the region. By roadsides and in waste places, in pastures and in cultivated fields, on dry mesas and hillsides, these pestiferous grasses abound. Both species are equally abundant, but they seldom grow intermixed, the one which first obtains a foothold appearing able to hold the ground against its companion, as well as against other vegetation. *B. rubens,* while attaining a more luxuriant growth in better soils, adapts itself also to those of the most sterile character. Dry and hard clay hillsides it coast with a close felt-like covering an inch or two high, that completely chokes out the native spring annuals that were their former occupants. . . . Both are nearly valueless for forage, as animals will eat only the youngest growth, and both are injurious in hay, notably *B. maximus* [*B. diandrus*], by reason of its greater height and long stiff awns.

The impact of these rapidly expanding bromes and their danger to the native flora was prophetically stated by Parish (1920), who wrote that both species,

> being vernal in their growth, are not as troublesome in cultivated grounds as in fallows and the dry soils of plains and hills, which they often occupy to the exclusion of the native vegetation. As a result, some delicate indigenous herbs, formerly abundant, are now rare. The two species seldom grow

together, the broncho grass (*B. diandrus*) usually monopolizing the better soils, while *B. rubens* can occupy the most arid hillsides, which it covers with a dense depauperate growth. Both species are sparingly eaten when young by stock, but are practically worthless as forage, and soon drying up they become a serious fire menace.

He found that red brome was "one of the most widespread, abundant and thoroughly naturalized grasses of the cismontane region; . . . rapidly overspreading arid plains and hills, and cultivated grounds."

Anstruther Davidson (1893a), a medical doctor from Scotland, wrote of "a great change our flora has undergone in the last thirty years" since the publication of Brewer and Watson's state survey in 1876–80, as "many introduced species that were rare at the survey are now common, and new species have become established." Davidson wrote that red brome was "fairly common about Los Angeles, and also at Santa Monica and Pasadena, and like the former (*B. diandrus*), [is] increasing rapidly." In retrospect, Davidson (1907) wrote that red brome was "rare and local in Los Angeles County in 1892, but now may be found in many parts of the County, even as far as the Mojave Desert." In his flora of Los Angeles, Abrams (1904) stated that red brome was already "common in sandy soils along the coast and in our interior valleys," while Millspaugh and Nuttall (1923) found it to be common everywhere on Santa Catalina Island. Davidson and Moxley (1923) made frequent observations of introduced species because Moxley was a meter reader in Los Angeles. He walked all over town, including in vacant lots from which he took botanical collections of the local weeds and where red brome was found to be "fairly common in sandy wastes." Smiley (1922) stated that "the red brome is one of the weedy grasses most characteristic of the interior valleys of the State." Many collections of red brome were taken in many areas of central California by 1900 (Appendix 3).

Parish (1890a) reported that both brome species had quickly reached the "borders" the desert. However, botanical collections in the desert record only red brome in the desert interior beginning in the 1920s (Table 4.1). Most early collections state that it grew near roads and fields in semi-arid parts of the Mojave Desert, including in the Antelope Valley and higher mountains of the northeastern Mojave. McGuire (1935) collected red brome in Clark County, Nevada, and noted in 1932 that "this troublesome weed [is] commonly distributed along road sides and in waste places in Washington, Colorado, Utah, Arizona and New Mexico."

The original collection of *Bromus diandrus* was made near Mission Dolores at San Francisco by Bolander in 1862, the botanical sheet record-

TABLE 4.1 EARLY BOTANICAL COLLECTIONS
OF *BROMUS MADRITENSIS* AND *BROMUS TECTORUM*
IN THE MOJAVE DESERT

Collection/Herbarium	Year	Location
Bromus madritensis		
Parish (RSA)	1887	Fort Tejon
Jepson 5468 (JEPS)	1913	New York Mountains
Epling (LA)	1925	Palmdale
Abrams 11709 (LA)	1927	head of Antelope Valley
Wolf 1546 (RSA)	1928	south side of Antelope Valley
Epling (LA)	1930	Argus Mountains (Death Valley)
Duran 3420 (RSA,UC)	1933	between Mojave and Palmdale
Duran M40 (UC)	1933	Silver Canyon, White Mountains
Clokey, Anderson 6518 (UC)	1935	N of Barstow
Axelrod 286 (UC)	1935	Deep Creek, San Bernardino Mountains
Jepson 75619	1935	Providence Mountains
Jepson 18348 (JEPS)	1937	Mitchells Cavern
Beal 299, 304 (JEPS)	1937	Providence Mountains
Yates 6496 (UC)	1937	Antelope Valley
Jepson 19692, 19754 (JEPS)	1940	Panimint Mountain
Bromus tectorum		
Peirson 4543 (RSA)	1924	Victorville
Abrams 11744 (RSA, CAS)	1927	Antelope Valley between Neenach and Fairmont
Jones (RSA)	1929	Hesperia
Jenson 375 (UC)	1934	Phelan
Axelrod 271 (UC)	1935	Deep Creek, San Bernardino Mountains

[a]. Collections/herbariums:
RSA, Rancho Santa Ana Botanical Garden, Claremont, California
JEPS, Jepson Herbarium, University of California, Berkeley
LA,   University of California, Los Angeles
UC,   University of California, Berkeley

ing that the plant was "likely to be met with as a weed about settlements" (Brewer and Watson 1876–80: 319), illustrating a historical pattern among introduced species in which they begin their expansion with disturbance and later generalize into the landscape. Davidson (1907) found it "already frequent in the waste grounds throughout the city and rapidly expanding" in 1893; it had become "common along streets and waste places of Los Angeles by 1904. Abrams (1904) also found it in waste places. After discovering ripgut brome in 1888, Parish (1920) wrote that "this grass spread with great rapidity, and is now as abundant as *B. rubens,* not only in the south, but throughout the State." At that time,

B. diandrus was "becoming a weed in open situations" on Santa Catalina
Island (Millspaugh and Nuttall 1923: 56). Ripgut was a widespread grass
in California and surrounding states by the 1920s (Jepson 1925; Abrams
1940), but was not documented in the California desert.

Bromus tectorum, or cheat grass, was never abundant in coastal California, at least in comparison with that in the Great Basin (Mack 1981,
1989). Abrams's flora of Los Angeles (1904) does not list it, and botanical sheets do not record it anywhere in California until 1899 (Appendix
3). Collections place it in the Mojave Desert by the 1920s, but normally
at higher elevations than *Bromus madritensis* (> 1200 m). Parish (1920)
stated that it was infrequent in southern California. Today it is occasional
in mountain pine forest and higher ranges of the Mojave Desert.

Bromus mollis (soft chess), a field weed that thrives with disturbance,
was first collected by Marcus Jones in 1882 at Oakland (Appendix 3).
It was first reported in southern California by Dr. H. E. Hasse in 1890
at Santa Monica. Davidson (1893b) found a few plants at Pasadena and
Los Angeles. Davidson's catalog notes it "on streets and wastes, not common" (Parish 1909). Abrams's flora (1904) has it "frequent in our coast
valleys along roadsides." In the following decade, Parish (1920) recorded
that "the spread of soft chess has been rapid, and it can be found abundantly in waste places, pastures, cultivated grounds, and by road sides
throughout the San Bernardino Valley. It has not yet made its appearance in the undisturbed soils of the mesas and hills." Soft chess still
avoids undisturbed soils in the arid interior valleys of southern California
today.

Two other invaders at the turn of the century were *Avena barbata* and
Brassica geniculata. Both species spread into the California interior well
beyond the ranges of their Franciscan congeners. Parish (1920) stated
that *A. barbata* was first recorded in 1885 (cited in Vasey 1885: 56), but
it was apparently not widespread until a decade later. Except for a collection by T. S. Brandegee in 1888 on Santa Cruz Island, most early collections were taken after 1895 (Appendix 3). Abrams (1904) called it
"bastard oats" and wrote that it was occasional in the valleys. All floras
from the 1920s report it to be widely distributed, even the most dominant species, in many localities (Parish 1920; Smiley 1922; Davidson and
Moxley 1923; Jepson 1925).

Botanical collections show that *Brassica geniculata* broke out across
California in the 1910s and was not mentioned in the botanical floras
until the 1920s. It was first recorded by Davidson in the streets of Los
Angeles in 1911. Over the next decade, summer mustard became widespread and abundant in the neighborhoods of Los Angeles, favoring the

light sandy soils of the interior (Davidson and Moxley 1923). Parish (1920) stated that it was rare at San Bernardino in 1914 and "little increased" in 1918. He said that it was an abundant street weed in San Francisco. Jepson (1925) found summer mustard along the coast from Berkeley to Los Angeles and spreading to the interior.

The new invasives settled into a landscape of preexisting exotics. Parish (1890a, 1920) noted that among introductions known to date to Franciscan missionaries, *Erodium cicutarium* and *E. moschatum* were the most extensive naturalized plants in coastal and inland valleys of California. *E. cicutarium* was abundant across the deserts. He speculated that their dispersal was accomplished by the carpels of filaree, which are "admirably adapted to transportation in the pelage of animals. In cultivated ground along the valleys and mesas or tablelands, it literally covers the ground in many parts" (Parish 1920). Most botanical floras treat *Erodium* spp. as interior species well adapted to drought. Abrams (1904: 228) called them "the prevailing species of the interior valleys and foothills." Likewise, Davidson (1893a) stated that "*E. cicutarium* is mostly confined to the foothill districts and to drier ground, where the other (*E. moschatum*) appears unable to maintain a hold." Similarly Jepson (1925) found it on barren hillsides, and Davidson and Moxley (1923) called it the interior species of the interior valleys and mountains. Smiley (1922) found the filaree "so completely naturalized that few people suspect their Mediterranean origin."

Burr clover was extensive throughout California. Abrams (1904) found it "everywhere common," and Davidson (1893a) stated it was abundant on the plains and lower foothills. Other botanical treatments and collections attest to its widespread extent, excluding the desert (Davidson 1893a; Parish 1920; Davidson and Moxley 1923; Appendix 3). Likewise, botanical floras indicate that *Trifolium gracilentum* and *T. tridentatum* were extensive across California, including in city lawns, while *Melilotus indica* (sweet clover) was restricted to damp grounds and moist places (Davidson 1891, 1893a; Abrams 1904; Davidson and Moxley 1923; Jepson 1925). Among the fileree and clovers, only *Erodium cicutarium* has extensive distribution in the deserts to this day.

The two *Erodium* species and the clovers were highly regarded as fodder plants. Lindley and Widney (1888: 159), in reference to agriculture in Pomona, called fileree "a wild grass that grows profusely throughout Southern California and furnishes food for all varieties of stock." The *Riverside Bee* (May 30, 1889) reported that "residents were more impressed with erodium than wild oats, . . . it may be found in equal abundance in the valleys, and on the table lands and mesas. In even the driest seasons it does not fail to mature seed, thus being prepared for a more favorable year." The *Bee* also reported that the burr clover "grows abundantly in many of our fertile valleys and canyons, and, like the alfilaria, it makes well cured hay in the sun. It produces an abundance of burr-like seed pods, from which it receives its name."[7] Davidson (1891) remarked, "The pretty storks bill . . . supplies, along with the burr clover, the principal grazing in the earlier summer. *Medicago denticulata* (Burclover) and *M. sativa* (Lucerne, alfalfa) are two of the most important fodder plants in California. *M. denticulata*, or burr clover, has by natural processes spread over the greater part of the lower country, and not only affords maintenance to stock in its green state, but also when matured its ripe burrs being greedily eaten by horses and sheep as they lie around the withered remains of the parent stem."

Brassica nigra was most abundant along the coast, especially in heavy soils, similar to what accounts from the mid–nineteenth century reported. Davidson's (1891) account of black mustard reads like Duhaut-Cilly's assessment of the plant in Los Angeles sixty-five years earlier (see Chapter 3). Davidson called it "one of the greatest pest[s] of cultivation, and detrimental to growing crops. In most lands and grain fields, where it has secured a foothold, it grows most luxuriantly to a height of four or eight feet [1.2–2.5 m] or more, with stems as thick as a walking cane, and forming with its interlacing branches thickets as impenetrable as brushwood. Popularly it is said to have been introduced at an early date by the Spanish monks. . . . However introduced, its natural fertility, aided by the blackbirds and finches, has spread it over the length and breadth of the country." Lindley and Widney (1888) described forests of mustard trees along the old stage roads. Parish (1920), found it "abundantly naturalized as a ruderal weed and in grain fields" at San Bernardino "[but] in the coastal district, in the rich adobe soil of the hills and mesa, it often covers wide areas with a close growth 5–10 feet [1.8–3.0 m] high, excluding all other vegetation." F. J. Reed (1916), a botanist residing in Riverside, stated that black mustard "was common in cultivated ground but did not extend into the wild." It is presently restricted to heavy soils in valley floors of the Riverside-Perris plain. Botanical treatments of areas near the coast consistently note an abundance of black mustard in cultivated fields and adobe soils (Abrams 1904; Davidson and Moxley 1923), and early botanical collections locate it near the coast (Appendix 3).

Parish (1920) stated that *Avena fatua* was "abundant throughout the state, except in the deserts and mountains." In the interior it was "frequent in cultivated grounds, especially as 'volunteers' in grain fields" (cf. Davidson 1893b; Abrams 1904; Davidson and Moxley 1923). Beetle

(1947) claimed that *Avena fatua* had invaded most of the Central Valley. However, he may have confused *A. fatua* with *A. barbata,* which was mapped as the dominant grass in the northern Sacramento Valley by the California Vegetation Type Map (VTM) Survey of 1929–34.

The common fescue (*Festuca [myuros]*) was a dominant grass across California at the turn of the century. Parish (1890a) stated that "it has a very extended range . . . from San Diego to Oregon. In southern California it presents no appearance of an introduced species, not occurring in cultivated grounds or about habitations, but abundantly through the foothills and mesas," a true invasive species. Davidson (1893b) found there was "no native grass so common on hills, plains and in cultivated grounds [as the fescue]; yet its foreign derivation cannot well be doubted." Yet, Davidson and Moxley (1923) also found it on city streets.

Malva parviflora and *Hordeum murinum* were viewed as ruderals dependent on chronically disturbed habitat. Davidson (1891) wrote, "Over the wastes, orchards, and cultivated grounds where it is more or less prevalent, and where unmolested attains a height sometimes of 6 feet [2 m]. Once established it is not easily eradicated, as it grows and matures fruit throughout the greater part of a season." Other botanical treatments find it as an abundant naturalized weed in cultivated grounds, neglected dooryards, gardens, and waste places near dwellings (Parish 1920; Smiley 1922; Davidson and Moxley 1923; Jepson 1925). *Hordeum murinum* was "a very common and troublesome grass," especially on pastured land (Abrams 1904: 61) and an invader of fields (Jepson 1925). Parish (1920) found it "widely and abundantly naturalized in cultivated grounds, notably in overgrazed pastures, and in waste places; mostly in dry, sandy soils. . . . In 1890 [it was] reported by Hilgard to be 'a fearful nuisance' in central California. It was as abundant thirty-five years ago as at present, and [was] probably an early introduction."

Important evidence in biological invasions is whether botanical collections objectively reveal the temporal pattern in the expansion invasion of introduced species. While it can be argued that botanists provide nonrandom sampling in their search for novelty, and have personal sampling bias, they do work independently of one another. Hence their collective reports give probabilistic outcomes on the timing of species expansions. An important question is whether the temporal pattern of botanical collections can separate Franciscan exotics from those that first proliferated in the late nineteenth and early twentieth centuries.

The arrival of a newly introduced species is a subliminal process in which the species enters new landscapes earlier than it is first noticed or collected. The "breakout" years of alien species, the time when they first appear frequently in herbarium collections, often with "instantaneous" widespread distribution, can be sensitive to sampling density. The first detailed collecting in the Pacific Railroad survey and the state survey under W. H. Brewer in the early 1860s was followed by few collections in the 1870s, when most of California was still lightly settled in a grazing economy. Collections then increased exponentially after 1880.

Exotic species dating to the Spanish and Mexican periods, which were widespread by Frémont's time, e.g., *Avena fatua, Brassica nigra, Erodium cicutarium, E. moschatum, Malva parviflora,* and the clovers in the genera *Medicago, Melilotus,* and *Trifolium,* were still poorly sampled in the late nineteenth century. Moreover, the frequency of early collections in Appendix 3 reveal that Californians had little interest in unwanted plants such as *B. nigra,* the ruderal *M. parviflora,* and *Hordeum murinum,* but highly desired fodder legumes and wild oats were better sampled. Franciscan exotics were collected by the state survey in the 1860s, with a second pulse of collections beginning in the 1880s. However, *Bromus madritensis, B. diandrus,* and *Avena barbata* were rarely collected before the early to mid-1890s. *B. tectorum* was first collected in 1899 and *Brassica geniculata* in 1911. The rapid increase in collections after 1895 may reflect the primal expansion of these annuals, because their appearance is phased within a period of high botanical sampling. The location of botanical collections suggest that second-wave exotics were rapidly expanding across California, but botanical sheets rarely produced information on their abundance.

The indigenous wildflower flora persisted in the interior valleys and deserts well into the twentieth century, a fact that may surprise the novice, professional ecologist, botanist, and naturalist alike. Brandegee (1892) described a mixture of native annuals and alien species in San Francisco. *Eschscholzia californica* was common in the western part of the city and bloomed the entire year, consistent with Crespí's observations of mallows in the cool summer climate of Monterey Bay. Brandegee also listed many other native forbs as common, including *Viola pedunculata,* as well as species in *Phacelia, Amsinckia, Cryptantha, Nemophila,* and *Orthocarpus,* but they were growing with exotics that included *Erodium cicutarium, E. moschatum,* and *Malva parviflora,* as well as many species of *Trifolium* and *Medicago. Brassica nigra* was abundant about fields and waste places, and *Avena fatua* was common in the hills. *Stipa eminens (Nassella pulchra)* was reported in Mission Hills, but no other bunch grass was recorded. Brandegee cited the early botanical record of *Bromus diandrus (maximus),* but did not discuss the presence of bromes at San Francisco. Pacific Grove, according to Charles Frances Saunders (1931: 225), comprised "windswept, turfy downs, bright with sea-daisies, California buttercups, and es-chscholtzias, and ending suddenly at the sea's edge in perpendicular cliffs."

As stated by Chapin Hall (1929) in *National Geographic*, "In the early spring California dons her party dress and looks here best. . . . Literally, all outdoors become one vast garden of flowers, until it seems there is no end to the colorful panorama." He also wrote that "a mountain side, per-haps 20 miles in extent, covered with a mass of yellow poppies, is an eye-filling experience." What the article fails to convey is that the enclosed pho-tographs show flower fields on sand dunes near the coast or flower carpets in the desert, as much of central California had already been occupied by invasive species. Photographs of cattle pastures in the Coast Range reveal sheets of bromes and oats.

The floras of Jepson (1925) and Abrams (1940) suggest that wild-flower species were widely distributed in northern and central Califor-nia (Table 4.2). Botanists saw extensive flower fields in the interior. Re-ferring to baby blue eyes (*Nemophila menziesii*), Alice Eastwood (1893) described the blossoms at San Emigdio, most likely *Lupinus*, as "bits of the sky drawn down to earth." Jepson (1925: 941) wrote that owls clover (*Orthocarpus erianthus*) was "very abundant on the plains of the Sacra-mento and San Joaquin valleys, in the Sierra Nevada foothills and on the low hills of the Coast Ranges, often coloring wide stretches with stream-like bands of yellow." Goldfields (*Lasthenia gracilis*) was "every-where abundant, often coloring leagues and leagues of interior hills." Smiley (1922) wrote that the California poppy (*Eschscholzia californica*) was "extremely abundant and known to all Californians as one of the floral gems of the state. . . . [A few lupine species] grew so abundantly over our mesas and plains as to dominate the landscape during flower-ing time."

Similarly, botanical floras of southern California suggest that wild-flowers were still an important component of the herbaceous cover in the foothills and the valleys (Table 4.3). Most forbs were found through-out the region, while a few were considered to be ruderal (*Amsinckia in-termedia, Heterotheca grandiflora*). Several were concentrated in the chaparral of the foothills, including *Castilleja foliolosa, Cryptantha in-termedia, Mirabilis californica, Penstemon spectabilis, Phacelia minor,* and *Emmenanthe penduliflora*, similar to that in central California. *Ca-landrinia menziessii, Eschscholzia californica,* and *Tropidocarpum gracile* grew mostly on valley floors. Phillip Munz confirmed the accounts of the Spaniards of the widespread distribution of "little chia" (*Salvia colum-bariae*), which he described as occurring "in great numbers from the coast to the Colorado River."[8]

TABLE 4.2 GENERAL DISTRIBUTIONS OF WILDFLOWERS
IN CENTRAL CALIFORNIA

Species	Jepson 1925 (Calif.)	Abrams 1940 (Pac. Coast)
Amsinckia intermedia	v	*
Baeria (Lasthenia) gracilis	*	v
Calandrinia menziesii	v	v,r
Castilleja foliolosa	f	c
Chaenactis glabriuscula	*	*
Cryptantha intermedia	*	*
Draba cuneifolia	*	—
Emmenanthe penduliflora	c	c
Eschscholzia californica	*	*
Eucrypta chrysanthemifolia	v	f
Filago californica	*	*
Gilia achilleifolia	*	–
Gnaphalium californicum	–	f
Hemizonia wrightii	*	*
Heterotheca grandiflora	v	r
Layia platyglossa	*	*
Lepidium nitidum	*	*
Lupinus truncatis	—	f
Lupinus hirsutissimus	f	*
Mirabilis californica	—	f
Nemophila insignis (menziesii)	*	*
Nemophila aurita (Philostoma auritum)	v	v
Orthocarpus purpurascens	*	*
Pectocarya linearis	v	*
Penstemon spectabilis	*	f
Phacelia distans	*	*
Phacelia minor	v	—
Plagiobothrys californicus	v	*
Platystemon californicus	*	*
Salvia columbariae	*	*
Senecio californicus	v,c	—
Stephanomeria virgata	*	*
Tropidocarpum gracile	v	v
Viola pedunculata	v	v

v = valleys/plains near coast
f = foothills/slopes
* = valleys/plains and foothills/slopes
c = chaparral
r = ruderal

Although differences of opinion on the distributions of native forbs
are apparent in these floras, perhaps an artifact of year-to-year variation
in the floras being collected, these botanical volumes suggest that most
forbs were widespread from the coast to the interior, although their abun-
dances were not indicated. Among those with restricted distributions,

TABLE 4.3 DISTRIBUTION OF COMMON NATIVE ANNUAL
FORBS IN SOUTHERN CALIFORNIA

Species	Abrams 1904 (LosAngeles)	Reed 1916 (Riverside)	Davidson and Moxley 1923 (Los Angeles)
Amsinckia intermedia (menziesii)	*	r	v
Baeria (Lasthenia) gracilis	f	v	*
Calandrinia menziesii	v	v	v
Castilleja foliolosa	f	f	f
Chaenactis glabriuscula	*	f	*
Cryptantha intermedia	*	f	f
Draba cuneifolia	*	—	f
Emmenanthe penduliflora	c	f	c
Eschscholzia californica	v	v	*
Eucrypta chrysanthemifolia	c	v	-
Filago californica	*	v	*
Gilia achilleifolia	*	*	f
Gnaphalium californicum	*	*	*
Hemizonia wrightii	v	f	—
Heterotheca grandiflora	r	—	r
Layia platyglossa	v	*	*
Lepidium nitidum	*	v	*
Lupinus truncatis	*	*	*
Lupinus hirsutissimus	*	*	f
Mirabilis californica	f	f	—
Nemophila insignis (menziesii)	*	*	*
Nemophila aurita (Philostoma auratum)	f	—	v
Orthocarpus purpurascens	*	v	*
Pectocarya linearis	*	v	v
Penstemon spectabilis	f	f	f
Phacelia distans	*	*	*
Phacelia minor	*	f	f
Plagiobothrys californicus	*	—	v
Platystemon californicus	*	v	f
Salvia columbariae	*	*	*
Senecio californicus	*	f	—
Stephanomeria virgata	—	*	*
Tropidocarpum gracile	v	v	v
Viola pedunculata	*	f	v

v = valleys/plains/near coast
f = foothills/slopes
* = valleys/plains and foothills/slopes
c = chaparral
r = ruderal

Castilleja foliosa, Senecio californicus, Emmenanthe penduliflora, Eu-crypta chrysanthemifolia, Gnaphalium californicum, Lupinus hirsutis-simus, Mirabilis californica, and *Penstemon spectabilis* were considered foothill or chaparral species.

THE "ALTAR CLOTH OF SAN PASQUAL"

Southern California newspapers and lay magazines annually chronicled California's spring wildflower displays. Newspaper accounts have special value because they informally reported the status of wildflowers at the same localities every year. At a result, the trajectory of wildflowers over time scales of decades can be assessed site-specifically. The *Riverside Press Enterprise* described flower fields about town starting in 1885. The *Los Angeles Times* documented the famous poppy field of San Pasqual of Pasadena beginning in the 1880s. After urban sprawl erased this flower patch upward to the base of the mountains, the *Times* maintained its tradition by reporting on flower fields farther and farther away from town, all the way to the San Joaquin Valley and California deserts.

Wildflowers, even the showy introduced black mustard, were the centerpiece of southern California culture from the time the first towns sprung up in the coastal plains and valleys in the 1880s. More than the symbolic palm tree, winter wildflowers symbolized southern Californian bragging rights to those living in the cold, harsh climates in the eastern United States. Introduced flowers in gardens, and especially the midwinter bloom of the 'Washington' navel orange orchards, were also praised, often in photographs with the classic backdrop of snow-capped peaks of the San Gabriel and San Bernardino mountains.

Like San Francisco, the southern California coastal plains were already invaded by black mustard, while the best flower fields grew in the mesas and foothills. In an 1896 editorial, the *Los Angeles Times* stated, "Some [flowers] are found along the ocean shore, then on the mesa, or table-land are many beautiful flowers, but the finest are found among the foothills . . . , and in the little valleys at the foot of the more elevated regions."[9] From a 1907 article: "All along the King's highway [El Camino Real] from San Diego to the Tehachepi [sic] [the poppy] will be lifting their cups of gold to the sunlit skies. They will flame from the uplands of Altadena and flash among the mustard fields far and near from the mountains to the sea."[10] The following year it was reported that poppies grew on the hillsides while mustards grew on the plains,[11] reminiscent of Emory's account a half century earlier and Duhaut-Cilly's in 1827. The earliest account of black mustard in the *Times* was made in a brief

1882 editorial on "large patches of native mustard in the country around Los Angeles."[12] Clearly, the editor did not recognize it as an introduced species. Some years highlighted wildflowers even along the coast. A railway trip from Los Angeles to San Diego in 1892 distinguished itself as "a ride by the sea and through fields of poppies, . . . down to the bay."[13]

Wildflowers were described in books on southern California for the purpose of attracting immigrants from the eastern United States. Lindley and Widney (1888: 159) wrote that in Pomona the filaree (*Erodium cicutarium*) was "relieved by bright orange rugs, . . . solid beds of brilliant poppies. . . . There were lavender-colored lilies, bright-red cardinal flowers, pretty crucifers, vast bunches of violets, cream-colored bell-flowers, and the delicately-shaded tulip." In nearby San Antonio Canyon, "For weeks [wildflowers] make brilliant patches of color in the valleys and on the hills, varying from pale yellow to deepest orange" (159). In the hills at Lordsburg (La Verne): "To the south, almost as if you could reach out and pluck the flowers from their rounded sides, is the not less picturesque San Jose hills."[14] In a 1960 interview two years before his death, Theodore Payne recalled his inspiration for establishing the Theodore Payne Native Plant Foundation, which practiced wildflower conservation and even developed a farm to propagate wildflowers. Payne stated that when he came to California in 1893 from his native England, the most vivid impression made on him was "the endless miles of wildflowers in the San Fernando, San Joaquin, and Antelope valleys."[15]

In *Under the Sky in California* (published originally in 1913), Saunders (1931) took a long horse-and-buggy ride across southern California to trace the route of the Ramona Pageant. He captured both native flower fields and exotic black mustards near the coast. In Santa Clara River valley, "all the earth seemed one great jewel sparkling in the bright sunshine. The fragrance of a myriad flowers sweetened the dewy air. . . . Now and then we drove through thickets of wild mustard, higher than a man, as in Ramona's time, and covered with golden bloom" (106). Moving south from San Juan Capistrano, apparently in San Mateo Canyon, "The road, after following the line of surf for a while, turned inland and upward upon a great mesa . . . they made their midday camp amid wild flowers and nodding mustard in yellow bloom" (116). Leaving San Jacinto across the plains to the northwest, he found "wide, treeless stretches, sandy and rock-strewn but carped in places with wild grasses, filaree, and myriads of wild flowers of such beauty and abundance as had had no existence for us previously except in dreams" (133–34). On another trip, he crossed the "misty peaks of the Santa Ynez Sierra," and the "way . . . was bordered with wild blossoms of varied

hues and fragrance—pitcher sage, and yucca, and yellow mimulus, bro-
diaeas, styrax, bells, and lupines of many colors—the San Marcos road
is of such rare beauty"(169).

On Santa Catalina Island, overgrazed by sheep and feral goats, Saunders saw wildflowers protected by cactus, noting:

> But of all the floral beauty of the island, nothing is capable of giving greater
> pleasure than the wild cactus gardens of the inland hills. The sheep that
> have had the run of the interior of the island for a generation, would long
> ago have cropped it flowerless, had it now been for the prickly pear cac-
> tus, which, growing luxuriantly on the sunny slopes has been as a nursing
> mother to the multitudes of wild flowers that have gathered under its spiny
> skirts for protection from the marauding browers. The great slab-like arms
> of a cactus clump stretch and sprawl about upon the ground in a way that
> makes a very effective hedge, and within their beneficent sphere of influence
> such a tangle of lovely wildings grows and flourishes as is worth a long
> climb to see. Here are misty clouds of galium and flaming spikes of Indian
> paint brush, wild four-o'clocks, magenta-hued, lavender-cupped phacelias,
> and the white trumpets of native morning glories; here the cheerful suns
> of the plebeian yellow ox-eye blaze by the side of the delicate Catalina
> mariposa tulips. Blue brodiaeas and bluer nightshades are here, vetches
> in varying shades of purple and in white, velvety-leafed hosackias (*Cryp-
> thantha*) with clustered blooms of orange and yellow. (180–81)

In 1890, the State Floral Society chose the California poppy as the state
flower. According to the *Los Angeles Times*, "The magnificent orange-
colored poppy, whose clouds of bloom bedeck with glorious yellow miles
upon miles of California hillsides and carpet her valleys in the spring,
has been chosen by the State Floral Society as the state flower. . . . It flourishes from one end of the State to the other."[16]

The *Riverside Press Enterprise* frequently published anecdotal brevities describing flower fields on the hillsides rising above the orange groves
of town (Table 4.4), as well as accounts of weekend neighborhood flower
parties. Citizens raved about the wildflowers seen on Sunday walks or
horse and buggy rides in the hills. In 1891, "the Box Springs Mountains
[east Riverside] were covered with masses of golden poppies." In 1893,
the hills had "solid beds of yellow and blue and white." The poppies were
so abundant in 1905 that school children gathered the bounties "by the
handful and armload." In a political rally in nearby Redlands, President
Theodore Roosevelt, an avid conservationist and confidant of John Muir,
received a bouquet of wildflowers (Morris 2001).

The reports of wildflowers at Riverside show the importance of total
rainfall in the flower displays. The best displays occurred in 1891, 1893,
and 1905, all wet years with heavy spring rains. Flowers were less con-

TABLE 4.4 FLOWER REPORTS NEAR RIVERSIDE, 1884–1905

	PON[a]	Comments
1884–85	89	—
1885–86	94	April 27. [There were] myriads of beautiful flowers . . . that bedeck our . . . fields.
1886–87	59	—
1887–88	117	April 14. Through the plains . . . [the] flower[s] perfuming the air everywhere meets the eye. On the abrupt rocky hills, are beds of yellow and blue flowers.
1888–89	156	March 9. The wildflowers have not been so abundant this winter as last. The rains were not continuous enough.
1889–90	185	February 23. Immense field of poppies . . . [at] the mesa lying at the base of Mt. Cucamonga.
1890–91	129	February 28. Vegetation of all kinds is springing up and hills are taking on a delightful greenish tint. . . . In a few days, the golden poppy will tint the hill-sides with a warm, rich yellow, and beautiful flowers of a hundred different varieties [species] and hues will deck the are plains [with] many colors. March 28. The Box Springs hills are covered with masses of golden poppies.
1891–92	65	—
1892–93	124	March 23. Wildflowers are becoming very abundant, especially the beautiful golden poppy. April 1. The drive down the valley, following the canal [east Riverside], is one of great beauty at this season of the year. The hills are covered with a carpet of emerald green bedecked with beautiful wildflowers. The air is sweet with the fragrance . . . and the exhalations of blossoms so numerous as to seem like solid beds of yellow, and blue, and white. April 17. People gather gorgeous wildflowers upon the broad expanse of the plains and the sloping mesas, or to wander among the charming nooks and corners of the many beautiful canyon retreats (*Los Angeles Times*).
1893–94	71	May 3. Dry weather has affected the wildflowers (*Los Angeles Times*).
1894–95	164	March 31. The hillsides were never so gorgeously beautiful with wildflowers [Murrieta] (*Riverside Press Enterprise*).
1895–96	75	—
1896–97	128	March 25. West Riverside seems to have quite an attraction for wild-growing flowers, especially poppies.

	PONª	Comments
1897–98	55	—
1898–99	48	—
1899–00	69	—
1900–01	122	February 26. Dainty wildflowers on the hills . . .
1901–02	70	—
1902–03	129	March 6, 1903. The golden poppies are again in bloom; also cream cups, baby blue eyes, and several other small varieties.
		April 3. [Corona] The mesa is covered with wildflowers.
1903–04	57	—
1904–05	168	March 6. Those going to Coldwater Canyon report . . . they found such a profusion of wild flowers as they had never seen before.
		March 18. In has been many days [years] since there has been such an abundance of wildflowers, as are now to be found in this neck of the woods. This morning, crowds of school children and tourists were out on Rubidoux Hill, where they were gathering the beauties by the handful and the armload.
		March 20. Spare the poppies. But if the crowds of people and children who have been engaged in pulling up these beautiful flowers do not show more discretion, the poppies will not be there next year. . . . Therefore, all persons who want Riverside to have these poppies in great profusion are urged not to pick the flowers now making Rubidoux Hill so attractive.

SOURCE: *Riverside Press and Horticulturalist,* unless otherwise indicated.
ª. PON, percent of normal precipitation. Mean 25.4 cm.

spicuous most other years. There were no reports in 1885, 1887, 1896, 1898–1900, and 1904, all years with below-normal precipitation.

Beginning in the 1880s, the *Los Angeles Times* frequently published "brevities" on what became a famous field of California poppies above Pasadena, called the "Altar Cloth of San Pasqual" this chapter's Elizabeth Grinnell epigraph, and the "Mesa de los Flores." The Altar Cloth of San Pasqual was named La Sabanilla de San Pasqual—the altar cloth of Saint Pascal, a saint who knelt among the wildflowers and prayed while he tended sheep in Old Spain. The poppy field lay at the end of a railway from Los Angeles. Above the flowers was the tourist mecca of the Echo Mountain incline—a railway ascending a 30-degree slope to the Mount Lowe Tavern, and the Mount Harvard Observatory on the sum-

Figure 4.1. Gathering of wildflowers at the "Altar of San Pasqual," the famous poppy field above Pasadena, ca. 1900. Note the rail car in the background. Courtesy of the Herald Examiner Collection/Los Angeles Public Library.

mit of Mount Wilson. The poppies were admired by rail, and seen from above. The Altadena poppy field was a microcosm of a broader spectacle (Figure 4.1).

Charles Fredrick Holder (1889: 3), a fresh immigrant from the wintery east and the inspiration behind the Rose Parade, wrote picturesque descriptions of the Pasadena landscape, inspired most by the fact that he could see wintery snows on the nearby mountains and fragrant flowers in the middle of winter in town: "We can see the flurries and watch the white wraiths tossed aloft by the wind on the upper range: yet where we stand, the odor of the orange, rose, violet and hundreds of flowers fill the air." He recounted the region's recent history: "Fifteen years ago, the San Gabriel Valley constituted several large ranchos, owned by the Bandini family, and several others, Spanish and American. Great live oak trees covered large tracts, and the bare spots were overgrown with sage brush and grease-wood, or carpeted with a variety of flowers found nowhere else in the world in greater beauty or profusion" (5).

He described flowers from many sites about town. From Millard Canyon he looked down on the Pasadena plain and saw "Pasadena . . .

at your feet, a vast crazy quilt, or checkerboard colored in many tints and hues." Los Flores Canyon had "its profusion of flowers, looking down on Poppy Land" (Holder 1889: 38). Wilson's Ranch, directly below Mount Wilson, had "huge sycamores on whose branches the great California condor rests. . . . and pasture . . . blooming with flowers in mid-winter" (62). Holder was member of the Valley Hunt Club that originally established the Tournament of Roses, now the Rose Parade. The name Tournament of Roses refers to the fact that the first event in 1890 emphasized a tournament of races, and other competitions, not the parade (Hendrickson 1988, 1989). Holder described the wildflowers in relation to jackrabbit hunting: "Instead of following the roads, . . . the hunts take you across country where your horse beats out a wide swath in a growth of flowers which you had never heard of or seen before. You plow through them, literally a mass of color, brilliant and remarkable" (Holder 1889: 91).

Holder recounted the usual sequence of flowers during the growing season:

> At the. . . . first rain, [the mesas] assume a change of raiment—first green with the rapid growing alfileria, so valuable as natural fodder. This covers the length and breadth of the land, marking the high and by ways with newborn tints. Soon after the advent of the alfileria, come the delicate bell-shaped cream-colored flower, daintily poised on a slender stock [cream cups, *Platystemon californicus* or *Plagiobothrys*, which is abundant early in spring]. On the upper slopes they mass like snow, changing the color of the fields. In February, possibly earlier, . . . the poppy splashes of color come here and there; the afterglow of the mountains seem to have been transferred to the plains below and finally the slopes appear a fire with the golden flowers. In certain places, near canyons, on the hills of San Rafael Ranch, and in Wilson's Canyon—the "shooting star," or American cowslip is found [*Dodecatheon clevelandii*]. Yellow violets [*Viola pedunculata*] now appear covering the ground in many places, nodding in clumps of grain fields, or forming a gorgeous border to the roadsides together with the masses of a little blue cup-shaped flower—the "baby bluetts" of the children [*Nemophila menziesii*], but now others follow. . . . The painters brush colors the hillsides and fields with vivid tints, resting on a matting of velvet greens, formed by clovers of many kinds. The iris rears its graceful shapes in brilliant masses [blue eyed grass, *Sisyrinchium*], while evening primrose [*Camissonia*], rock rose [*Helianthemum*], wild pea [*Lathyrus,* or *Lotus*], tulip [*Calochortus?*, vaguely resembling a tulip], and many more in the floral throng. Bells of blue and white, trumpets of purple and pink, the lovely Penstemon, the gorgeous silene, huge ox-eyed daises [an asteraceae], delicate crucifers, golden dandelions [?], blue and white snapdragens [*Antirrhinum nuttallianum*], make this Christmas greeting. (107–8; botanical interpretations by Andrew Sanders, pers. comm.)

Holder also described the distribution of the introduced black mustard. In the "miniature valleys" of the hills between Pasadena and Los Angeles, "the wild mustard grows, rising five or six feet [1.8–2.0 m] in slender shafts, topped with yellow blossoms, as if a shower of floss had fallen from the skies. From the hill up this is a veritable golden yellow sea, shimmering and gleaming in the cool trade winds. Then descend and rise in the golden masses, the flowers meeting your horse's head, see the bunches of violets, spring up at every step" (41). Van Dyke (1886: 36–39) described a similar floral calendar for the Pasadena region. The rainy season began with the "star-like" alfileria and was followed by "shooting stars, . . . clovers, . . . and the light of the poppy that flamed along the meadows and blazes on the northern hillsides [next to the San Gabriel Mountains]." Other species described by him to bloom later in spring were clovers, lilies, iris, phacelia, vetches, lupines, asters, and "bell flowers, white and blue."

The *Los Angeles Times* first referenced the San Pasqual poppy field on June 6, 1883, when Jeanne C. Carr , wrote a sketch for a forthcoming volume *A Southern California Paradise* (Lindley and Widney 1888). Carr wrote: "The [Spanish] fathers were excellent judges of the quality of land. And nowhere was the outburst of floral spender which follows the early rains more dazzling than upon the *mesas* of San Pasqual, where the homesteads of Pasadena stand to-day. The sun-steeped escholzia [sic] overspread the earth with its ruddy gold, lupins [sic] rolled their blue billows toward the uplands."[17] In 1888 the *Times* reported,

> The profusion of wildflowers will attract the most latent interest, and "poppy land" as the upper mesa or Altadena is called, and that seems aire with their glow, is a marvel to the eastern eye—a solid mass of golden yellow, so vivid and intense that it has been seen from the ocean 25 miles away, the old mariners calling the Altadena uplands at this time the land of fire, or Terra del Fuego. A detailed picture of the mesa reminds you of John Muir's picture of the Central Valley. The roadsides are lined with wild flowers of every hue, that succeed each other in endless variety as the days slip by.[18]

Nearly thirty years later, Saunders (1914: 103) wrote of this floral landmark: "At the base of the San Gabriel Sierra north of Pasadena is an elevated mesa tilted to the south. . . . It was known in the Spanish days by the name of La Mesa de las Flores (the table of the flowers), because of the exuberant growth of poppies which covered it like a golden table cloth, its folds dropping down into the valley. The story goes that flowery flame could be seen plainly in clear weather from vessels at sea." Indeed, Saunders captured the Altadena poppy field in a color-tinted black-and-white photograph.

The *Times* reported the status of the flowers at the end of the rail line at Altadena nearly every year, often as a brevity in the editorial section (Table 4.5; Figure 4.1). Comments on the Altadena poppy field were accompanied by statements of the flowers on the mesas between Los Angeles and Pasadena, and on black mustard fields below Los Angeles to the seashore. The frequency of reports gives every impression that wildflowers were seen throughout the region most years. Exceptional displays occurred in normal years with well-distributed rains and in nearly all the wet years. What follows are highlights from extensive coverage in the *Times* for days listed in Table 4.5.

In 1888 the *Times* reported, "On every hillside a garden of blossoming wildflowers; in every valley the colors of the rainbow spread in unnumbered blossoms," and "the roadsides are lined with wild flowers of every hue, that succeed each other in endless variety."[19,20] While the dry winter of 1888–89 elicited no remarks on wildflowers, the following year, the third wettest at Los Angeles to this day, produced at Altadena "deep copper rivers [that] wind away or spread out in well-defined patches against the green."[21] By late spring came the "profusion of wildflowers with which the hillsides, mesa and canyons in the neighborhood are adorned in all the colors of the rainbow."[22] During the wet year of 1892–93, the railway ride along Mission Canyon showed "everything is abloom, . . . the ground is carpeted with flowers; fields of wild mustard are yellow as the sunshine . . . clematis garlands the roadside, and sunflowers, butter cups and 'baby-blue-eyes' are everywhere."[23] In the dry spring of 1894, the poppies still made "the foothills a field of the cloth of gold."[24]

Flowers on the mesa during the rainy year in 1895 were an epiphany for a lady who had just come in from viewing the poppy-covered hillsides of Pasadena: "She was farily bubbling over with delight. 'It is a dream of beauty,' she exclaimed. 'It is as if one of the brightest sunset clouds had dropped down and wrapped the hills in its mantle.'"[25] For others, the view from the Echo Mountain summit down to Pasadena was "vast golden, radiant floods of California poppies." The next month it was reported that "bloom[s] run riot everywhere in garden and field. The hillsides are golden with the nodding poppies; the broad levels are a sea of color."[26]

Dry winters diminished the poppy fields from 1898 to 1900 as in Riverside (cf. Table 4.4). In February 1898 the *Times* reported that "the gay and golden poppy of the foothills is now in evidence and a right brave show she makes."[27] In 1900 the poppies faced "great crowds going up by the electric cars and returning laden with golden flowers . . . and it seemed as though the flowers must be exterminated by their onslaught but the supply seemed to stand the strain very well. . . . It is noticeable, however, that they are retreating toward the mountains, and it seems only a question of

	PONᵃ	Flower Abundanceᵇ	Precip. Pattern Observations (July to June)	Dates of Flower Reports
1886–87	94	**	February/April	April 27, May 24
1887–88	93	***	Winter/spring	January 2, February 23, March 11, 29
1888–89	129	—	Midwinter drought	—
1889–90	233	***	Fall/winter floods	December 6, January 28, 29, 30, 31, February 11, March 16, 22, April 12
1890–91	89	**	Heavy/February	February 7, March 9, 26
1891–92	79	**	Spring	March 12, 27, April 7
1892–93	176	***	Evenly distributed	February 11, March 10, May 21
1893–94	45	*	December	February 2, March 18, May 3
1894–95	108	***	Winter/late spring	March 10, March 28, 31, April 9, 25, 28, 26, July 22
1895–96	57	**	January/March	January 1, March 1, March 17
1896–97	113	***	Evenly distributed	February 2, 6, April 10
1897–98	45	*	Winter/spring drought	February 23
1898–99	37	*	January/March	March 20
1899–00	53	*	Fall/winter drought	January 10. Fall rains
1900–01	109	***	Evenly distributed	February 27, March 1, 3, 8
1901–02	71	*	Spring	March 1
1902–03	129	**	Evenly distributed	February 14, March 14
1903–04	58	—	Late spring	—
1904–05	131	**	Winter/spring	May 15
1905–06	124	***	Winter/spring	February 18, 25, March 15, 26
1906–07	129	**	Winter/spring	January 31, March 21, 29
1907–08	78	**	Winter	February 24, 26, April 1
1908–09	128	***	Evenly distributed	January 28, February 7, 23. March 11, 21, April 21
1909–10	84	*	December	February 13, March 8, 28, April 2
1910–11	108	***	Winter/spring	February 5, April 11

	PON[a]	Flower Abundance[b]	Precip. Pattern Observations (July to June)	Dates of Flower Reports
1911–12	77	**	Midwinter drought,	March 11, April 1, 19, 28
1912–13	89	***	January/February	March 13, 30, April 11, 13, 18, June 7
1913–14	158	**	Fall/winter, flooding	March 9
1914–15	114	**	Winter	March 8, April 3, 4, 19
1915–16	133	**	Winter, flooding	February 6, March 30
1916–17	93	*	Fall/winter	May 13
1917–18	93	*	Late spring	May 1

SOURCE: *Los Angeles Times*.
[a]. PON, percent of normal precipitation at Los Angeles downtown, July to June. Mean, 38.0 cm.
[b]. Flower abundance: — no report, * local, ** normal, *** unusually abundant.

time when under present conditions the poppy must become practically exterminated in this vicinity."[28] When the drought ended in 1901, the *Times* reported that "the poppy . . . [is] making great splashes of color like a running prairie fire in the lowlands."[29] In 1902, the Santa Ana Valley had a profusion of wildflowers in the canyons and on the foothills: "Poppy hunters were out in full force."[30] The following year a rare snow fell on the poppies,[31] and in 1904 people only wished for wildflowers, as drought had produced "the dull monotony of dust" from the grain fields.[32]

Drought-breaking rains in 1905 resulted in excellent late-blooming flowers, but school children found it difficult to sell flowers about town. In 1905 Alice Merrill Davidson wrote that "in spite of cultivation and its weeds, it is probably true that there are still fields of wild flowers accessible to every California school child, only in most towns we must support that the dime for the trolley-car ride is also accessible."[33] Flowers were "sufficient" in 1906, after another wet winter and spring, in spite of "predations by young boys. The blooms are being plucked at a fearful rate. Boys about town pick them carelessly so that in many instances the plants themselves are destroyed. The blossoms are sold for a few cents a bunch about town."[34] The *Times* also reported that "although the population of Southern California has been so largely increased during the past year, the poppy crop has also been augmented sufficiently to go around, and there will be an armful for us all." It was also reported that "when Mr. Henry Huntington returns to Los Angeles from New York he will look in vain for the beds of poppies that would be expected to adorn the roadbeds of his interurban [railway] lines. . . . Sad to say the poppies have not popped" that spring. [35]

Another bumper crop developed in 1909, an average rainfall year but with well-distributed rains. In February the *Times* claimed "the greatest crop of [poppies] that we have ever had since the original pepper tree was planted at San Juan Capistrano [in the 1830s]."[36] Weeks later: "The plains are spangled with poppies amid the emerald carpet that covers almost every square foot of the whole territory of southern California."[37] By late spring had appeared "broad patches of golden poppies, streaked with long lines of yellow mustard three feet tall [1 m], and blotched at a thousand points with bunches of purple larkspur."[38] In 1911 the *Times* reported that "the California poppy and the wild mustard blossoms are painting the canyons and the sides of the mountains until they look like fields of the cloth of gold, recalling the heraldic days."[39] The following year, aviators visiting the Griffith Aviation Park and "gathered arms full of poppies on the flying field."[40] Flower fields were scarce after 1915, and massive floods distracted Los Angeles citizens from flowers in 1916. A string of dry years beginning in 1917 diminished the flower fields. City officials and members of the Los Angeles business and civic organizations toured rapidly developing areas across the mountains and visited poppies not in Los Angeles, but at Palmdale and the Antelope Valley, a prophetic event that earmarks the retreat of wildflowers and wildlands to the desert.[41]

The flowering season lasted from January to as late as April in wet years (Appendix 4), which is similar to accounts by Muir, Cronise, and others from central California from the mid–nineteenth century. In *Trees, Shrubs and Wild Flowers of Southern California,* Jeanne C. Carr wrote,

> In February, the advance-guard of the poppy family (*Eschscholzia californica*) is observed taking possession of old furrows in fields and orchards. . . . Two weeks later rank patches, with open, bright-yellow flowers, appear in company with blue *Nemophilas,* nodding cream-cups (*Emmenanthe penduliflora*), purple Calandrinias and yellow violets, which have bronze linings and delicate penciling of black lines upon their petals (?) . . . *Dodecatheons* . . . cover the moist banks. White "forget-me-nots" exhale their delicate odor. But not for many weeks shall we reach the summit of the floral year, when, perchance, after a walk or ride through miles of poppies. . . . We touch an island of blue larkspur or lupine. . . . All through March, April and May, plants of *Layia platyglossa* (tidy tips) are scattered over the ground. It is nearly always found associated with the moss like *Gilia dianthoides.* The commencement of the vernal years varies greatly, but Christmas nearly always finds the earth thickly furred with the starry mats of *Feleri,* (*Erodium cicutarium*), with young grass and clover, and from the first rain to the end of April the floral display increases, until, at its height, distinct bands of color, blue or orange may be traced in the landscape for many miles (Lindley and Widney 1888: 327).

Holder (1889: 75), while not a botanist, captured both wildland and urban phenology around Pasadena: "In October, the weather grows slightly cooler, and soon the rains come. This gives renewed lift to all vegetation, and instead of frost, snow, and ice, the land blooms like a garden, the mesas are carpeted with a succession of flowers, the low hills taken on a vivid tint, and the air is filled with fragrant odors. . . . Yet how cold Pasadena really is, is shown by the fact that the roses, callas, and all flowers, wildland and otherwise, bloom throughout the entire winter."

Ultimately, the flowers and exotic annuals cure in the dry season. Van Dyke (1886: 37), in a description of autumn, stated that "the plains and slopes lie bare and brown; the low hills that break away from theme are yellow with dead foxtail [farmers foxtail, *Hordeum murinum*], or wild oats, gray with mustard stocks."

Until circa 1920, there were very few reports of wildflowers in the sparsely populated desert, where access was limited by poor roads. In a rare exception, the *Riverside Press Enterprise* reported that "all the desert is ablaze of bloom" in 1905, a wet year with heavy spring rains, which also breached the Coachella Canal, flooding the Salton Sea with waters of the Colorado River. The report also describes "verbenas and baby blue eyes [in] bloom when they are only a few inches high."[42] One of the earliest accounts of desert wildflowers in the Mojave Desert north of Los Angeles was reported by the *Times* in 1906, another wet year with heavy spring rains. The Board of Public Works, directed by William Mulholland, made an auto tour to the Sierra Nevada, source of the future Owens Valley aqueduct. They reported that "the Mojave Desert is carpeted with flowers that come up after the rains, but nowhere else is it so gorgeous with blossoms as in the westerly bend of Antelope Valley. Thousands of acres are covered with poppies of vivid orange, so thickly that one could pick a dozen blossoms from any square foot of land. Under the poppies are myriads of small golden yellow flowers [*Lasthenia gracilis*], and here and there a streak of purple runs through the gold and orange embroidered thickly upon a carpet of vivid green. Miles and miles of flaming prairie unrolled before the speeding cars."[43]

When the aqueduct was completed, land values and land clearing skyrocketed in Los Angeles, especially in the San Fernando Valley (McWilliams 1946). California's flora was even used to bring about its own destruction. During the spring of 1912 imaginative land developers at Van Nuys, with little regard for nature's floral legacy, "celebrated poppy day" to entice investment. The public was invited to gather flowers among "the fields at the new town of Van Nuys . . . aglow with thousands of yellow blossoms." A picnic barbecue was tendered for those who came, and automobiles took visitors to the poppy fields. The *Times* later reported, "These beautiful spring days are attracting many people from the city to the golden poppy fields surrounding the new town of Van Nuys. . . . Nearly every car returning has people laden with the flowers, which may be gathered anywhere in this vicinity, without restriction. Van Nuys is growing as rapidly as ever, and in the next few months bids fair to eclipse even its past enviable record in the building line."[44] The *Times* also listed the sale of hundreds of lots. Poppies were replaced: "Four thousand sacks of seed potatoes have been planted around Van Nuys this season, 100,000 fruit and walnut trees have been set out in new orchards, and 7500 acres (3,000 ha) planted in sugar beets."[45]

Bit by bit, wildflower fields were carved away from the plains of Los Angeles, but automobiles allowed people to drive farther and farther away to new flower fields distant from the expanding town. Charles Frances Saunders (1914: 99) perhaps best encapsulated the impending fate of wildflower fields around Los Angeles, quoting a professor acquaintance:

> A humiliating fact in connection with our California wild flowers . . . is the average Californian's indifference to them. Not only does he not know their names, he does not even see them, as he slashes right and left in his haste to subdivide the State into building lots and orange ranches. Why, man, the gardens of Europe are full of California wild flowers, and have been for three generations—raised from seeds carried there by English collectors—such flowers as clarkias, collinsias, lupines, gilias, eschschotzias, godetias, phacelias, mariposa tulips, penstemons, and a score more. Now here . . .

In eulogizing Los Angeles's wildflower fields, Saunders invoked Hesperides, the nymphs of classical mythology who, with the aid of a dragon, protect a garden of golden apples: "Year by year more and more of these Hesperian gardens of the wild are being broken up by the encroaching settlements of men; and only the other day I saw a plowman, knee deep in eschscholtzia, driving his furrows straight through five acres of them and quenching their sheeted fire with the upturned earth" (104).

FLORAL BRILLIANCE
AND COLLAPSE ALONG THE "CIRCLE TOUR"

In 1926 Phillip Munz, lead author of *A California Flora*, published by the Rancho Santa Ana Botanical Garden in 1959, wrote in the *Los Angeles Times* that in southern California "every canyon, every hillside, every valley, every ridge has its own quota of color and beauty."[46]

Within decades this vision became a distant memory, and this is well-documented in both the *Times,* the *Riverside Press Enterprise,* and in *The Desert Magazine.*

From the 1880s to 1920, the wildlands of the Los Angeles plains and interior valleys hosted an impressive pastel of color every spring, as first described by the Franciscans in the 1770s and by writers after the Gold Rush. The retreat of wildflowers due to urbanization and agriculture was furthered by the expansion of bromes, slender wild oat, and summer mustard into areas that had never seen the plow or bulldozer—first in Los Angeles, then in the southern California interior valleys and San Joaquin Valley, and later in the desert.

Beginning in 1920, wildflowers were documented virtually every spring by the *Los Angeles Times* in coordination with the Automobile Club of Southern California, on whose board *Times* editors served, to encourage the readership to spend days off work seeing the wildflowers in the country (Figure 4.2). Articles in the *Times* were more than casual records, because temporal trends were made robust by the newspaper's focus on a "circle tour" to favorite sites and flower festivals, a practice which continued into the 1960s. Articles by Lynn J. Rogers, the outdoor editor, chronicle the experiences of scouting trips along highways to specific areas. One trip struck north from Los Angeles across the San Fernando Valley to Gorman and Arvin in the southern San Joaquin Valley. The "tour" returned via Tehachapi Pass and southward along the western Mojave Desert to Palmdale, with occasional tangential surveys along Highway 58 eastward to Barstow. Another trip struck east from Los Angeles through Riverside and Beaumont to the Sonoran Desert at Palm Springs and returned via Anza-Borrego State Park and San Diego. Wildflowers were also frequently highlighted in the column "Lee Side o'LA" by Lee Shippey. The accounts from the *Times* were paralleled by the *Riverside Press Enterprise,* which frequently reported on the state of wildflowers on the Riverside-Perris plain, especially in articles on the play *Ramona* by Helen Hunt Jackson each year at San Jacinto. *The Desert Magazine,* under the editorship of Randall Henderson, annually recounted the desert wildflowers in the spring issues from 1939 to 1962.

Table 4.6 recounts the annual status of wildflowers in several regions beginning in 1918, including Los Angeles, Riverside, the southern San Joaquin Valley, the southern coastal interior valleys, Antelope Valley, the Mojave Desert, Coachella Valley, and Death Valley. The magnitude of the flower display was evaluated using a qualitative ordinal classification. Many articles after 1950 that anticipated potential blooms were not cited as direct observations of wildflowers. The dates of articles in the *Los An-*

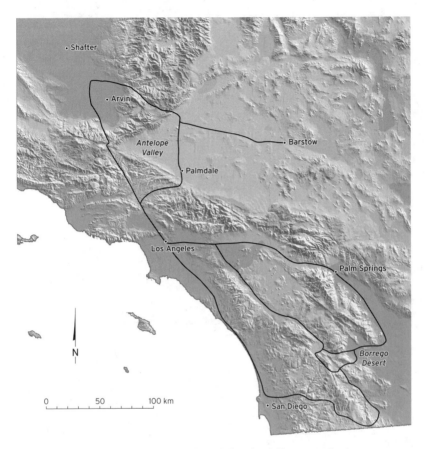

Figure 4.2. The "circle tour" taken by weekend wildflower enthusiasts
(*Los Angeles Times* May 14, 1922; April 7, 1929; May 1, 1938).

geles Times, the *Riverside Press Enterprise,* and *The Desert Magazine* are
given in Appendix 4.

The year-to-year abundance of flowers correlates with total annual
rainfall and generally fluctuates in phase across southern California.
Broadscale winter storms generally yield only minor regional differences
in annual precipitation departure, but some years produce extraordinary
rainfall in one district (e.g., 1949 in the Coachella Valley), typically as a
result of a single storm.

Before circa 1940, the annual frequency of wildflower displays was
highest in moist climates near the coast and decreased toward drier climates of the desert. An exception is the southern San Joaquin Valley,
which receives similar average annual precipitation as the Mojave Desert,
but soils are kept moist in winter by persistent tule fogs. Desert blooms

TABLE 4.6 REPORTS OF FLOWERS IN SOUTHERN CALIFORNIA SINCE 1918

	LA[a] PON[b]	LA	Riv PON	Riv	Bak PON	Kern	Plm PON	Ant	Moj	PS PON	Coa	DV PON	DV DV	Cen Calif
1918–19	57	—[c]	99	—	80	—	80	**	—	61	—	nd	—	nd
1919–20	84	*	119	*	85	—	102	—	—	148	—	nd	—	nd
1920–21	91	**	102	—	112	—	89	—	***	59	***	nd	—	nd
1921–22	131	**	203	***	142	***	133	***	***	224	—	nd	—	nd
1922–23	64	—	94	—	95	—	89	—	—	6	*	nd	—	nd
1923–24	44	*	84	*	59	***	44	—	*	24	—	nd	—	nd
1924–25	53	—	79	—	74	***	54	*	**	28	**	nd	—	nd
1925–26	117	—	140	—	81	***	89	***	—	177	***	nd	—	***
1926–27	119	*	142	*	100	***	121	**	—	132	*	nd	**	*
1927–28	68	*	116	—	95	—	60	—	*	83	—	nd	**	—
1928–29	85	*	80	—	72	***	76	***	—	109	*	nd	—	*
1929–30	77	**	130	**	77	***	78	—	*	152	—	nd	—	—
1930–31	84	—	118	—	93	*	74	*	*	118	**	nd	—	*
1931–32	113	**	150	***	151	***	130	***	***	242	*	nd	**	*
1932–33	79	**	91	***	113	*	80	*	—	84	*	nd	—	*
1933–34	97	**	47	***	36	*	71	**	***	47	*	nd	—	*
1934–35	145	**	120	***	135	***	134	***	***	160	**	128	**	***
1935–36	81	—	111	*	77	—	80	—	—	113	***	13	—	*
1936–37	143	*	218	**	152	***	163	***	****	225	**	171	—	—
1937–38	157	*	121	*	81	***	146	***	****	70	***	61	—	*
1938–39	87	*	108	**	110	***	137	***	****	130	***	207	*	*
1939–40	128	*	108	***	116	***	97	—	****	176	***	167	***	*
1940–41	219	**	230	***	186	***	232	—	*	221	***	156	*	*
1941–42	74	—	98	*	81	—	55	—	—	181	—	89	—	*
1942–43	122	—	152	—	169	*	163	—	—	223	—	75	—	—

Year						
1943–44	128	158	83	118	210	118
1944–45	77	96	135	57	75	57
1945–46	78	88	81	68	130	68
1946–47	84	91	82	159	55	159
1947–48	48	62	71	42	73	42
1948–49	53	77	65	105	111	105
1949–50	71	70	78	9	26	9
1950–51	55	56	88	58	30	58
1951–52	175	175	139	118	160	118
1952–53	63	96	99	49	137	49
1953–54	81	97	71	65	122	65
1954–55	80	83	74	90	83	90
1955–56	107	74	63	19	51	19
1956–57	64	98	75	86	75	86
1957–58	141	159	161	85	135	85
1958–59	37	43	39	65	45	65
1959–60	59	76	67	124	74	124
1960–61	32	29	65	33	29	33
1961–62	125	95	103	130	48	77
1962–63	58	55	73	36	50	59
1963–64	53	98	74	93	117	69
1964–65	91	80	92	44	64	108
1965–66	137	126	83	124	187	93
1966–67	147	120	114	89	107	36
1967–68	111	86	99	86	89	128
1968–69	183	193	141	145	143	122
1969–70	52	62	54	33	76	98
1970–71	82	63	107	73	42	55
1971–72	48	49	48	59	25	61
1972–73	142	120	128	103	80	167

TABLE 4.6 (continued)

LA[a] PON[b]	LA	Riv PON	Riv	Bak PON	Kern	Plm PON	Ant	Moj	PS PON	Coa	DV PON	DV DV	Cen Calif
1973–74 99	—	77	—	80	—	68	* *	—	75	—	87	—	—
1974–75 96	—	71	—	108	—	71	*	—	44	—	120	—	—
1975–76 48	—	77	—	70	—	37	—	—	86	* *	158	—	—
1976–77 82	—	82	* *	67	—	136	—	* *	129	* * *	125	—	—
1977–78 223	—	218		204	* * *	190	* *	* *	213	*	233	—	—
1978–79 132	—	152	—	107	—	143	—	—	139	*	99	—	—
1979–80 180	—	165	—	105	—	174	—	—	304	—	142	—	—
1980–81 60	—	59	—	77	—	60	—	—	47	—	45	—	—
1981–82 72	—	128	—	102	—	120	—	—	72	*	115	—	—
1982–83 209	—	184	—	156	—	190	—	—	164	—	155	—	—
1983–84 70	—	98	—	84	—	51	—	—	156	—	86	—	—
1984–85 86	—	75	—	65	—	88	*	—	56	—	90	—	—
1985–86 120	—	104	—	107	—	68	* *	—	112	—	50	—	—
1986–87 51	—	56	—	90	*	42	* *	*	85	—	90	—	—
1987–88 83	—	98	—	95	*	111	* *	* *	108	* *	265	* *	—
1988–89 54	—	66	—	60	—	50	—	—	32	—	31	—	—
1989–90 49	—	56	—	54	—	33	*	—	27	—	26	—	* *
1990–91 77	*	104	—	96	*	86	* *	—	92	—	81	—	—
1991–92 141	—	111	—	96	* *	158	* *	—	143	* * *	119	—	—
1992–93 183	—	208	—	150	* *	224	* *	—	243	* * *	162	*	*
1993–94 54	—	93	—	93	—	45	* *	—	48	* *	11	—	—
1994–95 164	—	189	—	149	—	116	* *	—	167	* *	157	—	—
1995–96 83	—	73	—	105	—	49	—	—	25	—	33	—	—
1996–97 83	—	113	—	103	—	46	*	—	26	—	44	*	—

Year														
1997–98	207	—	253	*	232	—	203	**	**	176	***	279	***	—
1998–99	61	—	64	—	110	—	33	—	—	15	—	57	—	—
1999–2000	77	*	64	*	82	—	54	—	**	32	*	52	—	*
2000–01	120	—	85	**	89	*	77	**	*	84	—	124	—	—
2001–02	29	—	34	—	56	—	28	—	—	6	**	21	—	*
2002–03	110	—	128	**	95	**	113	*	*	70	**	97	**	—
2003–04	62	—	70	—	73	—	64	—	—	73	***	101	***	**
2004–05	249	—	229	*	147	*	251	**	***	241	—	321	*	**
2005–06	88	—	78	—	112	—	98	**	—	47	—	86	*	—
2006–07	18	—	17	—	45	—	10	—	—	10	—	37	—	—

SOURCE: *Los Angeles Times, Riverside Press Enterprise* (Appendix 4); Feral Flowers, www.feralflowers.com/home.html.

a. LA, Los Angeles; Riv, Riverside; Bak, Bakersfield; Kern, Kern County; Plm, Palmdale; Ant, Antelope Valley; Moj, Mojave Desert; PS, Palm Springs; Coa, Coachella Valley; DV, Death Valley; Cen Calif, Central California.

b. PON, percent of normal precipitation. Means: Los Angeles, 38.0 cm; Bakersfield, 15.8 cm; Palmdale, 20.0 cm; Palm Springs, 13.6; Indio (Palm Springs column before 1927), 8.0 cm; Death Valley, 8.9 cm.

c. Flower abundance: nd, no data; — insufficient flowers to draw newspaper attention, or reports indicated flowers absent; * flowers present but reported as scarce or below usual abundance; ** "splashes" of flowers in a region, or reported as roughly average; *** unusually abundant.

occur only in the wettest winters, particularly in the arid Coachella Valley and Death Valley. By the turn of the twenty-first century, flower displays have declined throughout southern California, with the exception of the deserts. The decline coincides with the invasion of "second wave" invasive annuals.

Wildflower outbreaks, 1920–1934

In the 1920s and early 1930s, wildflowers occasionally painted the plains and hillsides across southern California, including plowed lands of Los Angeles. Although drought persisted from 1918 to 1921, flowers were still seen on the outskirts of Los Angeles. The *Riverside Press Enterprise* reported that "poppy fields at the foot of the Cajon pass attracted thousands of motorists."[47] According to the *Times,* "Not many miles from the heart of Los Angeles may be found one of the thickest and most beautiful poppy fields in southern California . . . at Cahuenga Pass until in runs into the Coast highway." It was "the prettiest poppy field within close range to this city."[48] Heavy rains in 1922 produced an outbreak of wildflowers across southern California. In the Sonoran Desert, where rainfall was twice normal, the *Times* reported "a sea of flowers, wind rippled, breathing a perfume that pervades the length and breadth of Coachella Valley."[49] Along the coast, the *Times* wrote that "poppy fields are glowing in almost every county in the southern portion of the State . . . as the desert is [in] bloom now like a natural garden . . . motorists are daily making up picnic parties to go out and enjoy the spring flowers."[50] Drought resumed through 1926, and the best displays were limited to Kern County. In 1925, the Tejon Ranch and areas north into San Joaquin Valley were "miles of gorgeous blooms, one flower touching another, an illimitable ocean of color and of riotous beauty. In one place was a stretch of blue lupines reaching off to the very horizon."[51] The following year, flowers colored the San Gorgonio Pass, where the *Times* wrote that from Banning "the valley sloping off to the east is like an inverted rainbow with its strips of blue, pink, yellow and purple blossoms."[52] Heavy rains in 1927 and 1928 triggered extensive blooms across southern California, especially in Kern County and the deserts. In the first annual flower festival at the town of Arvin, south of Bakersfield, thousands of visitors came from all over the state to see "its slopes aflame with poppies and its plains splashed with the purple of California lupins [sic]."[53] Residents of Kings and Fresno counties "saw the immensity of the valley could be realized by gazing east over the plains for twenty miles, a vast bed of flowers of all colors, while the hills to the west were a panorama of dif-

ferent colors from the flowers,"[54] recalling the descriptions of John Muir
and Clarence King sixty years earlier. Along the circle tour, land clearing had limited the poppy fields to the upper end of the San Fernando
Valley, near Chatsworth, to the hillsides of Santa Susanna Pass, and to
the fields around Simi and Moorpark. To the north of Los Angeles was
"a continuous mass of bloom on top of Gorman and down into the San
Joaquin Valley."[55] The desert was "a wonderland, a riot of color, a million rainbows lying side by side over sand dune, valley and on to the shoul-
ders of the mountains." West of San Jacinto was "a solid tract of several
hundred acres of yellow flowers [*Amsinckia, Lasthenia*], [and] over the
hillsides along the route to this city are myriads of Canterbury bells, blue
bells, Indian paint brushes, [and] California poppies."[56]

In 1928, flower fields near Palmdale were plundered by the plow, a
tragedy in hindsight because most agricultural lands ultimately failed to
yield crops in that region's semiarid climate. The *Times* reported that "although many fine fields of poppies were plowed up, and the fields planted
in grain there still remain approximately 600 acres (240 ha) . . . which
are well worth the trip to Antelope Valley."[57] Rains and flowers were
"scanty in the Coachella Valley as the sand verbena has not appeared in
the wonderful carpets that a wet year brings forth."[58] Light rains in 1929
produced blooms mostly near Arvin and Palmdale. At the Arvin Flower
Festival, wrote the *Times,* "The wildflower belt of the San Joaquin Valley . . . is most pronounced on the west side and on the gentle slopes of
the east portion of the coast range." The *Times* also reported that the
area had "many varieties of blooms, miles upon miles of flowery carpet
are reported covering the great level plains of the valley."[59] Wildflower
fields were reported in uncultivated areas of the southern California
plains. Sand verbenas were seen near Inglewood, and small flowers were
found on the mesas of San Diego. Large fields of introduced wild mustard grew along the alluvial fans of Foothill Boulevard east of Pasadena.[60]
Below-normal rains limited flower displays in 1930 and 1931, and the
spring of 1930 elicited one of the last accounts of wildflowers in the San
Fernando Valley: "Fields are now a blaze of yellow with California poppies, saffron-hued violet, and scores of other posies."[61] In 1931, the floor
of the southern San Joaquin Valley was covered with "limitless acres of
color spread like a Persian carpet."[62]

Heavy rains in 1932 produced "the most gorgeous display of wild
flowers that have graced the southland within the memory of the oldest
old-timer." Northern Death Valley was "ashimmer with gold, silver coral
and sapphire of desert bloom." Abundant rains brought "a wondrous
miracle . . . [the] bursting into blooms and casting a mantle of gorgeous

hue."[63] The Arvin Flower Festival was held "in the midst of 60,000 acres (24,000 ha) of wild flowers . . . with 50,000 people present. Crowds speeding in by train, airplane and motor were detained and charmed . . . by the variety of scents stretching in every direction." In the mountains above Arvin "blurs of orange poppies lured picnickers. . . . Acres of lupine and brodiaea flung a mantle of color onto every slope, and splashes of owls clover of deep magenta color, mingled with yellow sun cups and azure baby-blue eyes, invited sojourners into rainbow canyons."[64] At Riverside, "Gigantic masses of wild flowers, hundreds upon hundreds of acres in extent, are bursting into bloom in all parts of the valley . . . a field of yellow daisies, 1000 acres (400 ha) in extent, . . . [and] 500 acres (200 ha) of lupines in one great mass." Acres of color were seen in the Coachella Valley.[65]

In 1933, "the very dry condition" resulted in cancellation of the Arvin Flower Festival."[66] The *Times* reported in 1934 that the "Wild flowers of Kern thirsting for rainfall, [with] only dainty petaled gems appearing. The Kern County chamber reported "if no rain arrives immediately [the wildflowers] will wither and die."[67]

Expansion of bromes

Reports of flowers at Los Angeles became ever more infrequent, not only because of land development, but also as a consequence of the invasion of bromes, slender wild oats, and other European annuals that began choking out the native forbs. These invaders had advanced far beyond their initial proliferation described by Parish, Davidson, Abrams, and other botanists at the turn of the century thirty years before. G. B. Burtnett wrote in 1934,

> Before our towns were put on the map, the present vacant lot was part of a vast garden of blooms [that had] been sowing their seed and coming up again for thousands of years. With the settling and cultivation of land this natural cycle was destroyed. . . . Both those who mourn for the good old days when the San Fernando Valley was carpeted with alternate strips of blue and gold, when the cultivated fields of today were solid masses of both delicate and brazen colors, can see a ray of cheer ahead for lovers of the wild flower in natural settings. The enemies of the wild flower are weeds, foreign to California, the encroachment of agriculture and town sites. . . . Grain fields took the place of native plains, orchards and orange groves followed and many of these gave way for the town site. Then came the vacant lot. Cultivation ceased and weeds, mostly emigrant plants from foreign countries and, took possession of the ground. These weeds perpetuate themselves as once did the wild flowers. . . , . . . Vacant lots have been

growing weeds for many years. The ground is impregnated with their seed and their root systems.[68]

The abundance of brome was humorously described in the Lee Shippey's *Los Angeles Times* column, after a wet winter with remarkably well-distributed rains from October to April. Referring to foxtail (*Bromus madritensis, B. diandrus*) as a "cross between the wild oat and the seven-year itch," Shippey wrote,

> Of all California's bumper crops this year, none is more bumperish that the foxtail crop. In other years, foxtail used to get by, by posing as grass until, when we were off guard, it suddenly thrust up a plume of seed and ripened them. But still it was a lowly cousin of grass. This year each individual foxtail seems to think it is a tree. What used to be mere puddles of foxtail have expanded like the national debt and now you can, if you are fool enough, walk through tossing seas of foxtail which is waist deep. In many Southland towns and villages it has been necessary to hire mowing crews to harvest foxtail from the vacant lots lest children get lost in them.[69]

The best documentation of herbaceous vegetation was conducted by the California Vegetation Type Map (VTM) Survey of 1929–34 (Weislander 1938). The primary mandate of the survey was to publish maps by topographic quadrangle, an effort completed for most of coastal southern California. Legend information for California grassland clearly indicates that invasive species were dominant in the areas mapped, which included most of the uncultivated or urbanized plains of coastal southern California below the coastal sage scrub and chaparral belt. Grasslands extended inland to the edge of the Mojave and Colorado deserts. It is unfortunate that some quadrangles were mapped when herbaceous cover was too immature to identify the species composition. Maps generally show the symbol "Gr" without species-level information for individual polygons. Wildflowers are rarely indicated by the survey maps, although the legend designation "grass and herbs" in some of the quadrangles may in fact identify native forbs.

Legend descriptions of vegetation maps for coastal regions record grasslands dominated by *Avena, Bromus, Hordeum,* and *Erodium* (San Luis Rey, Corona, Triunfo Pass, Calabasas, Ramona quadrangles). In the San Luis Rey quadrangle, the dominant species rarely exceeded 1 foot (0.3 m) in height, but in large areas near the coast, oats (*Avena* spp.) formed almost pure stands, attaining heights of 2–3 feet (0.6–1.0 m). The legend for the Corona quadrangle states that grasslands were two distinct types near the coast. The most prevalent was composed of *Avena* spp. 2–3 feet tall (0.6–1.0 m), in association with another Franciscan invasive, clover (*Medicago hispida*). The second type was dominated by farmers foxtail (*Hordeum murinum*), which occupied valley floors. At San Juan Hill, grasslands of species of *Erodium, Medicago, Avena,* and *Bromus* had small colonies of native *Calandrinia menziesii* and *Baeria* (*Lasthenia*) *chrysostoma*. Black mustard (*Brassica nigra*), varying 4–6 feet (1.3–2.0 m) in height, was common over much of the grassland of the quadrangle. The coastal sage scrub had understory of *Erodium cicutarium, Bromus mollis, B. madritensis* (*B. rubens* on maps), *Avena* spp., and *Medicago hispida*. The Triunfo Pass and Calabasas quadrangles were mapped when the grassland was immature. Annual herbs were dominated by *Avena, Bromus, Hordeum,* and *Erodium. Hemizonia fasciculata* and *Brassica nigra* were also in abundance. Flats and southerly exposures in coastal sage scrub (termed "sagebrush") were covered with *Bromus rubens* (*madritensis*), *Hordeum murinum,* and *Medicago hispida*. Some northerly exposures had populations of native bunch grasses in *Stipa* (*Nassella*) and *Poa.*

The composition of grasslands in the interior valleys fronting the San Gabriel Mountains west of Pomona (Piru, Santa Susana, San Fernando, Pasadena, Pomona quadrangles) was similar to that on the coast, with grassland being dominated by *Avena, Bromus, Festuca, Erodium,* and *Medicago*. Most grassland occurred on deep-soil hills overlying Miocene or post-Miocene bedrock along the Santa Clara River valley, Oak Ridge, the Santa Susana Mountains, and from Hollywood through the Puente Hills to Corona. Except for the plains from Thousand Oaks to the western San Fernando Valley, virtually all grasslands in valley floors had long since been plowed. The VTM survey mapped mustards 4 to 10 feet in height (1.3–3.0 m, *Brassica nigra*) in the Whittier Hills and the San Jose Hills. Farther east, only "grass and herbs" were delimited from Pomona to San Gorgonio Pass. VTM workers stated that most grasslands in the San Bernardino quadrangle were "not natural," but represented agricultural clearings formerly cultivated that became occupied with "weeds and annual grasses."

The unplowed plains of the Elsinore quadrangle were also covered with "grass and herbs," dominated by *Bromus, Avena,* and *Erodium*. Lake Elsinore was dominated by *Distichlis spicata,* and Casa Loma by *Salicornia* spp. Herbaceous cover in coastal sage scrub consisted of *Erodium, Bromus rubens* (*madritensis*), and *Avena* spp. "Grasses and herbs" defined the grasslands of the San Jacinto quadrangle.

VTM maps along the edge of the Mojave Desert (Elizabeth Lake, Tujunga, Rock Creek quadrangles) rarely depict grassland except in disturbed areas. Legend information indicates that "grass and other herbs" embraced cultivated areas now grown over with *Bromus madritensis* and

Erodium cicutarium, the two most abundant invasives of the desert reported by Parish (1920) and seen presently. Coastal invasive species, including *Avena, Medicago, Festuca,* and *Brassica,* were not found this far into the interior. Two areas south of Palmdale were shown to have cover of the native desert needle grass, *Stipa speciosa,* in the Tujunga quadrangle, doubtless the same species seen by Frémont there in the 1840s. VTM workers reported that grasslands in the Mojave Desert consisted of abandoned cultivated areas dominated by *Erodium cicutarium* and *Bromus rubens (madritensis).*

Floral explosion, 1935–1941

From 1935 to 1941 came an unprecedented succession of spectacular flower years in which virtually all of southern California was almost annually decorated by extensive flower carpets, even the most arid deserts. The rainy winter of 1934–35 brought out wildflowers even in the Los Angeles region. Fields of lupine were seen in Carbon Canyon, Brea, La Habra Heights, and Yorba Linda Hills, and a solid mass of lupine was seen bordering Lake Elsinore.[70] The lower San Joaquin Valley was like "a vivid oil painting, with masses of color and dozens of varieties on hill and dale." At Death Valley, the *Times* wrote, "the entire section between Stove Pipe Wells and Furnace Creek is ablaze with desert marigolds."[71] The *Times* also reported that "this year on account of the gracious abundance and well-spaced distribution of the season's rains, the wild flower gardens . . . carpeting our so-called deserts have reached a perfection that few have ever witnessed."[72] A month later, flower lovers were "enjoying one of the greatest displays of wild bloom in the history of southern California."[73] In a retrospective article, the *Times* reported that the flowers of 1935 were the best display since 1921–22.[74]

After an off year, another great bloom colored southern California wildlands in 1937. Lynn J. Rogers wrote, "You can wander the length and breadth of the Southland today and catch nature in her pastel mood of bright color." Near Palmdale she saw "stretching before us like a flaming sea was a vast field of golden poppies" (Figure 4.3).[75] However, the flower pastures in Kern County began experiencing the same fate as those in the San Fernando Valley in 1912. According to the *Times,* "A warning was given not to attend the Arvin Flower Festival because 5000 acres (2,000 ha) of the formerly beautiful fields have been planted to potatoes and grain which might be destroyed if trampled upon." The festival moved to Shafter, 30 miles (50 km) northwest of Arvin.[76] In the Coachella Valley, flower fields near Indio were consumed by two bands of sheep,

Figure 4.3. Wildflowers in the Antelope Valley, 1936. WPA/Los Angeles Public Library.

totaling 1,000 head, from Phoenix, Arizona. Officials called on Riverside County authorities to put an immediate stop to the uncurtailed grazing that "threaten[s] destruction of hundreds of acres of beautiful desert wildflowers."[77]

The flooding rains of March 1938 brought "one of the most lavish displays of wildflowers since the great year of 1935."[78] The Shafter Exchange Club festival saw "brilliant spreads of flowers" throughout Kern County. At White Wolf grade in the Tehachapi Mountains, the *Times* wrote of "many acres of long-stemmed lupines and white popcorn flower while higher on the slopes are brodiaea and poppies. At the foot of the Grapevine are thousands of acres of lupines in a blue sea of blossom. Owl's clover and golden sunshine are to be found along the Taft highway." At the same time, "flowers were not so good the Mojave Desert" because the heavy rains had a late start: "Desert flowers here . . . will be found interesting only to those not expecting large expanses of color." Wildflowers "won the battle with man" at Arvin, as "two sections of the potato fields that were plowed over the previous year are now very brilliant with the golden glow of poppies." European weeds were winning the battle at Los Angeles, but had not made their way into Gorman. According

to the *Times*, "A few courageous blossoms here and there as you drive '
across the corner of San Fernando Valley gives no hint of the vast sea of
blue and purple that bursts before your eyes below the Grapevine." The
Mojave Desert, which had lighter rains, was "not literally one great garden, nor as abundant in blossoms, as in former springs."[79]

The spring of 1939 hosted the "best wildflower spectacle in many years
in the western Antelope Valley," which was a solid mass of bloom. The
blue lupine field at the Grapevine was "broken up by small patches of
golden poppies, evening snow, sun cups, owls clover and other smaller
varieties. The Mojave Desert was covered with millions of Coreopsis and
the route from Mojave to Barstow is covered with Indian paint brush,
lupine, Joshua blooms, [and] desert verbena." [80]

The year 1940 was unusual because wildflowers germinated early in
response to tropical storms in September 1939 and were nourished by
warm temperatures all winter. The desert was "thick with artists, for the
sunflowers are in bloom."[81] By March, Death Valley and the Mojave
Desert had "heavy carpets of flowers." At Death Valley, the *Times* reported that "the green and gold carpet of desert sunflowers continues to
spread its perfume throughout the floor of the valley." Flower fields in
Kern County were the best since 1935. By late March, wildflowers were
"the most profuse in 15 years."[82]On the Mojave Desert there was "a
matchless carpet of lupins, poppies, verbena, daisies, coreopsis, desert
primrose and other countless varieties."[83] Palm Springs had "one of the
greatest displays of wildflowers in recent years."[84] Wildflower displays
visited the Inland Empire around San Bernardino and Riverside. Sadly,
wildflower fields of the western Antelope Valley and San Fernando Valley were scarce due to plowing.[85] Late rains maintained the blooming
season from April to July.[86, 87]

The winter of 1940–41, the wettest year since 1889–90, produced a
once in a lifetime bloom across southern California. Expansive blooms
began unfolding at the Grapevine, Antelope Valley, and in the inland val-
leys near Riverside. The *Times* reported, "It is almost 50 years since the
Golden State has been as bountiful as it is today with wildflowers." In
Kern County "thousands of square miles of countryside [were] alive with
blooms of color . . . fields of poppies were seen even in Orange County."
Exotic black mustard still covered coastal southern California plains. The
Mojave Desert and Death Valley "offer[ed] another work of color."[88]
When columnist Lee Shippey descended the Grapevine, he wrote, "We
thought for a moment, that the ocean had moved into the valley below
us. There lay a sea of flowers, stretching away as far as eye could see."
He added, "I won't describe it. It is simply something that takes your

breath away and cannot be described."[89] The circle tour followed "a back road through the foothills to Beaumont [San Timoteo Canyon] and on every side colorful showings of poppies and other varieties of wildflowers were in evidence."[90] The growing season spanned into late spring. At Lake Elsinore, the *Times* reported on May 11 that while "some wheat and barley fields already are turning brown, . . . the rolling hills are a lovely green, and . . . flowers are everywhere." The *Times* also provides early evidence that invasive species, most likely wild oats, which thrive in high rainfall, were already mixed in the wildflower fields: "In some of the places where the flowers are blooming the grass has grown so high that visitors have to get out in the fields to see the blossoms to best advantage."[91]

World War II and long-term drought, 1942–1967

The string of almost annual wildflower shows that began in the 1930s ended with the 1941 landscape masterpiece, and the onset of World War II. If wildflowers proliferated during the war, it was not recorded for posterity. Gas rationing and required donations of spare tires to the government prevented normal reporting by the *Times*. Columnist Lee Shippey dwelled on "memories of loveliness."[92] People could not drive to the deserts, including the scouts, and some observers were drafted into the military. Shippey wrote that "minds were turned elsewhere . . . and people could not drive to the deserts."[93] Even *The Desert Magazine*, which gave outstanding annual accounts of desert flowers in its spring issues beginning in 1939, ceased reporting in 1943–45. Randall Henderson, the magazine's editor, was a public relations officer for the military in the Sahara Desert.

The deserts were also newly invaded by split grass (*Schismus barbatis*), a small desert grass native to southern Europe, northern Africa, and the Near East (Jackson 1985; Brooks 1999; Appendix 3). *Schismus barbatis* was first collected in California as early as 1935 at Kettleman Hills by Robert F. Hoover (Robbins 1940; Twisselmann 1967). It became a dominant annual in the deserts by the 1940s. By 1951, its range included hundreds of thousands of acres of arid California, Arizona, and Baja California. According to Ernest Twisselmann (1967: 185), in the San Joaquin Valley split grass expanded explosively during the great drought of 1945 to 1951, and he noted that "it is tenacious in extreme drought because a fraction of seed germinates each year, leaving most seeds for reserve when a cohort perishes before reproduction." Split grass also joined other invasives in coastal grasslands.

The flower displays were probably limited by deficient rainfall in 1942, but the following two winters yielded heavy rains in winter and spring. In 1943, "hillsides and lowlands of Arvin were carpeted with golden poppies, lupine and other varieties." The only evidence of wildflowers reported in *The Desert Magazine* were two photographs of sand verbena near Palm Springs. The *Times* reported a begrudging cancellation of the Arvin Flower Festival: "burning up gasoline and rubber to see the flowers is unpatriotic—if not unsafe—and the Arvin Booster Club, for one, has swallowed its pride over the display and has withdrawn its annual invitation to motorists."[94]

California experienced profound drought and poor flower displays in the postwar period. In 1946, Lee Shippey saw "exciting patches of wildflowers" between Santa Barbara and San Luis Obispo, but only "ankle high" miniature flowers grew in the Antelope Valley in 1947.[95] *The Desert Magazine* provides insight as to how the fickleness of climate led to variability in wildflower displays, as evidenced by reports of naturalists and botanists, in particular Mary Beal, a resident of the Mojave Desert who frequently wrote for the journal (Appendix 4). In 1946, cold and lack of rain in February stopped the growth of plants germinating in fall. Tiny wildflowers barely had enough soil water to reach bloom. The following year, the failure of spring rains forced annuals to reach early flower, leading to smaller plants and smaller blooms. In 1950 there was insufficient precipitation to initiate germination and the rains arrived too late the following spring.

Lavish displays covered the Mojave Desert in 1949 due to extraordinary snowfall. The snowpack, which lasted one to two months on the desert floor to as low as 3,000 feet elevation (900 m), led to abundant growing-season soil moisture.[96] Below the "snow belt," the Coachella Valley "was bathed in color."[97] Lynn Rogers took the "circle tour" of Kern County–Mojave Desert and wrote that she saw "one of the most beautiful flower displays Southern California has shown in more than nine years." Kern County had "some of the most vivid panoramic displays of wild-flowers in nine years, but old-time residents will find the districts that once supported lavish acres of wild flowers have been converted to agricultural uses. However, there still are great many sectors where the plow has not turned the earth."[98]

More and more land was converted to potatoes.[99] Columnist Lee Shippey focused on the resulting Kern County dust storms: "Forty-thousand acres (16,000 ha) which used to produce nothing but wildflowers, which tens of thousands annually drove to see, now are producing potatoes and cotton, and when high winds hit that broad, level sweep the dust is about as pleasant as smudge in the Pomona Valley."[100] The *Times* later reported that "a combination of agricultural encroachment and insufficient rain has all but wiped out the colorful picture that formerly greeted thousands of lovers of nature, botanists, and tourists in Kern County's valley fields and hills . . . the profusion of color that formerly carpeted Kern County's hills and dales is gone forever." An effort by the Kern County Chamber to effect legislative action for the establishment of a "wildflower preserve" was unsuccessful. "In recent years, Kern County's wild-flower season has been thought of more in terms of the Mojave Desert."[101] History repeated itself. Los Angeles handed its flowers over to Kern County and Kern County handed its flowers to the Mojave Desert.

Drought persisted through the 1950s, but excellent desert blooms were described in *The Desert Magazine* in 1952 and 1958, both El Niño years with above-normal, well-distributed rains. The *Riverside Press Enterprise* reported that 1952 brought the best flowering in a decade, whether on a 30-mile circle tour around Lake Mathews or a 350-mile tour of the Sonoran Desert.[102] Kern County hosted "the most brilliant mantle of glorious wildflowers . . . in 15 years."[103] Flowers did not proliferate in Los Angeles. Airplanes were used to plant wildflower seed in the Whittier Hills,[104] an effort that clearly did not recognize the seed bank adaptation of the native herbaceous flora, and more importantly the competition produced by invasive grasses. In 1958, Kern County had the best flower year in a decade, and Death Valley was "carpeted in brilliant yellows, whites, and purples."[105] "Old-timers" stated that the "flowers around Twenty-Nine Palms were more beautiful than they have been since 1938."[106] By late spring, Kern County flowers were seen in "massive spreads."[107] In retrospect, the *Times* wrote that 1958 was "one of the best in recent memory."[108]

Only local blooms were reported during most of the 1950s. Colorful hillsides in 1950 and 1951 were limited to Kern County.[109] Patches of lupine were seen in the Palos Verde Hills.[110] In 1953–56, *The Desert Magazine* made consistent reports of excellent mass germination from early winter rains, but wildflower failures resulted from poor spring rains. In 1953 a descent of San Gorgonio Pass toward the desert at Palm Springs brought notice of "shy desert wildflowers poking their heads out of the sand." Many flowers died without blooming.[111] Citizens in the Coachella Valley protested sheep grazing in the wild flower reserves.[112] The circle tour of Kern County brought "vistas of color."[113] Flowers were found only in "secluded areas" of Kern County in 1954.[114] In 1956 there were a few good blooms on the circle tour. Kern County had the "best panoramic displays of wildflowers since 1952." The foothills of the Antelope Valley had patches of wildflowers from Lancaster to Gorman.[115]

Wildflowers were virtually absent during extreme drought in 1959–61. In 1959, the *Times* reported that "wildflowers are a rare sight in this drought-stricken spring." One patch of poppies was found at the town of Tehachapi, and a few others were seen near Mojave.[116] The Tehachapi Mountains in the following spring had "many splashes of color," but photographs show only an ankle-high flower carpet.[117] The winter of 1960–61 was the driest according to instrumental records. Few annuals came up in winter except for sparse displays in Kern County, and spring was virtually rainless.[118] Good flowers required a long drive north to Coalinga.[119]

Drought-breaking winter rains in 1962 spawned a good flower display across southern California, even along the coast, for the first time in decades. On the steep hills between Bonsall and Pala, "poppies were bursting out . . . in places where they had been unseen for 15 or 20 years."[120] There were "spectacular appearances of wild flowers" along the Barstow freeway, but displays were "not up to their usual standard" in the Coachella Valley, where precipitation was below normal.[121] Columnist Matt Weinstock wrote that the hills of Gorman were a "mosaic of colors," adding that "year after year, there have been fringes of poppies and lupine. This year . . . millions of seeds, which have lain dormant all this time, felt the urge to sprout and did."[122] The hills around the *Ramona* play near Hemet were abloom with "poppies and wildflowers."[123]

Drought again diminished the flowers in 1963–64. In 1963, "splashes of color were seen in Kern County and poppies were seen along the highway at Temecula."[124] In April 1964, "miracle rains" brought out flowers along the coast, on the Riverside-Perris plain, and in the high desert.[125] Heavy rains in 1966 produced "wildflowers . . . from the desert to the sea in southern California."[126] The *Times* reported that even "vacant city lots show splashes of color that inspire us to hope that wild flowers will bloom again as in days of yore."[127] The following year the *Times* conceded that "wildflowers are nowhere near as prevalent in southern California as they were 50 and 100 years ago before the valleys were cultivated and cities and towns developed."[128]

Wildflowers were still important in Kern County when Twisselmann published his Flora of Kern County. He was keenly tuned to the year-to-year variability of blooms. He wrote that the California poppy ranged from "broad flaming beds that cover in spectacular fashion . . . the southeast slopes of the Tehachapi Mountains and the Antelope Valley" to, in dry years, the presence of "scattered plants consisting of a small inconspicuous tuft

TABLE 4.7 SELECTED WILDFLOWERS IN KERN COUNTY

Species	Comments
Eschscholzia californica	Common and colorful in all associations except the most arid open plains on the desert
Emmenanthe penduliflora	Occasional in the cismontane woody . . . associations. It is often abundant following fires.
Nemophila menziesii	A common and most attractive spring annual in sandy soil in cismontane upper Sonoran associatons. The flowers are bright blue of uncommon intensity.
Nemophila pulchella	Occasional in dense colorful bright violet-blue colonies on north slopes
Phacelia ciliata	Dense in heavy or fine clay on open slopes, frequently coloring large areas a rich blue. The often form spectacular patterns with the bright yellow of equally dense colonies of *Monolopia lanceolata*.
Phacelia distans	Common in all the cismontane associations to the chaparral
Amsinckia douglasiana	Fireweed is one of the spectacular sights in a good spring in the Temblor Range, where it often colors great expanses of broad slopes brilliant orange.
Cryptantha intermedia	Common in the Upper sonoran [ca. > 1200 m]
Plagiobothrys canescens	Valley popcorn flower. Common in the lower and Upper Sonoran [ca. < 1500 m].
Salvia columbariae	Common in all associations below the yellow pine forest. The seeds, which were a staple in the diet of the Piute Indians, are exceptionally nutritious; a tablespoon a day is said to be an adequate emergency diet.
Orthocarpus purpurascens	Common in broad colonies on open hillsides . . . in all the cismontane associations . . . often Coloring the slopes a warm rich pink
Hemizonia pallida	A late spring tarweed . . . often abundant on the higher plains and lower hill slopes, coloring them a light yellow after the spring annuals have dried
Chaenactis glabriuscula	Golden girls. Common and colorful . . . from the high plains around the valley to the edge of the yellow pine forest.
Lasthenia chrysostoma	Gold fields. Abundant in the valley and the surrounding hills . . . , often coloring broad expanses bright yellow in the spring.

SOURCE: Twisselmann (1967).

of leaves topped by one large flower, truly a Cinderella dress for a ball!" (Twisselmann 1967: 241). His colorful descriptions and range distributions of wildflower species attest to the flowers' abundance at the time of publication (Table 4.7).

Twisselmann assessed the primary invasive species. *Bromus diandrus* was "a fair forage plant when green, [but] the stiff awns sometimes cause severe damage to livestock when the plant is dry. Veterinarians report that awns lodged in dog's nostrils are a common problems in cities and towns" (176). Red brome (*Bromus madritensis*) was "common and often abundant in all associations below the yellow pine forest. . . . While often considered of secondary forage value, local livestock people praise red brome for its rapid growth and resistance to drouth and cold weather" (176–77). He added that while red brome was especially common after prolonged heavy grazing, "it also crowds out native plants, in light or sandy soils that are not grazed at all" (177).

Twisselmann passed away soon after his flora was published and, like Samuel Parish, did not see come true his own or Parish's prophetic insights concerning brome invasions and the conversion of wildflowers to brome fields.

Ascent of brome grassland in the interior, 1968–1987

The two wet bloom years of 1966 and 1967 were followed by two more rainy winters that culminated in the southern California floods of 1969, the wettest four-year period since 1937–41. Invasive bromes exploded in abundance across the interior valleys of southern California, including in Riverside and Kern counties (Table 4.6). For the first time, normal to wet years did not bear abundant wildflowers. A new pattern emerged and has persisted to the present in which wildflowers proliferate only in the first wet year after a long drought.

In 1969 several articles in the *Times* predicted great desert blooms that never came to pass.[129] Only in the Coachella Valley did "flowers pop up through the gravel."[130] The rains came too late for flowers in 1970, and the failure of spring rains kept fall germinates from maturing in 1971 and 1972.[131]

The wildflower crisis reached political levels. Governor Reagan urged that California save the poppies, asking citizens "to save the state flower—the golden poppy—by helping to pay for a proposed new park in the Antelope Valley." The governor stated, "Today, there are only a few areas left where our state flower, the golden poppy, still flourishes in sizable displays." He also said that "it could become extinct through neglect, just as did the State animal, the California grizzly bear."[132]

It is perhaps ironic that well-distributed heavy rains triggered abundant blooms throughout southern California in 1973, the very next year. Wildflowers broke out in Los Angeles city lots and coastal hillsides, as in 1962, after long-term drought.[133] The bloom in Death Valley was the best in thirty to forty years, and a "bumper crop" of poppies was photographed at Lake Mathews near Riverside.[134] Flowers fields in 1974 were limited to the Antelope Valley and the upper elevations of Kern County, where soils were moistened by a rare 2-foot (60 cm) snowfall in January.[135] Annual grasses, dominated by *Bromus madritensis,* were abundant enough to carry a fire from the San Jacinto Mountains into creosote bush scrub at Snow Creek, the first desert fire in recent memory (O'Leary and Minnich 1981).

In 1976, a 2,000-acre (800 ha) wildflower park was established in the Antelope Valley as a result of Governor Reagan's initiative.[136] State officials conducted a five-year study to find the most consistent poppy-bearing land in the region, concluding that the western Antelope Valley had the best poppy fields because of its steady annual rainfall, a high elevation of approximately 3,000 feet (900 m), and its undeveloped land. In 1976, officials dedicated the reserve, known as Poppy Park.[137] The reserve personnel subsequently became an excellent source of year-to-year wildflower status in the Antelope Valley.

Drought suppressed wildflowers in the mid-1970s. The only reports of blooms in 1975 came from Poppy Park and the Antelope Valley.[138] In September 1976 the eastern deserts were drenched by tropical cyclone Kathleen, which produced one year's rainfall in twenty-four hours (8–12 cm). The *Times* reported that "while most of California remains parched, some desert areas . . . are damper than usual and are producing prodigious displays of wild flowers."[139]

The winter of 1978 was the rainiest year of the twentieth century in much of southern California and followed a long drought. The *Times* reported that "after record-breaking rains," there was "a greater abundance of blooms than in many years past."[140] Many regions had "bumper crops of blooms," especially Anza-Borrego Desert State Park and in the Mojave Desert, where "the bright wedge of land known as the Antelope Valley [came] alive again with rich carpets of grass, California poppies and scores of other wildflowers."[141]

The reference to grass was an omen. Although rainfall for 1978–83 was the highest six-year running average in southern California since records began in the late nineteenth century, blooms virtually ceased throughout the region, even in the Mojave and Colorado deserts. Only the Antelope Valley poppy preserve had occasional showings (Table 4.6). Exotic grasses, especially *Bromus madritensis,* were extraordinarily productive, generating pastures in interior valleys and deserts, and were an extreme fuel hazard. The deserts also experienced rapid proliferation of a new invader, Sahara mustard (*Brassica tournefortii;* Minnich and Sanders 2000). This native of the African and Middle Eastern deserts had been seen locally in Riverside, Imperial, and southwestern San Bernardino counties in the 1950s (Munz and Keck 1959). The first collection from the Imperial Valley in 1927 was misidentified and was not reexamined until decades later by Andrew Sanders. It may have arrived with exotic date palms from the Middle East. After hurricane Kathleen, the hills near Palm Springs were bright green with Sahara mustard, especially in wind-blown deposits (Minnich and Sanders 2000). Like most mustards, a sticky gel forms over its seed case when it rains, permitting seed to disperse long distances by adhering to animals and the wheels of automobiles. By the 1990s, it ranged 1,000 kilometers from the Owens Valley to Arizona and south to Sinaloa, Mexico (Andrew Sanders, pers. comm.).

Above-normal rains in 1979 brought only predictions of wildflowers.[142] Even heavier rains fell in 1980, but only local blooms were reported in the Coachella Valley.[143] Red brome covered the Mojave Desert, and creosote bush scrub resembled a grassland more than a desert (Figure 4.4a). Reports of fires in the desert were rare until 1978. For example, in 1919 six miners "reported seeing a column of black smoke shoot up from the old volcano of Lavic, east of Ludlow."[144] They apparently saw a fire in the distance, because the volcano has not erupted in historic times. In Joshua Tree National Park, one of the desert lands with a long fire-history record, only small fires broke out after the wet winters in 1942, 1966, and 1967. In 1978, lightning triggered a 6,000-acre (2,500 ha) fire in pinyon-juniper woodland of Joshua Tree National Park, but fire records indicate the primary fuel was old-growth shrubs, pinyons, and wildflowers rather than introduced grasses (Minnich 2003). The herbaceous fuels of exotic grasses triggered an unprecedented outbreak of fires in the desert the following year. In July 5,200 acres (2,080 ha) burned in Lucerne Valley.[145] In 1980, lightning sparked two fires that covered 4,000 acres (1,500 ha) in the Mojave Desert (July 2), and a large burn in the San Gabriel Mountains spread into the desert near Palmdale (July 17).[146] A 30,000-acre (13,000 ha) fire in the San Jacinto Mountains moved across desert scrub to Palm Springs (July 29).[147] By late summer, the desert had experienced the "worst fire season on record," as

Figure 4.4. (a) Brome-dominated grassland in creosote bush scrub near Kramer Junction in the Mojave Desert, in 1980 after three wet years; (b) the same area with understory of native fiddleneck (*Amsinckia tessellata*) and other wildflowers after heavy rains in 2005, following multiyear drought. Photo by author.

brome-fueled fires had charred 33,000 acres (13,200 ha),[148] especially along the semiarid western margin of the Mojave and Colorado deserts (Minnich 1983). Red brome and cheat grass carried reburns through the 1978 fire in Joshua Tree National Park on 1982 and 1984 (Minnich 2003).

The great El Niño of 1982–83 produced the third-wettest winter of the century at Los Angeles, but predictions by wildflower fanciers of "one of the most spectacular blooming seasons in the Southland" again failed to pan out. Only "tiny bright flowers" were seen in the lower Colorado Desert.[149] My disappointing visit to Anza-Borrego Desert State Park brought home memories and photographs of flowering shrubs and dense understory of brome and Sahara mustard. In a study of desert fires near Palm Springs, red brome formed 40 percent cover at Blaisdell Canyon (Brown and Minnich 1986). Fires again broke out in the desert, with one burn spreading to the edge of Palm Springs and several others charring thousands of acres in the hills near Victorville.[150]

By the mid-1980s, Poppy Park was one of the few places in southern California where wildflowers were still seen in massive displays (Table 4.6).[151] The years 1985 and 1986 were considered to be good years for flowers on the reserve,[152] but the dry year of 1987 brought "an uncolorful spring in the California deserts. Only the red and violet petals of the exotic filaree attracted notice."[153] By late spring, California poppies were seen "on many a grassy slope."[154] Abundant rains in fall and late spring produced colorful flower displays in 1988. The *Times* reported that "during the past month, wildflowers have splashed their colors across the low desert. . . . The show was especially spectacular in the Mojave Desert, including Antelope Valley."[155] Death Valley had the "bloom of the decade."[156]

Brome "crash" and proliferation, 1988–1998

Newspaper accounts and interviews beginning in the late 1980s suggest a repeat of the year-to- year floral changes observed since 1967: proliferation of invasive annuals and suppression of native forbs in wet years, followed by brome "crashes" in drought and proliferation of wildflowers the first wet year after drought. After mass germination of bromes and oats, the failure of rains causes mass mortality of these grasses before they reach reproductive maturity, catastrophically reducing their seed banks (Salo 2004). This new pattern was captured since the late 1980s in serial photograph surveys of two desert sites (Blaisdell Canyon, Snow Creek) near Palm Springs, and from vegetation sampling at Riverside (Figure 4.5; Figure 4.6).

Figure 4.5. Creosote bush scrub on Blaisdell alluvial fan near Palm Springs (cf. Brown and Minnich 1986): (a) dense cover of invasive *Bromus madritensis* and *Brassica tournefortii* in 1988; (b) bare ground during two-year drought in 1990. The drought extirpated *B. madritensis* from this site and it has not returned. Photo by author.

Wildflowers failed during extreme drought in 1989–90 and were absent from the Antelope Valley poppy reserve, where "the last time the poppies carpeted the hills in a blaze of color was in 1988."[157] Neither were flower displays reported elsewhere in southern California. However, the abundance of red brome at Riverside fell from 800 individuals m[-1] in 1989 to 100 m[-1] in 1991, with a corresponding increase in native wildflowers with long seed life, including *Phacelia distans, Cryptantha intermedia,* and *Emmenanthe penduliflora.* Red brome was extirpated from Blaisdell Canyon (Brown and Minnich 1986; Table 4.8). The slender wild oat (*Avena barbata*) was extirpated at Snow Creek.

Drought ended with the "March miracle" rains of 1991 that salvaged yet another dry winter. The late rains produced a "great wildflower display" at Poppy Park, "the best in ten years."[158] Flowers even broke out in the Puente Hills of Los Angeles. A letter to the *Times* described "a breathtaking display of wildflowers. . . . The hills are velvety green, many carpeted with mustard,"[159] an invasive species and perhaps a sign of the times in public perception. In 1992, Anza-Borrego Desert State Park had a "bumper crop of wildflowers," the *Times* reporting that "colors are erupting . . . more wildly than they have in years."[160]

Heavy rainfall in California from 1993 to 1995 led to a reemergence of abundant cover of introduced species. According to the *Times,* wild mustard and grasses took over territory normally occupied by more varied flora. At the Huntington Garden in San Marino, "everything [was] just bursting . . . the weeds as well."[161] Red brome increased at the expense of wildflowers and mustards at Riverside and Snow Creek. The rains increased the cover of filaree and Sahara mustard at Blaisdell. Wildflowers were reported only at Poppy Park and the Coachella Valley. Poppy Park manager Jim Geary was perplexed by the poor flower showing despite heavy rains in 1992: "Early winter rains in December prompted a profusion of small poppy plants in January. Warm sunny days fuel their growth. But . . . constant, heavy rain and cool, cloudy days have set the bloom back."[162] The culprit appears to be exotic bromes, as the *Times* reported that "the grasses came up and choked the poppies."[163] Flowers were excellent in Anza-Borrego Desert State Park, perhaps because bromes were still in low numbers from previous drought. But in 1993, despite twice normal rainfall, the bloom at the park was "disappointing" because "hardy non-native species like mustard [were] overwhelming native plants in many areas . . . wild mustard (*Brassica tournefortii*) and grasses [had taken] over territory normally occupied by more varied flora."[164] The *Riverside Press Enterprise* reported that "belly flowers were

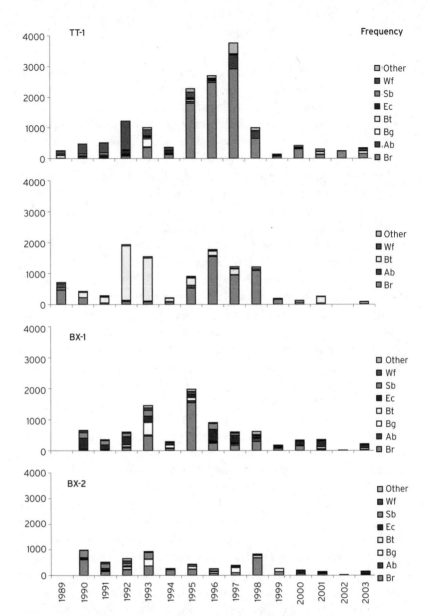

Figure 4.6. Annual herbaceous cover in spring at Box Springs Mountains of Riverside, 1989–2003: (a) density (m^{-2}), (b) biomass (m^{-2}). TT = Two Trees Canyon, BX = Box Springs Canyon. Wf, wildflowers; Sb, *Schismus barbatis*; Ec, *Erodium cicutarium*; Bt, *Brassica tournefortii*; Bg, *Brassica geniculata*; Ab, *Avena barbata*; Br, *Bromus madritensis*.

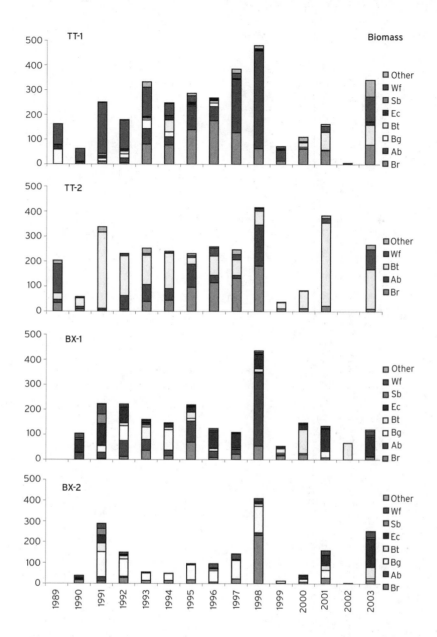

smothered by the heaviest intruder grass crop in years."[165] In 1994, Mary Snider of Poppy Park stated that "repeated downpours last winter provoked the furious growth of weeds, which deprive the poppies of critical moisture and sunlight."[166] Large grass fires broke out again in the Mojave Desert near Victorville (Brooks 1999).

In 1995, the Poppy Park began a program of controlled burns to reduce invasive grasses, with mixed results. Although rain was plentiful, the *Times* reported that "a lack of cold weather failed to eliminate non-

TABLE 4.8 ANNUAL HERBACEOUS COVER (%) AT SNOW CREEK AND BLAISDELL ALLUVIAL FAN

Snow Creek	1983	1988	1990	1991	1992	1993	1994	1995	1996*	1997	1998	1999	2000	2001	2002	2003	2004	2005
Bromus madritensis	4	5	1	2	5	5	5	6	1	2	3	0	0	0	0	0	0	0
Avena barbata	1	0	0	0	0	0	0	0	0	0	0	0	0	0	0	0	0	0
Brassica tournefortii	0	4	2	1	4	2	3	1	2	2	5	1	1	3	0	4	4	4
Erodium cicutarium	4	4	4	5	4	4	3	3	4	5	5	3	3	3	1	5	5	2
Schismus barbatis	3	2	2	2	2	2	2	4	1	3	3	1	1	3	1	4	2	6
Native annual wildflowers	2	1	1	4	2	2	1	1	2	2	1	1	1	2	0	3	2	1
Total	6	7	4	7	7	6	7	7	5	6	7	3	3	4	1	6	5	7

Blaisdell	1983	1988	1990	1991	1992	1993	1994	1995	1996	1997	1998	1999	2000	2001	2002	2003	2004	2005
Bromus madritensis	4	2	0	0	0	0	0	0	0	0	0	0	0	0	0	0	0	0
Brassica tournefortii	2	3	0	1	4	2	3	1	1	1	3	1	1	2	0	4	3	4

Species																	
Erodium cicutarium	3	2	1	4	4	3	2	2	3	4	3	1	2	0	3	4	4
Schismus barbatis	3	5	1	2	2	1	3	3	2	4	2	1	2	0	2	3	4
Native annual wildflowers	2	1	0	3	2	1	1	1	2	2	1	1	2	0	3	1	3
Total	6	6	1	4	5	4	4	2	4	5	3	1	3	0	5	4	5

NOTES: * years of fire

% cover:

o = absent
1 = <1
2 = 1–5
3 = 5–10
4 = 10–25
5 = 25–50
6 = 50–75
7 = >75

native weeds that compete with poppies for sunlight and water; reserve officials are unsure how last year's 15-acre controlled burn to kill off weeds will affect this year's bloom."[167] The reserve experienced a "once in five year bloom" later in the spring.[168] The park conducted another controlled burn to "enhance the following year's flower display by reducing the non-native plants and allowing native plants . . . to expand their range into the burned areas."[169] John Crossman, resource ecologist for the Mojave Desert state parks, summarized the grass invasion problem in the 1990s: "In '95, we had just over 20 inches [50 cm], and that was a good year. In '93 we had over 25 inches [63 cm], and that was not a good poppy year. In '91 we had only about 12 inches [30 cm], and that was one of the best years of all." While rainfall was modest in 1991, the large bloom of 1991 may have been linked to brome crashes from the 1989–90 drought. In an interview in 1998, reserve personnel said the problem was "a large population of non-native grasses growing . . . in the reserve. During heavy rain season, these grasses have a tendency to grow up very quickly and crowd out the poppies."[170]

Drought revisited the desert in 1996, when poppies were "allegedly hanging on some rolling hills called the Antelope Buttes,"[171] and the flowers were scarce at Poppy Park in 1997.[172] At Snow Creek, brome cover decreased to less than 5 percent by 1996, but light, well-distributed rains produced bumper crops of exotic annual grassland at Riverside. Studies in the Mojave Desert by Brooks (1999) and Brooks et al. (2004) document that the exotic species *Bromus madritensis, Schismus barbatis,* and *Erodium cicutarium* concentrated beneath shrubs and produced biomass of 0.3 ton ha^{-1} in 1997.

Heavy rains from the great El Niño of 1998 produced a breakout flower year, especially in the deserts. These areas were drenched by tropical cyclone Nora in September 1997. Frequent heavy rains fell over the desert and the remainder of southern California from November to April. The *Times* reported that "suddenly the driest parts of California's desert are exploding with wildflowers. Botanists said they have found flowers that usually only appear after fires—plants usually only seen in dry years. Some spring flowers appeared in fall. Still other flowers came up in winter. Most of this topsy-turvy mix is continuing to bloom as the spring regulars join in."[173] Anza-Borrego had "the best bloom in my 16 years," according Bob Anderson, a docent for the park.[174] In the Mojave Desert, a "40-mile stretch of Interstate 40 between Barstow and Needles [was] a carpet of gold."[175] In a trip to Las Vegas, I rarely encountered red brome for a 100-mile stretch along Interstate 15 from Hesperia to the California-Nevada border. Drought-induced die-off of bromes in 1996–97 apparently contributed to the wildflower outbreak. At a Death Valley park there were "more flowers, and larger," than most of the employees remember seeing before.[176] The more innocuous invasives were still abundant in the desert. The Blaisdell site was still covered by filaree, Sahara mustard, and split grass. Wildflowers produced an estimated 1–5 percent cover. Red brome increased at Snow Creek, but filaree and Sahara mustard continued to dominate the site. At Riverside, where brome had proliferated continuously since 1993, the El Niño rains supported bumper crops of slender wild oat and red brome, with annual productivity of 4 tons ha⁻¹. Wildflowers were absent throughout coastal southern California and Kern County in spite of heavy rains, continuing a trend since the 1960s.(Table 4.6). Two fires burned 15,000 acres (6,000 ha) in the high deserts of Joshua Tree National Park in May 1999,[177] the flames carried by red brome and cheat grass that flushed in 1998 (Minnich 2003).

The "perfect drought" and desert blooms, 1999–2005

Extreme drought prevailed over most of southern California from 1999 to 2001. Annual rainfall in the deserts was as low as 15 percent of normal, and blooms were rarely reported. Photo resurveys recorded only 10 percent cover of filaree and split grass at Blaisdell and Snow Creek (Table 4.8). Red brome was extirpated from Snow Creek. At Poppy Park, according to the *Times*, "fickle flowers have played hard to get, teasing with occasional patches of orange, but mostly hiding in the sandy soil."[178] Red brome "crashed" at Riverside. At Box Springs, densities as high as 2,700 stems m⁻¹ in December 1998 declined precipitously to less than 50 m⁻¹ the following month and remained less than 150 m⁻¹ until 2003 (Figure 4.7). While rains supported sparse blooms in the desert, the best flower displays occurred in the interior valleys of southern California. An "usually thick batch of wildflowers" made the best flower season in the Riverside-Perris plain in decades: "the hills and fields look[ed] as if God had dropped his water color set."[179]

The winter of 2001–02 was the driest in southern California since instrumental records began in 1849. Portions of the Imperial Valley had no measurable precipitation for an entire year. Dennis Schramm, assistant manager of the East Mojave National Preserve, stated, "Forget the wildflowers of spring that never arrived." Referring to the desert scrub community, he said that the "creosote and other hardy native plants are now in a state of indefinite wilt, their leaves curled up tight to conserve every drop of moisture. Burro bushes stand sere and brown. They never

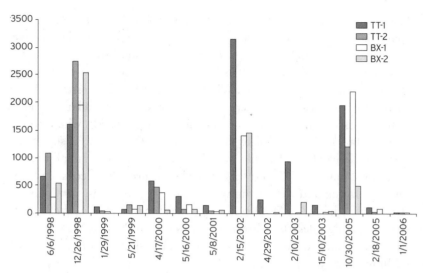

Figure 4.7. Frequency of *Bromus madritensis* at Box Springs Mountains of
Riverside (m^{-2}), 1998–2006. TT = Two Trees Canyon, BX = Box Springs
Canyon.

bothered to grow new leaves."[180] Even the coastal southern California
hillsides failed to turn green all winter.

El Niño rains in 2003 brought out abundant flowers in several areas
where bromes were diminished by drought. The *Times* reported, "After
failing to appear in bone-dry 2002, the California poppy has returned
with a fiery flourish at Poppy Park."[181] Hillsides in the Riverside-Perris
plain hosted poppies, fiddlenecks, phacelia, and goldfields.

The winter of 2004–05 was among the three wettest rainy seasons in
coastal southern California, and was the wettest year in instrumental
records in the eastern deserts. The Mojave and Colorado deserts, now
liberated from invasive bromes, split grass, and filaree by chronic drought
since 1996, supported what many called the bloom of the century. The
Times reported on March 8 that at Death Valley "the wettest year on
record . . . transformed this forbidding wilderness . . . into a vividly unfamiliar world of wildflowers and reflecting pools, triggering ecological
cycles not seen before on so large a scale. . . . It's our best bloom in history; . . . you may not see it this good again in your lifetime."[182] Extensive wildflower fields carpeted the Indio Hills and lower elevations of
Joshua Tree National Park (Figure 4.8). Death Valley was "in the midst
of the most spectacular bloom people here can remember," and even a
10-kilometer lake filled the basin at Badwater. Repeat photographs of
1980 brome pastures in the Mojave Desert revealed a solid mass of *Am-*

Figure 4.8. Once in a lifetime bloom in 2005. Carpet of lupines and desert dandelions at Joshua Tree National Park. Courtesy of Victoria Minnich.

sinckia tessellata and other wildflowers in 2005 (Figure 4.4 b). Blooms were not as spectacular in the Riverside interior valleys as in previous years, as red brome had increased in numbers since 2003. Travelers were "treated to emerald-hued hillsides and slashes of bright color as flowers and grasses bloom profusely." Andrew Sanders, herbareum curator at the University of California, Riverside, said that "the downside is [that] the weedy grasses overtake wildflowers in some areas." The deserts were "the best viewing," where the weeds were not prolific.[183] Wildflowers were sparse in Kern County and were not reported from exotic annual grasslands covering uncultivated regions of coastal California.

The historically unprecedented outbreak of desert fires that followed was also a "teaching moment" that native wildflowers can be as potent a fuel source as introduced species. In June 2005, a conflagration burned through 65,000 acres (25,000 ha) of native grasses (*Aristida, Bouteloua*) in the Providence and New York mountains of the northeastern Mojave Desert.[184] Volcanic-capped hillslopes at Pioneertown in the eastern San Bernardino Mountains sustained a 1,000-hectare fire,[185] carried by thick dried wildflowers of *Amsinckia, Phacelia, Chaenactis, Salvia columbariae,* as well as *Camissonia californica, Geraea canescens, Lepidium*

TABLE 4.9 ANNUAL SPECIES COVER (%) AND BIOMASS (TONS HA^{-1}) IN THE PIONEERTOWN FIRE AND MOJAVE DESERT, APRIL 2005

	Native Species										Exotic Species				Total
	At	Cc	Cs	Ep	Er	Gc	Pp	Pt	Srd	Sc	Bmd	Ec	Sb	Tot	Biomass
Pioneertown fire															
Pion-burn- SE	12.4	1.3	1.4	0.2	6.5	0.9		4.0		8.2	3.3	0.8	0.7	40.0	2.67± 1.87
Pion-burn-SW	0.2		2.9	0.4	0.2			1.5		5.4	3.8	3.3		17.8	1.06± .73
Landers-N	11.8	0.7	0.7				5.7		0.1			0.1	14.7	34.2	1.38± 2.06
Landers-S	32.5	0.3	7.8					0.1			0.7	28.3	0.8	70.9	2.15± 1.02
Mojave Desert															
Kramer-SE	36.9							0.3	1.0			9.4	1.3	47.7	–
Kramer-burn	23.6		0.7						2.0		0.6	3.5	15.1	46.5	1.97± 1.79
Kramer-W	38.3								0.2		0.1	0.3	31.5	71.6	3.90± 2.87
Hwy-14			3.3					1.6				10.2	4.9	26.1	1.14± 0.77
El Paso Peaks	14.4		0.5					0.1			1.2	13.0	2.0	31.9	1.54± 1.03

NOTES: At, *Amsinckia tessellata;* Cc, *Camissonia californica;* Cs, *Chaenactis stevioides;* Ep, *Emmenanthe penduliflora;* Gc, *Geraea canescens;* Pp, *Pectis papposa;* Pt, *Phacelia tenacetifolia;* Srd, *Salvia carduacea;* Bmd, *Bromus madritensis;* Ec, *Erodium cicutarium;* Sb, *Schismus barbatis;* Sc, *Salvia columbariae*
Other annuals present: *Eucrypta* spp., *Lepidium lasiocarpum, Lotus strigosus, Nemacladus* spp.

lasiocarpum, Lotus strigosus, Pectis papposis, and *Salvia carduacea* (Table 4.9). The same crop of wildflowers supported a 30,000-hectare fire the following summer.[186] Four fires burned portions of Joshua Tree National Park,[187] the flames carried by native grasses *Aristida* and *Bouteloua,* which grew in abundance after heavy summer rains in 2005. The Mojave Desert, which experienced numerous small fires, was covered with abundant dried forbs dominated by *Amsinckia tessellata,* with biomass ranging from 0.5 to 3.0 tons ha^{-1}. Herbaceous fuels persisted into 2005–06 because of limited decomposition in the dry winter of 2005–06.

"SECOND WAVE" INVASIONS
AND THE FADING OF CALIFORNIA WILDFLOWERS

Although Franciscan missionaries brought invasives in the late eighteenth century, how is it that wildflower flora has persisted through most of interior California well into the twentieth century? The recent collapse of the interior forb assemblage cannot be explained from the expansion of Franciscan exotics. While the transformation of coastal pastures by *Avena fatua* and *Brassica nigra* between 1780 and 1840 is largely unrecorded, the written record and botanical collections since Frémont suggests that Franciscan exotics have not expanded their ranges since the mid–nineteenth century. Alternatively, the wildflower collapse in the twentieth century is related to their displacement by "second wave" invasives (*Bromus madritensis, B. diandrus, Avena barbatis, Brassica geniculata*) that first proliferated at the turn of the twentieth century.

Frenkel (1970) and Blumler (1995) assert that all key invasives, including *Bromus madritensis, B. diandrus, Avena barbata,* and *Brassica geniculata,* date to the Franciscan and Mexican periods. This view is premised on two botanical collections of the state survey in the early 1860s (Brewer and Watson 1876–80) and on proposed botanical oversights or confusions by nineteenth-century botanists. While Brewer and Watson (1876–80: 553) list over one hundred persons who made botanical collections in California between 1791 and 1878, Frenkel (1970: 40) insists that "few of those botanizing California prior to 1860 paid any attention to introduced species." However, the tradition has always been that botanists are likely to record their types, i.e., the first collection of a species on a new continent will be collected and recorded because it represents a "novel" finding. It is highly unlikely that the hundreds of botanists combing California, including David Douglas, Torrey of the U.S.-Mexican boundary survey, Vasey of the Pacific Railroad survey, Fré-

mont, Green, Nuttall, Asa Gray, and Brewer, would have completely ignored invasive species. Frenkel (1970) and Blumler (1995) propose that
bromes, slender wild oats, and summer mustard were confused with Franciscan species: *Avena barbata* with *A. fatua,* and *Brassica geniculata* with
B. nigra.

An earlier-arrival hypothesis, of course, "buys time" to explain the
widespread distribution of "second wave" invasives. However, if these
exotics were already in California by the gold rush, why did they fail to
displace the indigenous flora long ago? The two collections of bromes
in the 1860s represent two possibilities: (1) that the bromes were already
present for some time, or (2) that the collections represent excellent observations of a species' first arrival. The rapidity at which both bromes
have raced through California since 1895 lends credence to the latter
hypothesis.

Informal landscape descriptions by those with botanical background
such as Frémont, Muir, King, Hittell, and Cronise, never recognized
bromes. Brewer's (1883: 961) summary of invasive weeds in California
states that the "most troublesome pest" of California was *Hordeum murinum* in overgrazed lands. He further stated that "plants other than true
grasses furnish a larger portion of the forage of those regions which have
rainless summers, some of which) alfileria, bur-clover, etc," (8) all introduced Franciscan species highly praised as fodder plants by botanists
and agriculturalists. Similarly, Hilgard's 1891 public lecture on "immigrant plants" in Los Angeles (reported in the *Times*) did not list any "second wave" exotics as potential pests in California.

Collections of "second wave" species were infrequent until 1895, when
botanists such as Parish, Davidson, Abrams, and others stated that
bromes and *Avena barbata* were expanding rapidly into new territory.
Parish explicitly detailed the arrival and expansion of *Bromus madritensis* and *Bromus diandrus* in the San Bernardino Valley beginning in 1888.
California Vegetation Type (VTM) maps show that second-wave exotics
led by species of *Bromus, Avena barbata,* and summer mustard (*Brassica
geniculata*) were already abundant or dominant in the hills surrounding
the Los Angeles basin by 1929–34. Salo (2004) examined nineteenth-
century botanical records and concluded that *Bromus madritensis* proliferated in the late nineteenth century. She speculates that bromes were
spreading in the Central Valley earlier than reported by Parish (1920)
because of vectored seed dispersal combined with disturbance from post–
gold rush livestock grazing, processes that also occurred in southern California. However, grazing intensities had already been high, if not greater,
in southern California since the early nineteenth century. The efficiency

of *B. madritensis* seed transport in the fur or feathers of native animals would appear to mitigate against an inclusive domestic livestock facilitation hypothesis (Schiffman 2000).

The decline of native wildflowers in the twentieth century went hand-in-hand with the expansion of second-wave exotics into more arid habitat far beyond the geographic ranges of *Avena fatua* and *Brassica nigra*, reaching the edge of the California deserts. *Bromus madritensis* spread across the California deserts in the 1920s and 1930s and was joined by *Schismus barbatis* in the 1940s and *Brassica tournefortii* in the late 1970s. Newspaper reports of wildflowers virtually ceased at Los Angeles in about 1940, at the southern California interior valleys and southern San Joaquin Valley in the mid-1960s, and at the deserts in the late 1970s.

Naveh (1967) posits that invasions are related to the open, unstable nature of annual grassland, but his model is premised on bunch grassland theory. Because precipitation is more unreliable than in Mediterranean Europe, California herbaceous vegetation can be destabilized more rapidly in the face of introduced species. After elimination of "original" perennial vegetation, unoccupied space became new habitat for invading pioneering species, all of which self-fertilize. A few well-adapted species quickly built up large, homogeneous, highly productive populations, encouraged by pastoral land use, first in Mediterranean countries, and again in California by the Spaniards and first settlers. California grassland ecologists have remarked at the phenomenal pace at which invasives have taken over the state over the past two centuries (Mooney and Drake 1986; Huenneke 1989). A simpler explanation is that California invasives come from similar climates in Mediterranean Europe, and hence were "preadapted" to California pastures (Jackson 1985; Sims and Risser 2000).

Generations of students of California grasslands have argued that introduced species spread across the state in response to human-induced disturbances, including agropastoralism, urbanization, and burning (Parish 1920; Biswell 1956; Frenkel 1970). But as pointed out by Blumler (1995), California valley grassland is comprised of 80–90 percent introduced species, whose abundance is unrelated to disturbance. He argues that the success of Mediterranean species in California is related to climate and nonconvergent evolution—the evolution of dissimilar adaptations in similar environments on one land mass (Europe) compared to another (North America). Exotics arrived with superior adaptations to native organisms, notably the capacity for rapid seed dispersal, the capacity for mass recruitment at high densities, and rapid early season growth rates that competitively displace or force into greater seed dormancy the indigenous, slow-growing forbs relying on seed dormancy as an adaptation to fluctuating interannual precipitation (Cohen 1966; Ellner 1985a,b; Venable and Brown 1988).

Introduced species are remarkably successful through mass, early season recruitment at suffocating densities at the expense of native forbs. But this trait also predisposes exotics to mass premature mortality (reproductive "crashes"; Salo 2004). In contrast, selection pressure for native wildflowers, whose recruitment patterns are cued to fire, carbon, rain, and floods, leads to adaptation for long seed life in environments of high climatic variability and recurrent disturbance. Variable and unpredictable moisture, temperature, and light at the time of germination and during the cool winter months cause extraordinary variation in productivity and floristic composition from year to year and disturbance (Talbot et al. 1939; Heady 1958; Naveh 1967; Frenkel 1970).

While invasive species theory predicts that introduced species will experience contagious steady advance in response to seed dispersal capacity in optimal habitats, Mediterranean invaders have recently been delivered setbacks by drought. Red brome became dominant in the desert during very high rainfall from 1978 to 1986, but was virtually extirpated from these areas by the late 1990s (Figure 4.4).

The fate of invasives is also mirrored in the dominance pattern of annual species. Time-series data as well as photo resurveys in Blaisdell and Snow Creek (Table 4.8) suggest a coherent pattern of annual dominance with fluctuating interannual precipitation. Exotic grasslands appear to alternate between two end-member states in which wet years favor bromes and oats and dry years favor *Erodium* spp., *Brassica geniculata, B. tournefortii, Schismus barbatis,* and wildflowers, due to "crashes" in bromes and oats (Talbot et al. 1939; Biswell 1956; Heady 1958; Salo 2004). In Kern County rangelands, Twisselmann (1967: 91–93) noted, using life zone terminology to organize his thoughts, that "the most successful plants in years of scant rainfall are the native *Festuca microstachys* var. *simulans, Lepidium dictyotum . . . , Lasthenia chrysostoma,* and the introduced Arabian grass (*Schismus arabicus [barbatis]*), red-stemmed filaree (*Erodium cicutarium*), red brome (*Bromus madritensis*), and common foxtail (*Hordeum glaucum*). In wet years the upper Sonoran associations were relatively common (*Bromus diandrus, B. madritensis, Avena fatua, Festuca megalura [myuros]*)."

Nineteenth-century writers appear to have captured these same trends with flood and drought, changes that were attributed to overgrazing by livestock. Among the best observers, Frémont saw wildflowers, *Erodium,* and clovers in the Central Valley in normal rainfall winters that followed drought in 1840–44 (Haston and Michaelson 1994). Wild oats and black mustard grew along the coast. Brewer (1883: 963) stated that wild oats were "most abundant between 1845 and 1855, when hundreds of thousands of acres were clothed with it as thick as a meadow." Wild mustard also was an important dominant over large areas at about the same time, particularly on better lands in the coastal valleys (Bryant 1848; Cronise 1868; Jepson 1925). Brewer witnessed extreme drought in 1862–64 and found that wild oats and mustard were succeeded by filaree, which increased in abundance between 1865 to 1870 (Brewer 1883; Hittell 1874; Perkins 1863; cf. Burcham 1957: 194). Muir, King, and many others saw local flower pastures in the Coast Range, and extensively in the interior valleys in the late 1860s. However, it is unclear whether *Avena fatua* was ever abundant in the Central Valley outside the floodplains. The historical descriptions of "barrens" after both wet and dry winters is inconsistent with the presence of wild oats, a grass capable of producing abundant cured hay that was the foundation of the late-Spanish and Mexican grazing system along the coast.

The contrast in the adaptive modes between native forbs and European invasive annuals is related to differences in the climate of the western Mediterranean basin and California (Jackson 1985). The Mediterranean climate of Europe has an early onset of cool-season precipitation and less interannual variability of total precipitation than in California. The early rainy season is caused by the southward expansion of the jet stream into the warm Mediterranean Sea, which encourages heavy precipitation episodes by October. In California, where the cold upwelling California current prevails year-round, cyclones of the jet stream bring heavy rains by November and December. The rainy season peaks in November-December in the Mediterranean, and in January-February in California (Table 4.10).

Modest interannual precipitation variability and short dry seasons in the western Mediterranean, compared to California, select for species characterized by mass germination of propagules with short seed life. Germinating rains in Mediterranean Europe, one to two months earlier than in California, select for prewinter mass germination and rapid growth in warm temperatures under high evapotranspiration demand. In California, native forbs germinate selectively with optimal conditions, the seed sometimes lying dormant in the soil for years. Germination normally occurs in winter with cold temperatures, the native wildflowers growing slowly in winter under low evapotranspiration demand and then growing rapidly to flower in spring.

California is a seasonal desert in summer and experiences highly vari-

TABLE 4.10 AVERAGE MONTHLY PRECIPITATION (MM) AT SELECTED STATIONS
IN CALIFORNIA AND WESTERN MEDITERRANEAN EUROPE

	Jul	Aug	Sep	Oct	Nov	Dec	Jan	Feb	Mar	Apr	May	Jun
California												
San Francisco	0	1	5	24	62	94	114	92	72	33	10	3
Los Angeles	0	2	7	11	33	61	81	87	62	26	7	2
Mediterranean												
Seville	1	4	20	66	70	84	64	62	57	59	39	9
Lisbon	4	5	33	75	100	97	95	87	85	60	44	8
Madrid	11	12	24	39	48	48	33	34	23	39	47	26
Rome	16	33	68	93	111	90	81	75	65	55	32	16
Athens	5	8	10	53	55	62	48	41	41	23	18	7
Casablanca	0	1	6	34	65	73	57	53	51	38	21	6

SOURCE: Europe, Euro Weather, www.eurometeo.com; California, Western Regional Climate Center, www.wrcc.dri.edu.

able interannual precipitation in winter. Selection pressures, regardless of specific mechanism, are for long seed life and seed banks as an adaptation to disturbance and precipitation variability. Annual wildflowers surviving on dormant and refractory seed are also better adapted to the long summer drought than are perennial bunch grasses that must maintain photosynthetic metabolism without growing-season precipitation (Naveh 1967). Indeed, on the basis of habitat and germination requirements, *Stipa (Nassella) pulchra* lacks the ecological character of a dominant species (Bartolome and Gemmill 1981).

Premature mortality of bromes and oats (death before production of seed) at Riverside occurred during winter rainless periods of at least one month and cumulative seasonal precipitation of less than 4 centimeters from December to February, and less than 15 centimeters in March and April. These rules recognize that herbaceous biomass, leaf area, and evapotranspiration increase cumulatively over the growing season. Hence it takes less-extreme dry spells to cause herbs to desiccate, perish, or to be placed into terminal inflorescence under warmer temperatures in spring compared to fall. If the total rainfall before the rainless period exceeds 15 centimeters, annuals reach flowering regardless of precipitation distribution.

The sampling array in the hills at Riverside documents an outbreak of wildflowers—*Phacelia, Cryptantha, Salvia (columbariae), Chaenactis, Layia, Eschscholzia, Emmenanthe*—following drought in 1989–91; a proliferation of *Bromus madritensis* and *Avena fatua* in wet years from 1992 to 1998; and a return of mustards and flowers in the drought from 1999 to 2003 (Figure 4.6). Similar trends were observed at repeat photo points in the Sonoran Desert at Snow Creek and Blaisdell Canyon near Palm Springs. Above-normal precipitation may have triggered the proliferation of bromes in the desert after 1978, but frequent drought-triggered brome retreat (increase in wildflowers) since the 1990s. Salo (2005) speculates that bromes expanded in two periods, 1930–45 and the late twentieth century, in association with higher rainfall from the Pacific Decadal Oscillation. While there is strong evidence for brome expansion since the late 1960s, the wet phase of 1930–45 was accompanied by spectacular native blooms in California, suggesting that the bromes had yet to dominate grasslands in interior California.

The modern dominance of "second wave" exotics has produced a new pattern in the interannual variability of flower abundance. While abundance correlated with precipitation in the early twentieth century, the best flower years now occur in the first wet years after invasive-depleting droughts. As a result, mass blooms have become ever more infrequent throughout southern California except in the hyperarid deserts, beyond the range of red brome. The future of desert biological invasions may be one of episodic expansion and retreat with precipitation variability.

The winter of 2001–02, the driest year since 1849, was the ultimate test of brome resilience against premature crashes. However, field data at Riverside suggests that individual winters are never dry enough to engender "perfect" crashes, i.e, total extinctions at regional scales, as seen recently in the desert. While drought may suppress bromes, rare survivors of the bottleneck will rapidly recolonize interior valleys within years. Bromes and oats will maintain a permanent presence in coastal California. With broad-scale extirpation, brome recolonization may require decades in the desert. *Bromus madritensis* has not been observed at Blaisdell since 1989 and at Snow Creek since 2001.

It was hypothesized that the displacement of indigenous herbaceous cover by Mediterranean annuals possessing different fuel characteristics increased fire hazard and reduced fire intervals in many areas, the grass/fire feedback hypothesis (D'Antonio and Vitousek 1992). However, the historical role of fire throughout California has been mitigated by the removal of fuel by livestock grazing. The relationship between changing California pastures, climate variability, and grazing pressure is poorly documented because comprehensive fire records for ranchlands do not exist before the 1960s, and the interannual status of herbaceous species composition and biomass is rarely sampled.

Moreover, the Spanish record demonstrates that indigenous flower pastures along the coast burned extensively before the invasion of European annuals. In a modern example, wildflowers dominated by *Amsinckia tessellata* with standing biomass of 1.0 to 2.0 tons ha^{-1} fueled the 30,000-hectare Pioneertown fire in 2006 in the Mojave Desert after unprecedented heavy rainfall in the 2004–05 winter (Table 4.9). It is problematic whether the displacement of indigenous herb cover by Franciscan wild oats and black mustard would engender change in burning rates in a positive feedback in coastal California (D'Antonio and Vitousek 1992), in view of the widespread burning in native forbfields recorded in Crespí's journal. However, the feedback model may have merit in the interior, where summer barrens are now covered with cured brome grassland. It would be expected that wet years would produce long growing seasons, high biomass, and fire outbreaks in the flower fields of interior California. Still, Brewer (1966) lamented the dust and barrenness of the Central Valley in the summer of 1862, after widespread heavy rain and floods the previous winter. A pattern of limited productivity and burning in both wet and dry years raises the importance of the brome invasions in increasing the rate of burning of inland regions in the 1960s.

The effect of fire on exotic annual grassland parallels the effect of drought. While the intrinsic flammability of bromes and slender wild oats may encourage burning, postfire successions often consists of increases of less flammable native and exotic forbs with long seed life, whose germination is fire-, charate-, or smoke-stimulated (Keeley 2000). Sims and Risser (2000) concluded that fires shift the annual grassland species from grasses to legumes and forbs. The effect of fire on population dynamics and species composition of annuals depends on the season of burning. Time-series data at Riverside show that spring burns convert brome grassland into forbfields (Figure 4.6). Spring or early summer burns, just after desiccation of the herb layer, select for seedbanking forbs and bunch grasses because nonshattered seed of bromes and oats in ambient position are killed by high flame temperatures. Seed survives on the soil with lower flame temperatures in fall burns after shatter (cf. Wills 2000). Postfire annuals including *Erodium*, legumes, and native flowers disarticulate and leave minimal fuel. Hence fires do not recur on these sites until after there is recolonization by bromes and oats, often years later (cf. Meyer and Schiffman 1999; Brooks et al. 2004).

The distribution of bunch grasses in botanical treatments at the turn of the twentieth century is similar to that in Brewer and Watson's (1876–80) state survey of flora, i.e., moist areas of the Sierra Nevada foothills and the Coast Range. Abrams (1904) found *Stipa (Nassella) pulchra* on mesas, grassy hills, and in open places in the chaparral belt, while Davidson and Moxley (1923) found it to be common on adobe banks on plains and in foothills from Los Angeles to San Diego in the Coast Range. In central California, Jepson (1925), Smiley (1922), and Beetle (1947) found it mostly in the Coast Range and the Sierra foothills.

HISTORICAL DEVELOPMENT
OF EXOTIC ANNUAL GRASSLAND

Historical accounts reveal that pre-European herbaceous cover in California consisted of ubiquitous dense fields of wildflowers from the coast to the foothills of the Sierra Nevada and southern California Coast Range, with thin cover across the desert. This is a far cry from the bunch grassland baseline accepted by many in the ecological community. First, Clements deduced vegetation change using space-for-time substitution based on climax and relict theory. The vegetation was originally bunch grassland, but cattle grazing selectively removed bunch grasses and led to proliferation of both wildflowers and European annuals, the expansion of introduced species eventually leading to modern grasslands. The Stanford school (Mooney et al. 1986; Huenneke 1989) concluded that annuals were "invasives" that spread without the help of livestock or other disturbances, but left unclear the status of bunch grassland as a baseline vegetation. This study concludes that the baseline vegetation of wildflowers was displaced by invasives.

The botanical vocabulary of Franciscan missionaries limits even speculation on spatial patterns of species composition, only showing that the Franciscans consistently observed the wildflower life-form along the coast from San Diego to San Francisco, often for days at a time. We can be certain only of the abundance of chia (*Salvia columbariae*) in southern California. The species composition is hypothesized to be the same suite of species collected or observed since the mid–nineteenth century, notably in the genera *Eschscholzia, Amsinckia, Nemophila, Lasthenia, Calandrinia, Chaenactis, Chorizanthe, Cryptantha, Emmenanthe, Eriogonum, Eucrypta, Gilia, Layia, Lepidium, Lupinus, Phacelia, Salvia, Senecio,* and *Viola,* to name a few. While the Franciscans never recorded wildflowers in the interior valleys of California, mid-nineteenth-century explorers, botanists, naturalists, and settlers saw extensive forbfields throughout this region.

Coastal flower fields produced cured *pasto* and *zacate* in summer. Spanish journals clearly show that burning was extensive in this assemblage, the fires mostly initiated by Native Americans to enhance local food resources. Several genera leave abundant cured biomass, including *Amsinckia, Phacelia, Chaenactis,* and *Lepidium,* and *Salvia (columbariae).* The "interior flower field" assemblage disarticulated into barrens in summer, which burned much less frequently than along the coast. Fire intervals were highly variable in response to fluctuating interannual precipitation. The extraordinarily widespread extent of burned territory recorded by Crespí in the summer of 1769, after a winter of normal precipitation days recorded in the Portolá expedition's journals, suggests that coastal forbfields are sufficiently productive to burn most years. Perhaps the interior pastures burned extensively in abnormally wet years. Most fires were reported in reeds and tule swamps in the Central Valley.

While the Franciscans were unflattering in their descriptions of the desert, it is reasonable to assume that this region was covered by thin but highly variable cover of wildflowers, as seen today. This botanical community has subtle floristic differences from those on the coast, with desert wildflowers often consisting of closely related species to coastal members of the same genera, including *Phacelia, Amsinckia, Cryptantha, Esch-*

scholzia, Lasthenia (western desert margin), *Chorizanthe, Chaenactis, Emmenanthe, Eriogonum, Eucrypta, Gilia, Layia, Lepidium, Castilleja, Lupinus, Malvastrum, Mentzelia, Mirabilis, Coriopsis, Pectocarya, Plantago, Salvia, Senecio, Malacothrix, Abronia, Camissonia,* and *Geraea* (Jaeger 1941). Desert lands were barren most years but could support dense flora pasture after extraordinarily wet winters, with sufficient biomass to carry fires such as in 2005. The bunch grasses *Hilaria rigida* and *Nassella (Stipa) speciosa,* first reported in the desert by Frémont, were abundant in the Antelope Valley and at higher elevations of the Mojave Desert. Dense growth of native annual grasses *Bouteloua* and *Aristida* carried fires across 50,000 hectares in the high deserts, including the Mojave National Preserve and Joshua Tree National Park in 2005–06.

By the mid–nineteenth century, coastal pastures of California were already covered with European species. Black mustard was extensive in coastal southern California by the 1790s and presumably in coastal central California by that time. Mission bricks reveal that wild oats began their expansion somewhat later, possibly as late as 1810, but had reached their modern range along the coast and Central Valley floodplains by the gold rush. The reports of Frémont, Muir, Brewer, Clarence King, and others clearly show that coastal flower fields were no longer ubiquitous. Interior California beyond the ranchos was still graced with indigenous flower fields in the mid–nineteenth century. Muir saw flower carpets and Hittell saw a "galloping rainbow" in his rail crossing of the Central Valley. Interior flower fields were invaded by *Erodium,* clovers, and fescue, which coexisted with native forbs. The rapid expansion of introduced annuals suggests that they dispersed into new territory independently of livestock grazing.

Reports of fire from the mid–nineteenth century pale in comparison with Crespí's remarkable account of California in 1769. Coastal oat-mustard prairies doubtless carried fire at high frequencies, but pastures were also being consumed by one million head of cattle and other livestock. Interior valleys burned rarely in summer because the new invaders of the interior—*Erodium,* clovers—left little cured fuel alongside the sparse remains of native forbs. Zalvidea, Muñoz, Frémont, Wilke, Darby, Muir, King, Brewer, Hittell, and many others lamented the barrenness of the Central Valley. Desert flower fields were little changed from pre-Hispanic times, as native forbs were joined only by *Erodium cicutarium.*

Beginning in 1880, a second wave of European invaders, led by *Bromus madritensis, B. diandrus,* and *Avena barbata,* began invading oat-mustard prairies along the coast and into interior flower fields. Franciscan oat-mustard prairie apparently reached its ecological range by the mid–nineteenth century. "Second wave" invaders were abundant and dominant constituents of exotic annual grassland by the time of the California Vegetation Type Map (VTM) Survey of 1929–34. Red brome moved across the Mojave Desert for the first time in the 1930s, and has since been joined by two annuals from Middle Eastern deserts: split grass (*Schismus barbatis*) in the 1940s and Sahara mustard (*Brassica tournefortii*) in the 1970s. Historical newspaper accounts indicate that bromes had displaced wildflowers at Los Angeles and other coastal regions by the 1940s, in the interior valleys and southern San Joaquin Valley by the late 1960s, and in the deserts beginning in the late 1970s. Wildflowers were suppressed even in the deserts until the 1990s, when the abundance of brome was reversed by "crashes" from drought, after 1988 in the Sonoran Desert and 1997 in the Mojave Desert. Wildflowers are presently most abundant in the California deserts, where brome crashes are frequent even under normal precipitation. "Second wave" exotics, especially *Bromus madritensis*, enhanced burning in the interior valleys and western margins of the desert.

Wildflowers have persisted in Franciscan exotic annual grasslands along the coast almost exclusively in nutrient-poor soils, such as in the coastal dunes around Monterey Bay as seen by Crespí, Frémont, and the *National Geographic* (Hall 1929), in beach dunes near Santa Monica (Mattoni and Loncore 1997), in serpentine rocks at Jaspar Ridge (Huenneke et al. 1990), and near Davis (Michael Barbour, pers. comm.).

Since 1960, bromes and slender wild oats have come to dominate all of interior California, with Franciscan oats and black mustard still prevailing along the coast. Both assemblages have invasive companions in *Erodium, Schismus, Medicago, Trifolium,* and *Brassica geniculata*. Tall-statured bromes and oats tend to dominate in years with high rainfall, while *Erodium, Schismus,* clovers, and summer mustard tend to dominate in drought. Wildflowers now persist sparingly in semiarid regions such as the Carrizo Plain, the southern San Joaquin Valley, and interior valleys of southern California, with their abundance increasing after invasive "crashes" from drought or spring fires. In the past, wildflower abundance was proportional to total annual precipitation and well-distributed rains. Today, wildflower splashes occur only rarely, in the first wet years following long-term drought.

CHAPTER 5

Lessons from the Rose Parade

There is one California wild flower that every Californian, however unobserving, knows and loves, as the Briton his daisy or the Irishman his shamrock, and that is the native poppy or eschscholtzia. Poets apostrophize it; artists paint it and craftsmen work it into their handiwork; it is sown in gardens and tradesmen employ it as a mark for their merchandise. Every spring millions of its blossoms are brought indoors and set in vases and bowls, where it illuminates the rooms of half of California with the glow of imprisoned sunshine. To a degree that can be said of no other State device, it is the floral emblem of the Commonwealth—not a token voted by a little knot of flower enthusiasts, but the spontaneous choice of a whole people, who love it and admit it into their daily life.

— Charles Frances Saunders (1914: 102–3)

In New York, people are buried in the snow. Here our flowers are blooming and our oranges are about to bear. Let's hold a festival to tell the world about our paradise.

— Charles Fredrick Holder in 1890 (Hendrickson 1989: 2)

Saunder's praise of the California poppy is truly historic, as this flower— that formed brilliant carpets throughout the state only two centuries ago, and annually drew weekend tourist crowds to the valleys and deserts only a half century ago—is so rare that Governor Reagan in 1972 began an initiative to set up a poppy reserve that eventually came to pass in the Antelope Valley. The near demise of the *copa de oro* and its floral compatriots was not only the product of land clearing, but also of displacement by invasive European annuals, transforming nearly all of California's landscape, plowed or not. Late–nineteenth century California

culture, which was only one century removed from the pre-Hispanic baseline, valued its wildflower heritage, and this is still expressed in modern
pastimes. Floral societies sprung up in all the local towns and hosted
weekend flower parties for the neighborhood. The New Year's Rose Parade in Pasadena was the institutional outcome of the combined forces
of southern California's floral societies. One rigorous parade requirement
was that the floats must be entirely covered with flowers.

The official story is that the New Year's Rose Parade (Tournament of
Roses) was established by the Valley Hunt Club, an exclusive private social club dominated by "an elite cadre of Pasadena's super rich," many
from the cold climates of the Midwest (Hendrickson 1989). Winter flowers and the mild California climate were the central concepts behind the
tournament. The idea was inspired by Charles Fredrick Holder in 1889,
who stated "in New York, people are buried in the snow. Here our flowers are blooming and our oranges are about to bear. Let's hold a festival to tell the world about our paradise" (Hendrickson 1989: 2). On January 1, 1890, there was a parade of flower-covered horses and buggies
and an afternoon of public games on the "town lot." The parade was to
resemble the West's version of the festival of roses in Nice, France. In
fact, the first tournament was devoted more particularly to a program
of sports, especially to recollection of the old Spanish California days of
horsemanship and the parade of flower-bedecked vehicles was an inci-
dental feature (Hendrickson 1989). The tourney of rings coupled with
the floral displays prompted Professor Holder, the first president of the
Tournament of Roses, to say, "Now we have the name we want—The
Tournament of Roses" (Hendrickson 1989).

But Holder's ideas, even the concept of floats in the parade, had prece-
dent in ladies' flower organizations that had already celebrated four
"flower festivals" in the 1880s.[1] In 1889, the city of Los Angeles put on
a grand exhibition at "Hazards Pavilion."[2] The ladies were "making all
the arrangements and a corps of carpenters and laborers . . . were not
given breathing time by their fair bosses." The highlight was a "floral
mount" of "Wilson's Peak, on which rests the observatory." The mountain was a "miniature" that was "an exact duplicate of Nature's original . . . [and] gives one an idea of the grandeur, the wildness of the mountain on which is to be placed the beacon light of science for southern
California [the Mount Wilson Observatory]." Floral societies of each city
had their own booths, just as each city now contributes individual floats
to the Rose Parade. The perishability of cut flowers in the festival was
solved by contributions from the local population. According to a *Times*

reporter, "a large quantity of fresh flowers had been received and the decorations of the different booths had been perfected." Flowers apparently came from the hillsides and people's yards. The pavilion was jammed with one of the most fashionable audiences that was ever assembled in the city. The visitors to California were astonished by the "variety of flowers." But even flowers can be "mind numbing" in time. According to the *Times*, "Those who have become accustomed to the infinite variety of this semi-tropic region, grow to ignore that which is so immediately observed by the strangers from the Atlantic-bound East." This forum was a microcosm of the cosmpolitanization of flowers. In addition to native wildflowers, floral favorites were "sunflowers, crocus, dahila, heliotrope, lily, dandelion, hollyhock, japonica, violet, mignonette, Touch-me-not, tulip, and nightingale." The European garden rose was a favorite in the decorations and the eventual inspiration for the Rose Parade. Thanks to Holder's proposal for a Rose Parade, "floral mounts" became "parade mounts" with wheels.

But even this heritage has withered as the Rose Parade has lost sight of its historical baseline. As Los Angeles grew to cosmopolitan status, transportation became more efficient, and its citizens consequently became detached from the neighboring landscape. An increase in the speed of technology resulted in an increase in the overall pace of life, leading to a decrease in the human ability to stop and observe and form bonds of attachment to local landscapes. The globalization of flowers paralleled the globalization of the world's biota (Coates 2006). Native flowers no longer dominantly covered the hills and, as a result, were quickly forgotten—by the untrained eyes of the masses and even by the few trained eyes of botanists and naturalists.

A plethora of evidence documents that the California landscape was coated by wildflowers. Pre-Hispanic indigenous cover of California was not grassland, not even prairie—a mixture of flowers and grasses—but forbland like that found in most deserts of the world. California is a desert half the year. This story has been overlooked because of a flawed hypothesis that bunch grasses were pervasive in the past. As a consequence, we take for granted the rapidly fading wildflower heritage because the perception of past vegetation among the scientific community and the public has been built upon this erroneous premise. This bunchgrass story has canalized us to perceive California ecosystems in a certain way, preventing us from observing, doubting, and searching for alternative evidence to construct alternative stories. After all, language is the way we model the world. The bunchgrass model attains even greater stature because wildflowers are no longer a reminder of their former vast presence in California wildlands. The bunchgrass theory diverted the attention of citizens and scientists from the state's wildflower heritage.

The bunchgrass-grazing hypothesis was created by range managers and scientists influenced by the Dust Bowl tragedy of the 1930s, to a point that it became an idée fixe that has kept blinders on us. Unfortunately, science is invariably compromised when it is interwoven with government policy (Grove and Rackham 2001). The fundamental finding that wildflowers are being displaced by pulses of invasive annuals demonstrates the importance of historical ecology in testing and constraining hypotheses on vegetation change over time, because information on past herbaceous vegetation in California cannot be tested against real-time observations. Historical ecology is treated as a "natural experiment" in which one must account for the evolutionary ratchet of time—the irreversibility of the past—and all we can do is accept that it has happened, accumulate as much evidence as possible, and learn from it in all future circumstances. The hope is that the known baseline can be pushed farther back to paleo time scales from palynological and macrofossil evidence. It is unclear whether pollen or phytolith records will be useful in the assessment of the forb flora, because these methods preferentially record wind-pollinated species, especially grasses. Perhaps packrat midden studies can be expanded into Mediterranean ecosystems from the California deserts. Just as primary historical evidence has been reanalyzed in this study, there is potential for large amounts of archaeological evidence to be reinterpreted from a historical ecological perspective.

Amazingly, in the modern era of endangered species protection virtually none of the native wildflowers, now only a fraction of their former abundance, are part of the agenda for species protection and recovery. Newly expanding star thistles garner higher priority as invasives than bromes and oats that have already suffocated the wildflower flora. People's views have changed from landscapes to species in an environment of increasingly specialized science and the legal framework of endangered species protection. What is striking about eighteenth- and nineteenth-century writings is the interest in holistic landscapes as broad-scale vegetation or as a pastoral resource. The Rose Parade exemplifies the misdirection of modern concern: people are looking at the individual flowers instead of the float. Must we continue to pursue science as framed by existing government policy, or instead can we use science to question, challenge, and change government policy?

Legal structures are tunnel-visioned to the "protection" of rare species, not biological landscapes. Species protection is irrational, because the salvation of one species usually comes at the expense of its neighbors. Isolation of one species is close to impossible, if not futile, and it is vital to deal with the entire floristic package. Species have multiple and complex relationships with other species and with the abiotic elements of landscapes. Moreover, the first assumption in the implementation of National Environmental Policy Act and the Endangered Species Act is that "total protection" is the only way to save a species. Natural disturbance processes are largely ignored and disregarded thanks to the early twentieth century Clementsian legacy that fire and disturbance are "unnatural." In fact, grazing, fire, and floods, which have always been a part of the landscape dynamic, may be beneficial to species protection. Grazing removes thatch, reducing the competitive advantage of exotic species over natives. Fire brings out native forbs in abundance, a relationship that was well understood by Native Americans at the time of European contact.

A "David and Goliath" opportunity for restoration of California's wildflowers rests on the observations that the damage to the region's herbaceous vegetation has been accomplished by only a few species. Exotic annuals came without their pathogens and have not been selected by local conditions, while native annuals in all their diversity have a history of adaptation and evolution over geologic time scales. Exotics have already shown their vulnerability, indeed local extirpations, to such factors as drought and early season fires. As with *Eucalyptus* (Coates 2006), natural predators will eventually catch up with the bromes and oats, encouraged with human intervention through biological control. Invasive species that dominate California's landscape are few in number. This fact alone brings promise, because land managers can reduce or extirpate a few huge "birds" with one "stone."

The restoration of California's wildflower flora will require management strategies involving the entire landscape, with a historical perspective. The greater we understand the past, at all spatial scales, the better we can manage landscapes for today and into the long-term future. Local research studies have restricted application because local, high-precision field measurements with limited time scales of data collection have often lead to dangers of extrapolation to landscapes. Small-scale findings have reduced the probability of addressing broad-scale ecosystems. As stated by Jackson et al. (2001), it is important to increase the breadth of data collection in order to see the big picture of landscapes over long time scales. Another approach is to accept a "less than total" understanding of ecosystems—sacrifice a level of precision, quality, and completeness of evidence collected—and to employ an "adaptive management" perspective. Based on the assumption that the system is chronically changing, and that our knowledge of the system is chronically changing, one must be open to new ideas and tinkering with the system of study, followed by revision (or overhaul) of management plans.

Potential avenues for effective management and conservation to preserve some degree of history in our landscape include spring burning, use of pathogens as biological controls of invasives, dedication of new wildflower reserves, and encouragement of seasonal grazing by domesticated livestock. California ecologists, botanists, and the public at large should redirect their attention to the state's wildflower flora, which has great potential renewal at extensive scales. The wildflower flora still exists as a long-lived seed "banked" in the soil of millions of acres of degraded California lands. Few wildflower species have become extinct, and during rare years and in scattered locales they still appear in abundance, reminding us of the past while directing us in the future.

Notes

CHAPTER 1

1. *Los Angeles Times* (October 31, 1954), Our wildflowers . . . for so little they offer so much, p. L-14; Seasonal wildflowers, Total Escape, www.totalescape.com/tripez/wildfl.html.
2. *Los Angeles Times* (March 3, 1903), Poppy day in capitol, p. 4.

CHAPTER 2

1. One league is a hour's ride on a horse, a variable quantity depending on the adversities of terrain, usually 2–5 kilometers.
2. The Santa Margarita Mountain spur extends west to Camp Pendleton only 10 kilometers from shore.
3. The use of "sheer soil" in Crespí's journal requires explanation, as this phrase is used repeatedly. It was translated from *pura*, meaning "pure" soil, which implies that it was likely a loam, with a high weathered clay fraction. Crespí frequently associated herbaceous cover with "pure soil" throughout California.
4. The wording "as before" reveals a writing style common in the Spanish journals, meaning that the vegetation of one day was similar to that encountered on previous days' marches (i.e., the English "ditto").
5. The widespread extent of Indian burning from San Diego County northward along the coast is suggested by a mission questionnaire published in 1798. While Native Americans were "granted ½ the year to gather their wild seeds, [they] were punished for setting fires," especially when burning was viewed as a "transgression against the common good, like killing cattle and sheep, or firing pasture" (Timbrook et al. 1982).
6. In Mexico *pajon* often refers to coarse straw or a tall grass. After floods it leaves perennial straw like a stalk when it dries. As with other sites, bunch grasses along the San Gabriel River grow in riparian habitats.

7. "Oy contéque subimos y baxamos 27 lomas. Entre la infinita variedad de flores, como tulipanes, y otras, de muy diversos colores, y muy lindas, con que desde ahora se empiezan á vestir aquellos camos, montes, y valles, de aquellas tierras, vi varias como en España; y entre ellas unas muy hermosas y pequeñitas de cinco hojas, que parece forman una cara, totalmente parecidas á las que vi en Cataluña en algunos jardines llamadas alli Pensamientos, solo con la diferencia que aquellas son amarillas, y algo moradas en los extremos de las ojas, y estas son todas amarillas y no huelen" (Web de Anza Archives, Font expanded diary, February 11, 1776).

8. "Tiene este río mucha chia buena para refrescar. . . . Tiene esta grandíssima llanura toda la campiña llena de chia mui buena de refrescar, y es tanta, que aunque la gentilidad sea mucha, me pareció impossible que puedan cojer ni la mitad de ella. A la actualidad estava en flor; y flor morada" (Brown 2001: 686–88).

9. In 1792, as part of a survey of the northwest coast of America, John Sikes made a sketch of the coast near Santa Barbara Point. It shows an indentation of coast at Santa Barbara with coastal plain and the Santa Ynez Mountains in background (Brown 1967). The Santa Ynez Mountains contain a few coast live oaks, but are otherwise featureless. An oak grove covers a steep coastal bluff in the foreground right, and another oak grove lies behind an Indian village. There are oak groves in canyons of steep coastal bluff on the right. Unfortunately, the pasture on the coastal plain and vegetation on the slopes are indistinguishable.

10. "Todo el terreno que ocupan es tan fertil, y hermoso, como los demas independientes de esta canal, en donde tiene, recrearse la vista, especialmente la que ha registrado la suma esterilidad del Golfo de California, donde no se ve, ni Arboles, ni Yerbas esteriles, y aqui por el contrario están tocando las aguas del mar, campos tan verdes, como floridas" (Web de Anza Archives, Anza diary, April 14, 1774).

11. "A las seis y media de la manaña salimos de esta ranchería y punta de San Juan Bautista de los Pedernales, siguiendo el rumbo del norueste por tierra llana cerca de la playa, mui empastada. A poco pasamos esta punta y divisamos otra retirada que formava otra encenada. Luego de salidos empesamos a encontrar el campo mui florido de varias especies de flores de todos colores, que aunque en todo el camino y en la canal encontramos muchas flores, no con la abundancia que aquí, porque todo es una flor, mucha abundancia de blancas, amarillas, encarnadas, moradas, azules: muchas violetas, amarillas o alelías que se suelen sembrar en jardines, mucha espuela de cavallero, mucho cardosanto florido, mucha chía florida, y lo que más agraciava el campo, el ver de todos los colores de mui diferentes species (Brown 2001: 710–11).

12. "En la Punta de la Concepcion ramata la serrania, que venimos dexando á la derecha; y desde ella muda el terano mucho de aspecto: toda la tierra muy florida y verde con mucha variedad de zacates buenos pastos, y hierbas olorosas y utiles; y oy desde que salimos hasta parar vi mucho hinojo marino totalmente el mismo en la hoya, y en el gusto como el que se cria en las peñas y en las paredes en la marina de Cataluña con la diferencia, que se laventa el tronoco de la tierra como media vara, tiene la oja mas delgada, y estaba ajora muy florido con unas flores amarillas á modo de mirasoles pequenos, que con la abundancia

causaban una hermosa vista en todo el camino" (Web de Anza Archives, Font expanded diary, February 28, 1776).

13. "El camino, como todo, de lindo pais, verde, frondoso, florido, fertil, hermoso, y lucido" (Web de Anza Archives, Font expanded diary, March 10, 1776).

14. A member of the Buckeye family, Hippocastanaceae, California buckeye (*Aesculus californica*), is an abundant species in the Santa Cruz Mountains.

15. "Los campos son tan verdes de yervas, y tupidos de diversas flores campestres, como los de mas atras" (Web de Anza Archives, Anza diary, April 1, 1776).

16. In his journal Font recorded twenty-five days of rain during the 1775–76 Anza expedition (an asterisk [*] denotes observations of storm runoff): November 3; December 12, 13, 14, 15, 23, 26; January 2, 21, 23; February 5, 6, 7, 8, 9*, 10*, 11, 12, 13, 14*, 15, 17; March 9, 10, 30; May 4 (Web de Anza Archives, Font diaries). Additional evidence that 1775–76 was not a drought year was Font's observation that the road along Bautista Canyon in the San Jacinto Mountains was damaged by floods (May 5). He also recorded on May 4 that he still saw snow in the canyons of the sierra, a normal occurrence in late spring.

CHAPTER 3

1. The term "Californios" evolved from the early nineteenth century until statehood to describe a population of resident Californians of Hispanic and non-Hispanic origins. They were in alliance against Frémont and others in the California conquest.

2. *Los Angeles Times* (April 15, 1923), California's composite wild flowers (Frances M. Fultz), p. X-17.

3. *Los Angeles Times* (March 2, 1952), For the first time in years: Will we see wild flowers like these again? (Theodore Payne), p. F-24.

4. *Los Angeles Times* (March 21, 1932), Visitor views poppies named for his ancestor, p. 8; *Los Angeles Times* (April 4, 1991), Rain's golden payoff poppies: Spring is expected to bring an abundance of wildflowers in the Antelope Valley (David Colker), p. 7-A.

5. *Los Angeles Times* (January 30, 1890), p. 7.

6. *New York Times*, (July 15, 1880), n.p.

7. New York Times (October 21, 1872), Homeward journey—Colorado revisited (anonymous), n.p.

8. Frémont's crossing of the Mojave Desert is perhaps best evoked by his description of the people with him: "and still forced on south by a desert on one hand, and a mountain range on the other; guided by a civilized Indian, attended by two wild ones from the sierra; a Chinook from the Columbia [River]; and our own mixture of American, French, German—all armed; four or five languages heard at once; above a hundred horses and mules, half wild; American Indian and Spanish dresses and equipments intermingled—such was our composition. Our march was sort of a procession. Scouts ahead and on the flanks; a front and a rear division; the pack animals, baggage, and horned cattle, in the centre; and the whole stretching a quarter of a mile along our dreary path. In this form, we journeyed, looking more like we belong to Asia than the United States of America" (Frémont 1845: 259).

CHAPTER 4

1. *Los Angeles Times* (March 29, 1891), poem by Elizabeth Grinnell, p. 12.
2. *Los Angeles Times* (Aug 3, 1891), Weed, p. 5.
3. *Los Angeles Times* (May 15, 1905), Save the wild posies, from a paper by Alice Merrill Davidson read before the State Federation of Women's Clubs, p. II3.
4. *Los Angeles Times* (July 5, 1890), Brevities, p. 7.
5. *Los Angeles Times* (May 17 1896), Pomona brevities, p. 28.
6. *Los Angeles Times* (December 3, 1922), State Highway Commission asked to help fight weeds (T. C. Holt), p. IX9.
7. *Riverside Bee* (May 30, 1889), n.p.   .
8. *Los Angeles Times* (September 19, 1926), n.p.
9. *Los Angeles Times* (July 16, 1896), Our native flora: Some flowers of Los Angeles County (G. W. Dunn), p. 11.
10. *Los Angeles Times* (January 31 1907), Poppy time is coming, p. II4.
11. *Los Angeles Times* (April 1, 1908), Clean up, dress up, p. II4.
12. *Los Angeles Times* (May 20 1882), About town, p. 3.
13. *Los Angeles Times* (April 9, 1892), News and business, p. 8.
14. *Los Angeles Times* (May 24, 1887), The town of Lordsburg: Fertile in soil, salubrious in climate, delightful in scenery (Pasadena Union), p. 3.
15. *Los Angeles Times* (December 20, 1960), The Salvation of our beauties (Ed Ainsworth), p. B6; cf. *Los Angeles Times* (May 5 1968), Farm planned to propagate wild flowers, p. B4; *Los Angeles Times* (April 15, 1973), Wild flower power, photographs, p WS1.
16. *Los Angeles Times* (December 18, 1890), The state flower, p. 8.
17. *Los Angeles Times* (June 6, 1883), Graphic sketch of a rich and charming spot (Jeanne C. Carr), p. O2.
18. *Los Angeles Times* (February 23, 1888), A midwinter day—A true land of flowers, p. 3.
19. *Los Angeles Times* (January 2, 1888), Are you weary and heavy laden? Come and take a rest, p. 3.
20. *Los Angeles Times* (February 23, 1888), A midwinter day—A true land of flowers, p. 3.
21. *Los Angeles Times* (January 30, 1890), Editorials, p.7.
22. *Los Angeles Times* (April 12, 1890), Los Flores y las Niñas—A pretty wildflower festival, p. 3.
23. *Los Angeles Times* (March 19, 1893), p. 8.
24. *Los Angeles Times* (March 18, 1894), Times eagle, p. 19.
25. *Los Angeles Times* (March 10, 1895), The Saunterer, p. 21.
26. *Los Angeles Times* (April 28, 1895), The Saunterer, p. 23.
27. *Los Angeles Times* (February 23, 1898), Editorial, p. 6.
28. *Los Angeles Times* (January 10, 1900), All along the line, p. I9.
29. *Los Angeles Times* (March 8, 1901), Beautiful Santa Ana Valley, p. A5.
30. *Los Angeles Times* (March 17, 1902), Poppy field day, p. 13.
31. *Los Angeles Times* (February 14, 1903), Snowflakes fall on sunshine land, p. A1.
32. *Los Angeles Times* (January 17, 1904), The Eagle, p. 5.

33. *Los Angeles Times* (March 10, 1905), Poppies fail to pop: Golden dream in limbo, p. II9; *Los Angeles Times* (May 15, 1905), Save the wild posies, from a paper by Alice Merrill Davidson read before the State Federation of Women's Clubs, p. II3.

34. *Los Angeles Times* (February 18, 1906), Vandals strip poppy beds: Flower fields are despoiled about Pasadena, p. I9.

35. *Los Angeles Times* (February 18, 1906), Pen points, p. II4.

36. *Los Angeles Times* (February 7, 1909), Editorial pen points, p. II4.

37. *Los Angeles Times* (March 11, 1909), Where everlasting spring abides, p. II4.

38. *Los Angeles Times* (April 21, 1909), The land of beauty, p. II4.

39. *Los Angeles Times* (April 2, 1911), Pen points, p. II4.

40. *Los Angeles Times* (April 1, 1912), Auto parties, p. II2.

41. *Los Angeles Times* (May 13, 1917), Large party wild visit Antelope Valley, p. V1.

42. *Riverside Press Enterprise* (March 29, 1905),

43. *Los Angeles Times* (March 30, 1906), An auto tour in Owens Valley: Board of Public Works finds plenty of water in Owens River and views the line of the conduit. Much seen in trip, p. II7.

44. *Los Angeles Times* (April 19, 1912), Poppy day at Van Nuys, p. II8.

45. *Los Angeles Times* (April 28, 1912), Fields are ablaze: Hundreds seek Van Nuys district to gather golden poppies—Building activity in new town unabated, p. VI5.

46. *Los Angeles Times* (September 19, 1926), Wildflowers of southern California (Philip A. Munz), p. K23.

47. *Riverside Press Enterprise* (February 6, 1920), p. 5.

48. *Los Angeles Times* (April 17, 1921), Golden poppy may be found not far away, p. VII1.

49. *Los Angeles Times* (May 7, 1922), Wildflowers riot of bloom (Mary E. Walter), p. II1.

50. *Los Angeles Times* (May 14, 1922), Carpet of wild flowers: A riot of bloom, p. VI6.

51. *Los Angeles Times* (April 5, 1925), Ocean of color and riotous beauty in Central Valley's vast fields of wild flowers, p. 35.

52. *Los Angeles Times* (Feb 21, 1926), Flowers of the garden and desert in Banning Pass (Mary Henderson), p. J2.

53. *Los Angeles Times* (March 14, 1927), Wild flower fiesta, p. 10.

54. *Los Angeles Times* (March 15, 1927), Wild-flower day observed in Porterville, p. 7; Storm-washed land now in full bloom (San Fernando), p. 7.

55. *Los Angeles Times* (April 3, 1927), Blossoms intrigue motorist: Desert country in bloom to present picture with riotous color, p. G14.

56. *Los Angeles Times* (April 27, 1927), Pathway to pageant garden of wild flowers: Road to Hemet lined with blossoms to greet visitors to annual Ramona Play, p. 13; *Los Angeles Times* (April 17, 1927), Flowers of desert at height: Winter's heavy downpour brings wealth of color to carpet earth (Charles H. Bigelow), p. G1.

57. *Los Angeles Times* (March 24, 1928), Poppies at full glory in valley, p. 6.

58. *Los Angeles Times* (April 22, 1928), Rare flowers in full bloom, p. G1.

59. *Los Angeles Times* (March 17, 1929), Annual flower pageant begins: Many areas report number of varieties in bloom . . . , p. G7.

60. *Los Angeles Times* (March 24, 1929), Gay array of flowers greets eye: This year's outdoor show of wild blooms becoming more beautiful daily, p F1.

61. *Los Angeles Times* (March 28, 1930), Valley area ablaze with wild flowers, p. 13.

62. *Los Angeles Times* (March 8, 1931), Flower fiesta day announced: San Joaquin spectacle to be at best in a week; Kern County invites public to feast its eyes, p. A1.

63. *Los Angeles Times* (March 27, 1932), Valley blossoms lure (Alma Overholt), p. F4; *Los Angeles Times* (April 10, 1932), p. E4.

64. *Los Angeles Times* (March 28, 1932), Arvin flower fete thronged: Thousands visit festival at Historic Tejon Rancho; Vast panorama of glorious bloom greets visitors, p. 10.

65. *Los Angeles Times* (April 9, 1932), Ramona pageant's scene blooming, p. 4; *Los Angeles Times* (April 3, 1932), Acres of color: California's wildflower show (Helen W. King), p. J12.

66. *Los Angeles Times* (April 16, 1933), Myriads of acres of varied hued blossoms greet motorists throughout Southland, p. D2.

67. *Los Angeles Times* (March 18, 1934),Wild flowers of Kern thirsting for rainfall, p.29.

68. *Los Angeles Times* (March 4, 1934), Eternal life for the wildflower (G. B. Burtnett), p. H8.

69. *Los Angeles Times* (May 20, 1935), No faddist for collecting but thorough when I go in for foxtail, Lee side o'L.A (Lee Shippey), p. A4.

70. *Los Angeles Times* (March 3, 1935), Flowers now blooming, p. E2.

71. *Los Angeles Times* (March 10, 1935), Blooms now carpeting fields and hillsides (Lynn J. Rogers), p. E1.

72. *Los Angeles Times* (March 11, 1935), Wildflower festival, p. A4.

73. *Los Angeles Times* (April 7, 1935), Vari-colored blooms transforms thousands of acres in southern California: Great show is at best, p. E7.

74. *Los Angeles Times* (March 19, 1936), Along El Camino Real (Ed Ainsworth), p. 14.

75. *Los Angeles Times* (April 18, 1937), Southland's fields aflame with flowers (Lynn J. Rogers), p. F1.

76. *Los Angeles Times* (March 11, 1937), Arvin flower warning issues: Kern Chamber states poppy fields now are planted in potatoes, p. 14.

77. *Los Angeles Times* (January 12, 1937), Indio aroused as sheep destroy desert flowers, p. A5.

78. *Los Angeles Times* (April 2, 1938), Wild flower festival ready: Kern blooms at peak for event tomorrow at Shafter, p. 9.

79. *Los Angeles Times* (April 17, 1938), Wild flowers furnish spectacular display: Hills and dales of California garbed by nature in dress of brilliant hues, p. E1; *Los Angeles Times* (April 10, 1938), Flowers of brilliant hue carpet Southland: California hills and valleys present nature's pageant of spring blooms (Lynn J. Rogers), p. F1; *Los Angeles Times* (May 1, 1938), Desert areas garbed in colorful dress of California wildflowers (Lynn J. Rogers), p. F1.

80. *Los Angeles Times* (April 9, 1939), Southland in full bloom: Wildflowers dress countryside in spectacular beauty, p. F4; *Los Angeles Times* (April 16, 1939), Southland cloaked in riot of color (Lynn J. Rogers), p. F1.

81. *Los Angeles Times* (February 13, 1940), What goes on? (Chapin Hall), p. A. *Los Angeles Times* (February 26, 1940), Lee side o'L.A. (Lee Shippey), p. A4; *Los Angeles Times* (March 30, 1940); *Los Angeles Times* (March 21, 1940), It looks like a big year for wild flowers, p. 14.

82. *Los Angeles Times* (March 25, 1940), Death Valley's wild flowers most profuse in 15 years: Warm winter and rains speed desert carpet, p. 16

83. *Los Angeles Times* (March 21, 1940), It looks like a big year for wild flowers, p. 14.

84. *Los Angeles Times* (March 31, 1940), Wild flowers glorify Southland (Lynn J. Rogers), p. F1.

85. *Los Angeles Times* (March 25, 1940), Along El Camino Real (Ed Ainsworth), p. A2.

86. *Los Angeles Times* (April 7, 1940), Rain to hold wild flowers, p. F6.

87. *Los Angeles Times* (April 14, 1940), Hills wear floral garb, p. F8; *Los Angeles Times* (July 24, 1940), Along El Camino Real (Ed Ainsworth), p. 8.

88. *Los Angeles Times* (April 20, 1941), Floral paradise unfolds (Lynn J. Rogers), p. E1.

89. *Los Angeles Times* (April 10, 1941), Lee side o'L.A. (Lee Shippey), p. A4.

90. *Los Angeles Times* (April 27, 1941), Floral pageant reaches climax (Lynn J. Rogers), p. F1.

91. *Los Angeles Times* (March 16, 1941), Southland now abloom (Lynn J. Rogers), p. E1. *Los Angeles Times* (May 11, 1940, Pastoral trails of yore beckon motor travelers (Lynn J. Rogers), P. F1.

92. *Los Angeles Times* (April 29, 1942), Lee side o'L.A. (Lee Shippey), p. B4; *Los Angeles Times* (February 13, 1942), What goes on (Chapin Hall), p. 8.

93. *Los Angeles Times* (March 17, 1943), Wild flowers present greatest array, but—, p. 8.

94. *Los Angeles Times* (March 17, 1943), Wild flowers present greatest array, but—, p. 8.

95. *Los Angeles Times* (May 28, 1946), Lee side o'L.A. (Lee Shippey), p. A4; *Los Angeles Times* (March 15, 1947), Wild flowers bloom in Antelope Valley, p A3.

96. *The Desert Magazine* (March 1949), Gorgeous flowering season is forecast, p. 25.

97. *Los Angeles Times* (March 20, 1949), Coachella Valley bathed in color (Lynn Rogers), p. E8.

98. *Los Angeles Times* (April 17, 1949), Kern wild flowers out in profusion (Lynn Rogers), p. E1.

99. *Los Angeles Times* (March 26, 1949), Poppies crowded out by potatoes, p. 7.

100. *Los Angeles Times* (February 16, 1949), Lee side o'L.A. (Lee Shippey), p. A5.

101. *Los Angeles Times* (March 26, 1949), Wildflowers crowded out by potatoes: Kern County's once vast flower fields bow to agricultural advance, p. 7.

102. *Riverside Press Enterprise* (April 4, 1952), Spring "busts out" in wild-flowers all over, p. 9.

103. *Los Angeles Times* (April 20, 1952), Wild flowers color Kern County Hills (Lynn Rogers), p. A22.

104. *Los Angeles Times* (January 22, 1952), Planes plant wildflowers on Whittier Hills, p. 21.

105. *Los Angeles Times* (March 22, 1958), Desert expects banner crop of wildflowers, p. 9; *Los Angeles Times* (March 1, 1958), Season comes early to Mojave Desert: Death Valley has carpet of flowers, p. A5.

106. *Los Angeles Times* (April 4, 1958), Wild flowers blooming richly in Kern County, p. B7.

107. *Los Angeles Times* (April 6, 1958), Kern wild flower displays intrigue (Lynn Rogers), p. C8.

108. *Los Angeles Times* (March 1, 1959), Wildflower season is now under way (Woodward Radcliffe), p. K16.

109. *Los Angeles Times* (March 17, 1950), Spring flowers carpet Kern fields and foothills, p. 23; *Los Angeles Times* (March 26, 1950), Tour guides thousands to flower areas (Lynn Rogers), p. D12; *Los Angeles Times* (March 25, 1951), Wild flowers of Kern County attract many (Jimmy Radoumis), p. E4.

110. *Los Angeles Times* (April 7, 1950), Flowers cover hillsides at Palos Verdes, p. A7.

111. *The Desert Magazine* (February 1953), Wildflower shoots bear promise of brilliant early spring display, p. 29; *The Desert Magazine* (March 1953), Cactus blossoms to be most conspicuous desert flower in May, p. 26.

112. *Los Angeles Times* (March 5, 1953), Wildflower damage seen from sheep: Coachella Valley citizens protest grazing in reserves, p. A7.

113. *Los Angeles Times* (April 1, 1953), Timely showers bring crop of wild flowers, p. A6.

114. *Los Angeles Times* (March 11, 1954), Wild flowers color Kern County secluded areas, p. A6.

115. *Los Angeles Times* (April 1, 1956), Kern County tour holds attractions: Fort Tejon, wild flower fields and Isabella Lake intrigue visitors (Lynn Rogers), p. C1.

116. *Los Angeles Times* (April 22, 1959), Wild flowers bloom in Tehachapi Valley, p. C7; *Los Angeles Times* (April 26, 1959), Kern County flowers in lush display, p. C11.

117. *Los Angeles Times* (March 22, 1960), Spring wild flowers bloom in Kern County, p. 27.

118. *Los Angeles Times* (February 17, 1961), Kern County unveils wild flower preview, p. C9.

119. *Los Angeles Times* (March 7, 1961), Coalinga hillsides and canyons alive with color of wild flowers, p. 16.

120. *Los Angeles Times* (March 19, 1962), On the move (Ed Ainsworth), p.28.

121. *Los Angeles Times* (April 1, 1962), Outstanding wild flower displays anticipated throughout Southland, p. C11.

122. *Los Angeles Times* (April 19, 1962), Fine time for those wild about flowers (Matt Weinstock), p. A6.

123. *Los Angeles Times* (April 29, 1962), San Jacinto and Hemet heed call of *Ramona*, p. H1.

124. *Los Angeles Times* (May 13, 1963), Wild flowers splash colors in Kern County, p. 28; *Los Angeles Times* (March 10, 1963), On the move (Ed Ainsworth), p. 30.

125. *Los Angeles Times* (April 20, 1964), Big wild flower year seen for Kern County, p. 26.

126. *Los Angeles Times* (April 13, 1966), Wildflowers in full blossom from ocean to desert areas, p. B1.

127. *Los Angeles Times* (April 10, 1966), After the rains, an explosion of wild flowers, p. 30.

128. *Los Angeles Times* (March 20, 1967), Wildflowers making Southland a riot of color (Charles Hillinger), p. A6.

129. *Los Angeles Times* (March 2, 1969), Flowers to bloom in Antelope Valley (George Lowe), p. J10; *Los Angeles Times* (March 30, 1969), Wildflowers due in Kern County (George Lowe), p. H4. *Los Angeles Times* (April 27, 1969), The desert is blooming, p. Q78.

130. *Los Angeles Times* (April 20, 1969), Flowers abound in the desert (Ed Barnum), p. K10.

131. *Los Angeles Times* (April 11, 1971), Trip of the week: "Hi" to Hi Vista (George Lowe), p. L3; *Los Angeles Times* (March 12, 1972), Antelope Valley for blossoms (George Lowe), p. I2.

132. *Los Angeles Times* (May 21, 1971), Save poppies, Reagan urges, p. B2; *Los Angeles Times* (September 17, 1972), Preserve for poppies to be sought, p. J17.

133. *Los Angeles Times* (April 15, 1973), p. WS1.

134. *Los Angeles Times* (March 27, 1973), Southland says it with wildflowers (Ellen Shulte Rodman), p. F1; *Riverside Press Enterprise* (March 17, 1973), Bumper crop of poppies at Lake Mathews, photograph, p. B1.

135. *Los Angeles Times* (March 31, 1974), Antelope Valley in bloom, p. J17; *Los Angeles Times* (April 11, 1974), p. WS13.

136. *Los Angeles Times* (May 2, 1976), California poppy gets protected area (Tendayi Kumbula), p. D1.

137. *Los Angeles Times* (March, 31, 2001), Around the valley the experts agree: There's just no predicting poppies (Cristina Sauerwein), p. B1.

138. *Los Angeles Times* (April 21, 1975), Wildflowers 'n' gold mines (Michele Hannan), p. F13; *Los Angeles Times* (April 27, 1975), Extra adit attraction (Jack Smith), p. D1.

139. *Los Angeles Times* (March 13, 1977), It's time to view desert flowers (Bart Everett), p. H7.

140. *Los Angeles Times* (April 16, 1978), It's nice, Mother Nature, p. G4.

141. *Los Angeles Times* (April 25, 1978), High, wide and handsome: The Antelope Valley—Wandering through the wildflowers (Mark Jones), p. H6.

142. *Los Angeles Times* (March 25, 1979), It's flower power in the valleys (Michele and Tom Grimm), p. D11; *Los Angeles Times* (March 20, 1980), It's not blooming time for desert wildflowers, p. SD A5.

143. *Los Angeles Times* (April 6, 1980), Where to look at wildflowers (Michele and Tom Grimm), p. F7.

144. *Los Angeles Times* (September 23, 1954), In other times, p. A4.

145. *Los Angeles Times* (July 10, 1979), Blazes blacken 5200 acres, burn 4 homes (Tom Paegel and George Ramos), p. B3; *Los Angeles Times* (July 10, 1979), Officials hope to contain Lucerne fire by tonight, p. A3.

146. *Los Angeles Times* (July 2, 1980), Fires in Elysian Park, Mojave extinguished: Lightning and arson blazes—plus a "phantom"—pop up, p. A24; *Los Angeles Times* (July 17, 1980), Wind-whipped fire south of Palmdale fully contained (Eric Malnic), p. A3; *Los Angeles Times* (July 7, 1980), Some fires still burn but heat of weekend eases (Kristina Lindgren and Roger Smith), p. B3.

147. *Los Angeles Times* (July 29, 1980), Crews struggle to control fire above Palm Springs: 7700 acres scorched (Lorraine Bennett and Dorothy Townsend), p. A3.

148. *Los Angeles Times* (August 22, 1980), Fires in desert worst on record (Lorraine Bennett), p. B25.

149. *Los Angeles Times* (March 13, 1983), The season's early for wildflowers (Michele and Tom Grimm), p. I4.

150. *Los Angeles Times* (July 18. 1983), Four blazes char 1,000 acres (Ted Thackrey Jr. and John Oaks), p. C1.

151. *Los Angeles Times* (April 6, 1986), Lancaster Poppy Park is in bloom (Michele and Tom Grimm), p. 10.

152. *Los Angeles Times* (April 4, 1991), Rain's golden payoff poppies: Spring is expected to bring an abundance of wildflowers in the Antelope Valley; Tours are available at the reserve outside Lancaster (David Colker), Valley Edition, p. 7A.

153. *Los Angeles Times* (March 28, 1987), Wildflowers when the California deserts are in bloom, a particular species of nature lover faces its greatest challenge (Bob Sipchen), Home Edition, p.1.

154. *Los Angeles Times* (April 11, 1987), Time to wake up and smell the Antelope Valley poppies (John McKinney), Home Edition, p. 6.

155. *Los Angeles Times* (March 27, 1988), A spectacular show of wildflowers (Michele and Tom Grimm), Home Edition, p. 7.

156. *Los Angeles Times* (April 2, 1988), Life in Death Valley, Home Edition, p. 6.

157. *Los Angeles Times* (April 9, 1990), Drought drains poppy preserve of its bloom (Lynn O'Shaughinessy), Valley Edition, p. 3.

158. *Los Angeles Times* (March 29, 1992), Bloom' good poppies . . . soon the Lancaster poppy festival is just one excuse to celebrate spring's blush (Ellen Clark), Home Edition, p. 5.

159. *Los Angeles Times* (April 18, 1991),Wildflowers (Gloria Gray), Home Edition, p. 6.

160. *Los Angeles Times* (April 8, 1992), photograph and caption, San Diego County Edition, p. 3.

161. *Los Angeles Times* (March 28, 1993), Travel insider winter showers have brought April Flowers sightseeing: Here is one writer's pick of six wildflower-viewing destinations in southern California (Christopher Reynolds), Home Edition, p. 2.

162. *Los Angeles Times* (March 29, 1992), Bloom' good poppies . . . soon the Lancaster poppy festival is just one excuse to celebrate spring's blush (Ellen Clark), Home Edition, p. 5.

163. *Los Angeles Times* (April 19, 1992), They're back! In the Antelope Valley, the hills are alive with this year's colorful carpet of wildflowers, even if the poppy reserve is a disappointment (Robin Abcarian), Home Edition, p. 1.

164. *Los Angeles Times* (March 28, 1993), Travel insider winter showers have brought April Flowers sightseeing: Here is one writer's pick of six wildflower-viewing destinations in southern California (Christopher Reynolds), Home Edition, p. 2.

165. *Riverside Press Enterprise* (April 16, 1993), Intruder grass crop, n.p.

166. *Los Angeles Times* (March 19, 1994), Desert bloom park's brilliant poppies herald the arrival of spring (Phil Sneiderman), Valley Edition, p. 1.

167. *Los Angeles Times* (April 2, 1995), Popular native poppies crown Antelope Valley hills with gold, Valley Edition, p. 2.

168. *Los Angeles Times* (April 21, 1995), Golden opportunity to see state flower: Brilliant poppy blooms will draw thousands to the two-day event, featuring crafts, a carnival and races (Sharon Moeser), Valley Edition, p. 32.

169. *Los Angeles Times* (June 10, 1995), Santa Clarita/Antelope Valley controlled burn planned for poppy field Lancaster: Tuesday's experiment aims to destroy plants that impede the flower's growth (Phil Sneiderman), Valley Edition, p. 22.

170. *Los Angeles Times* (March 10, 1998), Predicting poppies is impossible; Wildflowers: It's hard to say whether the fields will pop with color, experts say; But throngs of visitors are expected at the Lancaster reserve anyway (David Colker), Valley Edition, p. 1.

171. *Los Angeles Times* (April 18, 1996), Valley weekend: The bloom is off the poppy this year; Due to a warm winter, visitors will find few flowers at the California Reserve in the Antelope Valley (Bill Locey), Valley Edition, p. 1B.

172. *Los Angeles Times* (March 20, 1997), Wild things, we think we love you: Its wildflower season, but better act quickly—it could be one of the shortest in recent years (Angie Chuang), Home Edition, p. 50.

173. *Los Angeles Times* (March 16, 1998), Wildflowers transform State's desert; Nature: Normally arid area is exploding with color; Some call it the bloom of the century (Diana Marcum), Home Edition, p. 3.

174. *Los Angeles Times* (March 22, 1998), Anza-Borrego: Blooms day; Desert brightens with flowers and new B&B (John McKinney), Home Edition, p. 4.

175. *Los Angeles Times* (March 1, 1998), Desert flowers go wild this year, Home Edition, p. 3.

176. *Los Angeles Times* (April 13, 1998), It's a jungle out there, Orange County Edition, p. 1.

177. *Los Angeles Times* (May 31, 1999), Joshua Tree to remain open despite fire; Blaze: Although 13,000 acres have been blackened, firefighters expect full containment by Tuesday; Officials say lightning sparked flames, confined to northwest area of park, Home Edition, p. 3.

178. *Los Angeles Times* (March 31, 2001), The experts agree: There's just no predicting poppies, Valley Edition, p. B1.

179. *Los Angeles Times* (April 30, 2001), Painted ladies are leaving their mark; Insects: Wildflowers blossoming after heavy rains provide a banquet for migrating butterflies, which often end up splattered on cars (Scott Gold), Home

Edition, p. A3; *Riverside Press Enterprise* (April 18, 2001), Colors of spring: Late rains and sunshine combined to provide what is being the best flower season since the El Niño years of the mid-1990s (Thomas Buckley), p. B1.

180. *Los Angeles Times* (June 23, 2002), Chain reaction of thirst in California desert dry spell: Wildlife, scores of plant and animal species are suffering in the longest drought on record (Janet Wilson), Home Edition, p. B8.

181. *Los Angeles Times* (March 12, 2003), For Lancaster's nature enthusiasts, its poppy love once more: Some Antelope Valley dwellers are giddy about the return of the state flower, who absence last year was a serious blow to local tourism (Richard Fausset), Home Edition, p. B3.

182. *Los Angeles Times* (March 8, 2005), Brief, beautiful rebirth: Desert is teeming with wildflowers after record rainfall (Louis Sahagun), Home Edition, p. B1; *Los Angeles Times* (March 31, 2005), Lush with flowers, lousy with tourists: Death Valley's stunning bloom draws throngs, straining roads, motels and park services (Louis Sahagun), Home Edition, p. A1.

183. Riverside *Press Enterprise* (March 10, 2005), Wildflowers are blooming: Resurrection of beauty (Pat O'Brien), p. A7.

184. *Los Angeles Times* (June 26, 2005), Crews making headway against fires raging in Mojave National Preserve: The five blazes are 65% contained after burning 65,200 acres and razing five homes, six trailers (Louis Sahagun), Home Edition, p. B3; *Los Angeles Times* (July 18, 2006), High humidity, rain help damp fires' spread: Officials expect the San Bernardino County blazes—which have burned over 85,000 acres over nine days—to be contained tonight (Maeve Reston), Home Edition, p. B7.

185. *Los Angeles Times* (June 19, 2005), Wildfire threatens nearly 300 homes, Home Edition, p. B11.

186. *Los Angeles Times* (July 21, 2006), Lightning sparks several new desert fires, Home Edition, p. B3; *Los Angeles Times* (July 25, 2006), Low wind helps crews fight to contain new brush fires, Home Edition, p. B7.

187. MODIS Rapid Response System, Real Time, July 12, July 21, 2006, http://rapidfire.sci.gsfc.nasa.gov.

CHAPTER 5

1. *Los Angeles Times* (April 16, 1889), Flower festival: Preparations for the grand opening tonight; A small army of decorators at work—additional list of ladies who are to preside over the booths—Advance notes, p. 4.

2. *Los Angeles Times* (April 18, 1889). Flower festival: The second night at hazard's pavilion; A vivid description of Wilson's Peak—New members added to the various committees—An excellent programme carried out last night, p. 2.

Location of Franciscan campsites, Franciscan place names, and modern place names

PORTOLÁ EXPEDITION (1769–1770), SAN DIEGO TO MONTEREY
AND SAN FRANCISCO AND RETURN TO SAN DIEGO, DIARIES OF
CRESPÍ, COSTANSÓ, AND FAGES

	Place name in Franciscan diary or landscape description	Modern place name or landscape description	Latitude (° n)	Longitude (° w)
Prior to July	Misión San Diego	Mission San Diego	32° 47'	117° 06'
July 14	Los Pocitos . . . San Diego	La Jolla	32 50	117 13
July 15	San Jácome . . .	Carmel Valley	32 55	117 15
July 16	Cañada de San Alejo	San Elmo Lagoon	33 01	117 17
July 17	El Beato Simón . . .	Carlsbad	33 10	117 19
July 18–19	San Juan Capistrano	San Luis Rey	33 13	117 20
July 20	Santa Margarita	Santa Margarita Creek	33 17	117 22
July 21	Los Rosales de Santa Práxedis	SW base of Santa Margarita Mtns.	33 20	117 25
July 22	San Apolinario	San Mateo Canyon	33 25	117 33

Place name in Franciscan diary or landscape description	Modern place name or landscape description	Latitude (° n)		Longitude (° w)	
July 23	Santa Maria Magdalene	San Onofre	33	30	117	40
July 24–25	San Francisco Solano	San Juan Capistrano, El Toro	33	37	117	42
July 26	San Pantaleón	N of El Toro	33	41	117	45
July 27	Santiago Apostol	Santiago Creek	33	47	117	49
July 28	Rio de los Temblores	Santa Ana River, Placentia	33	51	117	52
July 29	Santa Maria	S base of Puente Hills	33	57	117	53
July 30	Rio La Puente . . . San Miguel	San Gabriel River, La Puente	34	02	118	01
July 31– August 1	Rio La Puente . . . San Miguel	San Gabriel River, La Puente	34	04	118	04
August 2	Rio Porciúncula, volcanoes of pitch	Los Angeles River, La Brea tar pits	34	05	118	15
August 3	San Esteban	Ballena Creek	34	05	118	20
August 4	San Rogerio	UCLA	34	04	118	28
August 5–6	wide valley, Santa Catalina plain	Encino, San Fernando Valley	34	10	118	30
August 7	hollow, N end of Santa Catalina plain	Sylmar, San Fernando Pass	34	19	118	30
August 8–9	San Largo . . .	Santa Clarita, Valencia	34	25	118	35
August 10	3 leagues [W]	Santa Clara River	34	25	118	44
August 11	Santa Clara	Santa Clara River	34	25	118	50
August 12	San Pedro Moliano	Fillmore	34	24	118	55
August 13	Los Santos Mártires . . .	10 km WSW of Santa Paula	34	21	119	07
August 14	La Asunción de María . . .	Ventura	34	16	119	16
August 15	Santa Cunegundis	Pitas Point	34	19	119	23

Date	Name	Location	Lat °	Lat ′	Lon °	Lon ′
August 16	Santa Clara de Monte . . .	Rincon Point	34	23	119	28
August 17	La Carpintería	Carpinteria	34	24	119	31
August 18	San Joaquín	Santa Barbara	34	25	119	39
August 19	La Laguna de San Joaquín	Santa Barbara	34	25	119	40
August 20	Puebla Islado	Golita, Isla Vista	34	25	119	52
August 21–22	San Luis Obispo	Dos Pueblos Canyon	34	27	119	58
August 23	San Gűido	Refugio State Beach	34	28	120	05
August 24	Gaviota	Gaviota	34	28	120	14
August 25	San Ceferino Mártir	15 km W of Point Conception	34	28	120	20
August 26	La Concepción	5 km W of Point Conception	34	28	120	25
August 27	La Concepción	5 km N of Point Conception	34	29	120	29
August 28	San Juan Bautista,	10 km E of Point Arguello	34	33	120	34
August 29	Pedernales, Santa Rosalia	Point Pedernales	34	36	120	38
August 30	Los Santos Mártires	Santa Inez River at beach	34	41	120	35
August 31	San Ramón Nonnato	San Antonio River at beach	34	48	120	37
September 1	Los Santos Mártires	Guadalupe oil field	34	58	120	38
September 2–3	Del Oso Flaco	Oso Floco Lake	35	02	120	37
September 4	San Ladislao, El Buchón	Pismo Creek	35	08	120	37
September 5–6	Santa Elena	San Luis Obispo Creek	35	11	120	42
September 7	Llano de los Osos	Los Osos Valley	35	18	120	47
September 8	high round island rock	Morro Bay, Morro Rock	35	23	120	51
September 9	Santa Serafina	NW of Cayucos State Beach	35	28	120	59
September 10	pine groves, San Benvenuto	Cambria	35	33	121	06

Date	Place name in Franciscan diary or landscape description	Modern place name or landscape description	Latitude (° n)		Longitude (° w)	
September 11	San Nicolás	San Simeon	35	37	121	12
September 12	San Juan de Dukla	Point Piedras Blancas	35	40	121	17
September 13–15	Santa Humiliana	Ragged Point	35	46	121	19
September 16	two tall mountain ranges	San Carpoforo Creek	35	47	121	16
September 17–19	top of mountain	Upper San Carpoforo Creek	35	49	121	16
September 20	Los Piñones	Burro Mountain	35	52	121	16
September 21–23	River	Nacimiento River	35	53	121	15
September 24	Las Llagas . . .	San Antonio River	35	58	121	11
September 25	several leagues north	Jolon Valley	36	02	121	12
September 26	Rio Carmelo, San Elzeario	King City	36	12	121	08
September 27	5 hours [N] over level land	Greenfield	36	21	121	17
September 28	4 leagues NW	Soledad	36	26	121	22
September 29	4 leagues NW	Chular	36	34	121	32
September 30	view of *Point of Pines*, *Monte Rey*	Salinas	36	39	121	42
October 1–6	Santa Delphina	Salinas River at beach	36	42	121	47
October 7	2 leagues N, Santa Brígida	Castroville	36	46	121	44
October 8–9	Rio del Pájaro	Pajaro River, Watsonville	36	53	121	45
October 10–14	Rio del Pájaro	Pinto Lake	36	57	121	46
October 15	Santa Teresa	Corralitos Creek	36	57	121	47
October 16	El Beato Serafino . . .	Soquel	36	58	121	57

October 17	Rio San Lorenzo, full flowing river	Santa Cruz, San Lorenzo River	36	58	122	01
October 18	stream named Santa Cruz	10 km W of Santa Cruz	36	59	122	09
October 19	San Pedro de Alcántara	Scott Creek	37	03	122	14
October 20–22	San Luis Beltrán	Waddell Creek	37	06	122	17
October 23	*Punta Año Nuevo*	Point Año Nuevo, Gazos Creek	37	10	122	22
October 24–26	San Pedro Regalado	San Gregorio State Beach	37	20	122	24
October 27	stream of San Ivo	5 km S of Half Moon Bay	37	24	122	25
October 28–29	Las Almejas	Half Moon Bay	37	28	122	26
October 30	Punta del Ángel Custodial	Moss Beach	37	31	122	30
October 31–November 3	bay of San Francisco	Pacifica	37	35	122	29
November 5	close to a lake	Crystal Springs Reservior	37	30	122	20
November 6–10	bay of San Francisco	Palo Alto	37	26	122	13
November 11	N 2 leagues	Redwood City	37	27	122	15
November 12	kept through same hollow	San Andreas Lake	37	38	122	27
November 13	near end of bay	Sweeney Ridge, San Pedro Creek	37	37	122	30
November 14	Punta del Ángel Custodial	Moss Beach	37	31	122	30
November 15–16	Las Almejas	Half Moon Bay	37	28	122	27
November 17	deep arroyo, 3 leagues S	Tunitas Creek	37	21	122	25
November 18	3 leagues S, San Ibón	Pescadero State Beach	37	15	122	25
November 19	*Año Nuevo*	Point Año Nuevo	37	07	122	19
November 20	San Luis Beltrán	Scott Creek	37	02	122	14

	Place name in Franciscan diary or landscape description	Modern place name or landscape description	Latitude (° n)		Longitude (° w)	
November 21	stream of San Lucas	W of Santa Cruz	36	59	122	09
November 22	Rio San Lorenzo	Santa Cruz, San Lorenzo River	36	58	122	01
November 23	Paraje del Corral	Corralitos Valley, Pinto Lake	36	57	121	51
November 24	Ranchería del Pájaro	Pajaro River, Watsonville	36	53	121	47
November 25	we rested	Pajaro River, Elkhorn Slough	36	50	121	45
November 26–27	Santa Delphina	Salinas River	36	43	121	45
November 28	*Monte Rey*	Monterey	36	36	121	53
November 28–December 10	*Monte Rey*	Monterey, Carmel	36	33	121	55
December 11	Santa Delphina	Fort Ord	36	41	121	45
December 12	4½ leagues ESE on R. San Elzeario	Chualar	36	34	121	32
December 13	3½ leagues ESE on same river	Soledad, Camphora	36	25	121	20
December 14	4 leagues SE	Greenfield	36	20	121	15
December 15	San Elzeario	King City	36	12	121	08
December 16	Las Llagas . . .	Jolon Valley	36	02	121	12
December 17	3 leagues S in high Santa Lucia Mtns.	Nacimiento River	35	53	121	15
December 18	Los Piñones	Burro Mountain	35	52	121	16
December 19–20	Santa Lucia de Salerno	upper San Carpoforo Creek	35	49	121	16
December 21–22	Santa Humiliana	near Point Piedras Blancas	35	41	121	20
December 23	El Laurel, San Juan de Dukla	San Simeon	35	37	121	12
December 24	El Osito	Cambria	35	33	121	06

December 25	Ensenada del Morro	Cayucos State Beach	35	28	120	59
December 26	high mountain rock	Morro Bay	35	23	120	51
December 27–28	El Buchón	San Luis Obispo	35	17	120	41
December 29	San Ladislao village	Pismo Beach	35	08	120	37
December 30	two lakes	Oso Flaco Lake	35	02	120	37
December 31	long lake	Guadalupe oil field	34	58	120	38
January 1, 1770	San Ramon Nonnato	San Antonio River at beach	34	48	120	37
January 2	large river of Santa Rosa	Santa Ynez River at beach	34	41	120	35
January 3	Pedernales	Point Pedernales	34	36	120	38
January 4	*La Concepción*	NW of Point Conception	34	32	120	33
January 5	*La Concepción*	E of Point Conception	34	28	120	20
January 6	Gaviota, San Luis Rey de Francia	Gaviota	34	28	120	13
January 7	San Gűido	Refugio State Beach	34	28	120	06
January 8	3 leagues, town of San Luis Obispo	Dos Pueblos Canyon	34	27	119	57
January 9	island town	Isla Vista, Santa Barbara	34	25	119	40
January 10	Carpintería	Carpinteria	34	24	119	31
January 11	La Asunción de María . . .	Ventura	34	17	119	17
January 12	San Hipólito, shortcut to LA	Camarillo	34	13	119	05
January 13	Los Santos Reyes	Newbury Park	34	11	118	56
January 14	El Triunfo	Thousand Oaks, El Triunfo	34	10	118	52
January 15	Santa Catalina de Bononia	Encino	34	10	118	30
January 16	Los Santos Mártires . . .	North Hollywood	34	08	118	28
January 17	La Porciúncula	Los Angeles River	34	05	118	15

January 18	Rio . . . los Temblores	Santa Ana River, Placentia	33	51	117	52
January 19	San Pantaleón, 4½ leagues SE	N of El Toro	33	41	117	45
January 20	San Francisco Solano	San Juan Capistrano	33	30	117	40
January 21	Stream of Santa Inés	San Mateo Canyon	33	25	117	34
January 22	San Juan de Capistrano	San Luis Rey	33	13	117	20
January 23	San Alejo	San Elmo Lagoon	33	01	117	17
January 24	Misión San Diego	Mission San Diego	32	47	117	06

SOURCES: Dates and Franciscan-diary place names from Brown 2001; Bolton 1927; cf. Minnich and Franco-Vizcaíno 1998.
NOTE: Italic place names are derived from the 1602–03 Vizcaíno voyage.

SECOND JOURNEY OF CRESPÍ (1770), SAN DIEGO TO MONTEREY

	Place name in Franciscan diary or landscape description	Modern place name or landscape description	Latitude (° n)		Longitude (° w)	
April 17	Santa Isabel Val., 4 leagues NNW	Soledad Valley	32°	55'	117°	14'
April 18	Beato Simón	Carlsbad	33	05	117	19
April 19	San Juan de Capistrano	San Luis Rey	33	13	117	20
April 20	Los Rosales de Santa Praxedis	Camp Pendleton	33	21	117	27
April 21	Santa Maria Magdalena	San Onofre	33	23	117	34
April 22	San Francisco Solano	San Juan Capistrano	33	30	117	40
April 23	Rio . . . los Temblores	Santa Ana River, Placentia	33	51	117	52
April 24–25	San Miguel River	San Gabriel River	34	02	118	00
April 26	Porciúncula, Los Santos Martires . . .	Los Angeles River, La Brea tar pits	34	05	118	21
April 27	Santa Cata-lina . . , S side	Encino	34	10	118	30
April 28	El Triunfo	El Triunfo	34	09	118	50

April 29	San Higino	Ventura plain	34	14	119	07
April 30	La Conversión River . . .	Ventura River	34	15	119	14
May 1	Carpintería	Carpinteria	34	24	119	31
May 2	4 leagues W, San Joaquin town	Goleta, Isla Vista	34	25	119	52
May 3	near San Gűido	Dos Pueblos Canyon	34	27	119	57
May 4	Gaviota, San Luis Rey de Francia	Gaviota	34	28	120	13
May 5	*Punta la Concepción*	near Point Conception	34	27	120	22
May 6	Pedernales	Point Arguello, Pedernales	34	34	120	38
May 7	San Berardo River	Santa Ynez River at beach	34	41	120	35
May 8	El Buchón	San Antonio River at beach	34	48	120	37
May 9	lakes and dunes	Guadalupe oil field	34	58	120	38
May 10–11	3 leagues past the 3 lakes	Oso Flaco Lake	35	02	120	37
May 12	San Ladislao	San Luis Obispo	35	17	120	41
May 13	San Ladislao, bear level	Osos Valley	35	18	120	47
May 14	San Benvenuto pinewood	Cambria	35	33	121	06
May 15	San Juan de Dukla	San Simeon	35	37	121	12
May 16	Santa Humiliana	Ragged Point	35	46	121	19
May 17–18	Santa Lucia Mtns., Los Piñones	Burro Mountain	35	52	121	16
May 19	Las Llagas . . .	Jolon Valley	36	02	121	12
May 20	San Elzeario	King City	36	12	121	08
May 21	kept along river 5 leagues	Soledad	36	25	120	20
May 22	embayment, Point of Pines	Salinas River at beach	36	40	120	48
May 23	Santa Delphina	Fort Ord	36	38	121	48
May 24	*Monte Rey*	Monterey	36	36	121	53

SOURCE: Dates and Franciscan-diary place names from Bolton (1927); Brown (2001).
NOTE: Italic place names are derived from the 1602–03 Vizcaíno voyage.

JOURNEY OF CRESPÍ AND FAGES (1772), MONTEREY
TO SACRAMENTO RIVER, DIARIES OF CRESPÍ AND FAGES

	Place name in Franciscan diary or landscape description	Modern place name or landscape description	Latitude (° n)		Longitude (° n)	
March 19	*Monterey*	Monterey	36°	36'	121°	52'
March 20	Santa Delfina	Salinas River, near Salinas	36	40	121	38
March 21	San Benito	San Juan Bautista	36	51	121	30
March 22	two broad valleys, 8 leagues	Las Llagas Creek, Morgan Hill	37	05	121	39
March 23	pass, arroyo w. water	Milpitas	37	27	121	55
March 24	parallel with . . . head of estuary	Fremont	37	33	122	04
March 25	8 leagues from point of estuary,	Hayward, San Lorenzo Creek	37	42	122	07
March 26	arm of estuary	Oakland	37	45	122	11
March 27	immense plain, *Farallons*	Oakland, Berkeley	37	49	122	20
March 28	great lake, 8 leagues extent	Richmond, San Pablo Bay	38	00	122	20
March 29	stretch of water ¼ league wide	Concord/Martinez, Carquinez Strait	38	01	122	07
March 30	great plain as level as palm of hand 3 arms of great river	Pittsburg, Antioch, Sacramento River	38	00	121	45
March 31	SSE, pass, beautiful valley	Walnut Creek, Danville	37	50	121	57
April 1	S, same valley 10 leagues	Pleasanton	37	40	121	52
April 2	S, valley of great estuary, SE 5 leagues	Fremont, Milpitas	37	33	122	00
April 3	hills between bay and Val. San Bernardino	Gilroy, Morgan Hill	37	05	121	38

| April 4 | San Benito | San Juan Bautista | 36 | 51 | 121 | 30 |
| April 5 | *Monterey* | Monterey | 36 | 36 | 121 | 52 |

SOURCE: Dates and Franciscan-diary place names from Bolton (1927).
NOTE: Italic place names are derived from the 1602–03 Vizcaíno voyage.

JOURNEY OF PALÓU (1774), MONTEREY TO SAN FRANCISCO

Place name in Franciscan diary or landscape description	Modern place name or landscape description	Latitude (° n)		Longitude (° w)	
November 21	*Monterey*	Monterey	36° 36'	121° 53'	
November 22–23	Santa Delfina River	Salinas River, near Salinas	36 40	121 38	
November 24	San Benito	site of Mission San Juan Bautista	36 52	121 30	
November 25	NNE, Pajaro R., Las Llagas	Las Llagas, near Morgan Hill	37 05	121 39	
November 26	S edge of large plain	San Jose	37 17	121 58	
November 27	arroyo with water	Los Gatos Creek, near Cupertino	37 17	122 03	
November 28	Crespí site of Dec [= Nov] 6–10	Palo Alto	37 25	122 10	
November 29	narrow lake, Val. San Andres	Crystal Springs, San Andreas Lake	37 33	122 22	
Nov 30– Dec 3	followed lake, near high hill	San Bruno Mountain	37 40	122 27	
December 4	lake, explored S. Francisco bay	Lake Merced	37 43	122 29	
December 5	canyon S of a lake	Laguna Salada, San Pedro Valley	37 36	122 30	
December 6–7	avoids cliff coast, arroyo	Half Moon Bay	37 28	122 26	
December 8	steep descent, two streams	Pescadero Beach	37 15	122 25	
December 9	*Punta Año Nuevo*	Point Año Nuevo	37 07	122 19	
December 10	2 arroyos a league apart, plain	near Davenport	37 00	122 10	
December 11	Laguna del Corral, camp of Crespí	Corralitos Valley NW of Watsonville.	36 50	121 41	

| December 12 | Santa Delfina R. | Salinas River | 36 | 40 | 121 | 38 |
| December 13 | *Monterey* | Monterey | 36 | 36 | 121 | 52 |

SOURCE: Dates and Franciscan-diary place names from Bolton (1926).
NOTE: Italic place names are derived from the 1602–03 Vizcaíno voyage.

FIRST ANZA EXPEDITION (1774), YUMA TO MISSION
SAN GABRIEL AND MONTEREY AND RETURN TO YUMA,
DIARIES OF ANZA, DÍAZ, AND GARCÉS

	Place name in Franciscan diary or landscape description	Modern place name or landscape description	Latitude (° n)		Longitude (° w)	
January 9	Colorado River, rivers unite	Colorado River, near Yuma, Arizona	32°	46'	114°	35'
February 10	Colorado River	Pilot Knob	32	43	114	45
February 11	small lake near river	Ciudad Morelos, Baja California	32	37	114	50
February 12	Santa Olaya	Benito Juárez, Baja California	32	36	115	00
February 13	7 leagues W	Queretaro, Baja California	32	35	115	08
February 14	one league W	Queretaro, Baja California	32	35	115	10
Febuary 15	dunes, returned to L. Olaya	Benito Juárez, Baja California	32	37	115	00
March 3	near camp of February 15	Mexicali	32	38	115	26
March 4	6 leagues W	Cerro Prieto	32	38	115	30
March 5	dunes, estuary of the sea	N end of Laguna Salada	32	34	115	45
March 6	Santo Thomás	Pinto Wash	32	40	115	49
March 7	4 leagues	Yuha Wash	32	43	115	55
March 8	Santa Rosa de las Lajas	Yuha Well	32	46	115	58
March 9	4 leagues N	N of Plaster City, Carrizo Wash	32	55	115	55
March 10	very large marsh	near Harpers Well	33	03	115	57
March 11	W 2 leagues	Fish Creek	33	05	116	03
March 12–13	San Gregorio	Borrego Valley	33	10	116	15

March 14	Santa Catharina	Coyote Canyon	33	23	116	26
March 15	Puerta Real de San Carlos	Terwilliger Valley	33	30	116	37
March 16	El Principe	Anza Valley	33	33	116	40
March 17	San Patricio	Tripp Flat	33	36	116	45
March 18	Valle San José	Hemet/San Jacinto Valley	33	44	116	57
March 19	lake, geese, San . . . Bucareli	Mystic Lake	33	47	117	04
March 20	Rio Santa Anna	Santa Ana River, Riverside	33	57	117	26
March 21	Los Osos	San Antonio Creek	34	02	117	40
March 22– April 9	Misión San Gabriel	San Gabriel	34	07	118	06
April 10	Porciúncula River	Los Angeles River	34	05	118	17
April 11	Carpintería	Carpinteria	34	24	119	31
April 12	Mextletlitán	Santa Barbara	34	25	119	40
April 13	Pedernales	Point Pedernales,	34	36	120	38
April 14	*Point Concepcion,* Rio Santa Rosa	Point Conception, Santa Ynez River	34	41	120	27
April 15	Misión San Luis	San Luis Obispo	35	17	120	41
April 16	Nacimiento R.	Nacimiento River	35	48	120	53
April 17	Misión San Antonio	Mission San Antonio	35	53	121	03
April 18–19	*Monterey,* Precideo	Monterey	36	36	121	53
April 20	Misión Carmel	Carmel	36	33	121	55
April 21	*Monterey*	Monterey	36	36	121	53
April 22	Los Ossitos	Soledad	36	26	121	32
April 23	Misión San Antonio	Mission San Antonio	35	53	121	03
April 24	R. Nacimiento	Nacimiento River	35	48	120	53
April 25	Misión San Luis	San Luis Obispo	35	17	120	41
April 26	Rio Santa Rosa	Santa Ynez River	34	41	120	27
April 27	met Fray Junipero Serra	near Point Conception	34	29	120	27
April 28	Dos Rancherias	W Santa Barbara plain	34	26	119	50
April 29	La Cuesta	Camarillo	34	13	119	03
April 30	R. Porciúncula	Los Angeles River	34	05	118	17
May 1–2	Misión San Gabriel	Mission San Gabriel	34	07	118	06

May 3	Arroyo de los Olisos	San Antonio Creek	34	02	117	40
May 4	Lake of San ... Bucareli	Mystic Lake	33	47	117	04
May 5	San Patricio	Tripp Flat	33	36	116	45
May 6	Santa Catharina	Coyote Canyon	33	20	116	25
May 7	San Gregorio	Borrego Valley	33	15	116	20
May 8	approaching Santa Olaya	near Mexicali	32	40	115	30
May 9	Santa Olaya, Colorado R.	Benito Juárez, Colorado River	32	36	115	00

SOURCE: Dates and Franciscan-diary place names from Web de Anza Archives, Bolton (1930a,b, 1933).
NOTE: Italic place names are derived from the 1602–03 Vizcaíno voyage.

SECOND ANZA EXPEDITION (1775–1776), YUMA TO MISSION
SAN GABRIEL AND MONTEREY, SAN FRANCISCO TO ANTOCH,
AND RETURN TO YUMA, DAIRY OF FONT

	Place name in Franciscan diary or landscape description	Modern place name or landscape description	Latitude (° n)		Longitude (° w)	
November 30–December 3	Colorado River, 3 branches	NE of Yuma, Arizona	32°	50'	114°	30'
December 4	pond	Yuma	32	46	114	35
December 6–8	Lake Ollaya	. Benito Juárez, Baja California	32	36	115	00
December 9	7 leagues WNW	Queretaro, Baja California	32	35	115	08
December 10	7 leagues WNW	Puebla, Baja California	32	35	115	20
December 11	Pozo de Santa Rosa	Mexicali	32	38	115	26
December 12	3 leagues N	Plaster City, Coyote Wash	32	45	115	55
December 13–17	7 long leagues NNW, San Sabastián	Superstition Mountain, Harpers Well	33	04	115	57
December 18	valley extends to Sierra Madre	San Felipe Creek, Ocotillo Wells	33	09	116	08
December 19	San Gregorio	Borrego Valley	33	15	116	20
December 20–22	Santa Catharina	Beatties Ranch, Borrego Valley	33	20	116	25

December 23	canyon crosses the Sierra Madre	Coyote Canyon	33	22	116	26
December 24–25	mountains of rock	Fig Tree Valley, upper Coyote Canyon	33	27	116	31
December 26	Puerto de San Carlos	Terwilliger Valley	33	30	116	37
December 27–28	Valle San Patricio	Tripp Flat	33	36	116	45
December 29	Valle San Joseph	Hemet, San Jacinto Valley	33	44	116	57
December 30	lake full of geese	Mystic Lake	33	47	117	04
December 31	R. Santa Ana	Santa Ana River, Riverside	33	57	117	26
January 1, 1776	R. Santa Ana	Santa Ana River, Riverside	33	57	117	27
January 2	Arroyo de los Alisos	San Antonio Creek	34	02	117	40
January 3	arroyo that joins San Gabriel R.	San Gabriel Valley	34	04	117	57
January 4–6	Misión San Gabriel	Mission San Gabriel	34	07	118	06
January 7	Santa Ana River	Santa Ana River, Placentia	33	51	117	52
January 8	Arroyo Santa Maria Magdelena, La Quema	San Juan Capistrano	33	33	117	40
January 9	San Juan Capistrano River	NW of San Luis Rey	33	17	117	23
January 10	La Soledad	Soledad Valley	32	55	117	14
January 11–February 8	Misión San Diego	Mission San Diego	32	47	117	06
February 9	Agua Hedionda	Carlsbad	33	08	117	19
February 10	near La Quema	Capistrano	33	30	117	40
February 11	Rio Santa Ana	Santa Ana River, Placentia	33	51	117	52
February 12–20	Misión San Gabriel	Mission San Gabriel	34	07	118	06
February 21	Porciúncula River	Glendale	34	10	118	19
February 22	Agua Escondida, little lake	Encino	34	10	118	30
February 23	Islands of Santa Barbara-	Newbury Park	34	10	118	54

	Place name in Franciscan diary or landscape description	Modern place name or landscape description	Latitude (° n)		Longitude (° w)	
February 23	Santa Clara River	Oxnard, Santa Clara River	34	15	119	10
February 24	La Rinconada	Rincon Point	34	23	119	27
February 25	Village of Mescaltitán	Santa Barbara	34	25	119	40
February 26	Ranchería Nueva	Las Llagas Canyon	34	27	119	59
February 27	Village of El Coxo	Point Conception	34	28	120	28
February 28	*Punta Concepción*, Pedernales	Santa Ynez River at beach	34	41	120	37
February 29	Laguna Graciosa	San Antonio Creek at beach	34	48	120	37
March 1	El Buchón	Pismo Beach	35	08	120	39
March 2–3	Misión San Luis Obispo	Mission San Luis Obispo	35	17	120	41
March 4	Monterey R., N of Santa Margarita R.	Atascadero	35	30	120	40
March 5	San Antonio River	San Antonio Dam	35	48	120	53
March 6–7	Misión San Antonio	Mission San Antonio	35	53	121	03
March 8	Monterey R. at San Antonio R.	King City	36	12	121	08
March 9	Los Correos	Soledad	36	25	121	20
March 10	*Monterey*	Monterey	36	36	121	53
March 10–21	Misión Carmelo	Mission Carmel	36	33	121	55
March 22	*Monterey*	Monterey	36	33	121	53
March 23	other side of Valley, Santa Delfina	15 km NE of Salinas	36	46	121	38
March 24	Paxaro R. Val., Val. San Bernardino	Morgan Hill	37	05	121	40
March 25	Arroyo San Joseph Cupertino	Cupertino	37	18	122	04
March 26	Arroyo San Matheo	Palo Alto	37	26	122	10
March 27–28	San Francisco	San Francisco	37	44	122	27
March 29	Arroyo San Matheo	San Mateo	37	33	122	23

March 30	Rio de Guadalupe	Fremont	37	30	122	03
March 31	Arroyo San Salvador	San Leandro/ Oakland	37	45	122	10
April 1	14 leagues NNW	Richmond	37	58	122	18
April 2	Puerto Dulce, straight of water	Concord, Carquinez Strait	38	01	122	05
April 3	level plain of immeasurable extent	Antioch, Central Valley	38	00	121	45
April 4	6 leagues SE, S	Central Valley, E of Livermore	37	40	121	40
April 5	Cañada de San Vicente	Diablo Range NE of Morgan Hill	37	20	121	30
April 6	out of the sierra	Diablo Range at E of Gilroy	37	04	121	29
April 7	well near Pajaro R.	N of San Juan Bautista	36	56	121	33
April 8–13	Misión Carmelo	Mission Carmel/ Monterey	36	32	121	56
April 14	Buena Vista	near Salinas	36	38	121	39
April 15	Valle San Bernabé	Jolon Road, S of King City	36	06	121	12
April 16	Misión San Antonio	Mission San Antonio	35	53	121	03
April 17	Cañada de Robles	San Antonio Dam	35	48	120	52
April 18	Rio Nacimiento, San Antonio	Atascadero	35	27	120	38
April 19–22	Misión San Luis Obispo	Mission San Luis Obispo	35	17	120	41
April 23	Laguna Graciosa	San Antonio River at beach	34	48	120	37
April 24	*Punta Concepción*	Point Conception	34	27	120	24
April 25	Mescaltitán	Santa Barbara	34	25	119	40
April 26	Rio de la Assumpta	Ventura	34	17	119	17
April 27	Agua Escondida	Agoura Hills	34	09	118	43
April 28	Porciúncula River	Los Angeles	34	05	118	17
April 29– May 2	Misíon San Gabriel	Mission San Gabriel	34	07	118	06
May 3	San Gabriel River	San Gabriel River	34	04	117	59
May 4	Santa Ana River	San Jacinto Valley	33	57	117	10
May 5	Cañon San Patricio	Bautista Canyon	33	33	116	43

| May 6 | Puerta San Carlos, Santa Catharina | Fig Tree Valley, Coyote Canyon | 33 | 27 | 116 | 31 |

SOURCE: Dates and Franciscan-diary place names from Web de Anza Archives, Bolton (1930a,b, 1933).
NOTE: Italic place names are derived from the 1602–03 Vizcaíno voyage.

FOURTH JOURNEY OF JOSÉ JOAQUIN ARRILLAGA (1796), FROM U.S.-MEXICO BORDER TO SAN DIEGO

	Place name in Franciscan diary or landscape description	Modern place name or landscape description	Latitude (° n)		Longitude (° w)	
October 24	large indentation in the sierra	U.S.-Mexico border, ca. 20 km SE of Ocotillo	32°	39'	115°	46'
October 24	Point San Sabastian	Carrizo Wash	32	53	116	05
October 25	San Felipe	San Felipe Valley, Agua Caliente	32	56	116	13
October 26	heights of sierra	Banner Canyon, Julian	33	05	116	34
October 26	laguna, meadow	Cuyamaca State Park	33	02	116	37
October 26	timber of San Diego, Arroyo San Luis	San Diego Canyon	32	50	116	45
October 27	Misión San Diego	Mission San Diego	32	47	117	06

SOURCE: Dates and Franciscan-diary place names from Robinson (1969); cf. Minnich and Franco-Vizcaíno (1998).

JOURNEY OF JOSÉ ZALVIDEA (1805), FROM MISSION SANTA BARBARA TO MISSION SAN GABRIEL

	Place name in Franciscan diary or landscape description	Modern place name or landscape description	Latitude (° n)		Longitude (° w)	
July 18	Misión Santa Barbara	Mission Santa Barbara	34°	25'	119°	40'
July 19	Misíon Santa Ines	Mission Santa Ynez, Solvang	34	36	120	07
July 20	11 leagues N	Sisquoc Creek	34	50	120	08

July 21	north, broken mountains	Sierra Madre Mountains, Cuyama Creek	35	03	120	05
July 22–23	plains, stream	W Cuyama Valley	35	00	119	50
July 24	east 4 leagues	E Cuyama Valley	34	55	119	30
July 25–26	8 leagues north	Buena Vista Lake	35	10	119	15
July 27–30	4 leagues north	San Joaquin Valley SE of Bakersfield	35	12	118	50
July 31	4 leagues north	near Lamont	35	15	118	55
August 1	5 leagues north	Kern River at Bakersfield	35	25	119	00
August 2	3 leagues south	S of Bakersfield	35	20	119	00
August 3	1 league, south of lake	Kern Lake bed	35	10	118	57
August 4–5	Grapevine Canyon	Grapevine	34	55	118	55
August 6–7	east [=SE] through canyon	Castaic Lake, Tejon Pass	34	50	118	50
August 8	wide valley	Antelope Valley	34	50	118	40
August 9	east across 15 league valley	Palmdale	34	35	118	05
August 10	passed hills of Misíon San Gabriel	N of San Gabriel Mountains	34	25	117	40
August 11–12	swamp full of water	Silverwood Lake	34	22	117	19
August 13	Misíon San Gabriel	Cajon Pass, Mission San Gabriel	34	07	118	06

SOURCE: Dates and Franciscan-diary place names from Cook (1960).

MORAGA EXPEDITION (1806), INTO CENTRAL VALLEY,
DIARY OF MUÑOZ

Place name in Franciscan diary or landscape description	Modern place name or landscape description	Latitude (° n)	Longitude (° w)
September 20	Misíon San Juan Bautista	Mission San Juan Bautista	36° 50'	121° 30'
September 21	east along plain, 1½ leagues	San Benito Creek, Hollister	36 50	121 24

	Place name in Franciscan diary or landscape description	Modern place name or landscape description	Latitude (° n)		Longitude (° w)	
September 22	E 8 leagues, W edge Tulare Plain	San Luis Reservoir, W Central Valley	37	05	121	03
September 23–24	tremendous swamps	W of San Joaquin River	37	10	120	54
September 25–26	fine meadows of good land	San Joaquin River swamps	37	12	120	45
September 27–28	4 leagues north	near Merced	37	17	120	30
September 29–30	3 leagues ENE	Merced River	37	27	120	38
October 1	NW 7–8 leagues, big river	Tuolumne River	37	37	120	40
October 2–3	N to river	Stanislaus River	37	48	120	50
October 4–5	NW 6 leagues	Cosumnes River	38	15	121	20
October 6	camp of October 1	Tuolumne River	37	37	120	40
October 7–8	camp of September 25	Merced River	37	27	120	38
October 9	8 leagues east	foothills E of Merced River	37	27	120	25
October 10	7 leagues east [=SSE]	Chowchilla River	37	10	120	05
October 11–13	8 leagues east [=SSE]	San Joaquin River, E of Fresno	36	50	119	50
October 14–19	east [=SSE] to Kings R.	Kings River	36	45	119	30
October 20–25	same direction	foothills W of Sequoia National Park	36	35	119	10
October 26	8 leagues	Tule River	36	05	118	55
October 27	6 leagues in same direction [S]	Poso Creek	35	30	118	55
October 28	river of Zalvidea journey	Kern River	35	25	118	55
October 29–30	3 leagues downstream	Kern River	35	23	119	00

| October 31 | many wild grapes | Grapevine Canyon, Tejon Pass | 34 | 50 | 118 | 55 |
| November 2 | Misíon San Fernando | Mission San Fernando | 34 | 16 | 118 | 27 |

SOURCE: Dates and Franciscan-diary place names from Cook (1960).

Spanish plant names
for California vegetation

ABETO fir (*Abies concolor*) in southern California

ABROJOS cactus

ÁLAMO, ALAMEDA, ALAMILLO poplar, mostly Fremont cottonwood (*Populus fremontii*)

ÁLAMO NEGRO black cottonwood (*Populus trichocarpa*)

ALISO traditional meaning in Mexico is alder (*Alnus*); in California it refers to sycamore (*Platanus racemosa*); Brown (2001) and Bolton (1927, 1930a,b, 1933) erroneously translate aliso to "alder"; Roberts (1989) and Minnich and Franco-Vizcaíno (1998) translate aliso to California sycamore (*P. racemosa*)

AMARANTH *Amaranthus* spp.

ARBOL tree

ÁRBOLES CORCHO cork tree, very likely *Quercus agrifolia* (term used by Pedro Fages in Priestly 1937)

ARBOLILLO shrub

ARBUSTO shrub, bush

AVELLANAS hazelnuts, California buckeye (*Aesculus californica*)

AVENO wild oat

BOSQUE, VOSQUE thicket, woodland, wood

BOSQUE CHAPARRO a shrubby growth

BOSQUE ESPINOSO literally, "spiny brush" or "wood" in chaparral, likely dominated by *Ceanothus*

BREÑALES "brambles," in reference to dense chaparral of the Santa Monica Mountains

CACHANILLA unknown

CACOMITES a species of *Iris*

CALABASAS wild gourd (*Cucurbita foetidissima*)

CARDOS SANTOS prickly poppy, holythistle

CARRIZO a large cane grass, reed grass, probably tule (*Scirpus* spp.)

CASTAÑOS chestnuts, California buckeye (*Aesculus californica*); could be *Chrysolepis chrysophylla*, but the only stands of this tree grow far from the Spanish routes in the redwood belt of the Santa Cruz Mountains

CAVALLERO larkspurs (*Delphinium* spp.)

CEBOLLIN wild onion (*Allium* spp.)

CEDRO coast redwood (*Sequoia sempervirens*) in the Santa Cruz and Santa Lucia mountains, incense cedar (*Calocedrus decurrens*) in southern California and the Sierra Nevada

CHAMIZO, CHEMIZO thicket; used ca. 1790–1890 in reference to chamise-dominated chaparral (*Adenostoma fasciculatum*); in the desert it refers to thickets of other species

CHAPARRAL used by Crespí in San Diego County as a "thicket" of shrubs, like traditional usage in Mexico; in the Spanish period "chaparral" was never used in its modern usage

CHIA *Salvia columbariae*

CHOYA pencil opuntia, mostly *Opuntia acanthocarpa* and *O. echinocarpa* of the Sonoran Desert, *O. littoralis* in coastal southern California

CIENEGAS marshes, swamps

CIPRES cypress, in reference to *Cupressus macrocarpa* at Carmel, and *C. sargentii* in the Santa Lucia Mountains

CLAVELES marigolds

COBEÑAS, COVENAS something that gives much shade and shelter, perhaps cottonwoods

COGI ESPLIEGO lavender (*Salvia apiana, S. mellifera*)

DATIL little date, Mojave yucca (*Yucca schidigera*)

DESIERTO desert, wasteland, area free of herbaceous cover or pasture

EMBOSCADO thicket of (small) trees

EMPASTADA pasturage, herbaceous cover useful to livestock

ENCINILLOS small live oaks

ENCINO evergreen oak tree, almost exclusively coast live oak (*Quercus agrifolia*), but refers to canyon live oak (*Q. chrysolepis*) in the high mountains, mostly the San Jacinto and San Bernardino mountains in southern California

ENCINOS CHAPARROS a growth of low brush or chaparral, very likely dominated by scrub oaks in the *Quercus dumosa* complex

ENCINOS DE POCO small live oaks

ENEA rushes

ENMONTADO DE MATORRALES thickets of chaparral

ENMONTADOS brush

ESPINAS thorns

ESTERIL sterile or barren land, desert

ESTERO estuary, translated by Brown (2001) as "inlet," of the sea into terrestrial stream drainage

FLORES flower

FRESNO ash tree (*Fraxinus latifolia*)

GALLETA the perennial grass *Hilaria rigida*

GOBERNADORA the "governor," or a plant that rules (i.e., dominant species), creosote bush (*Larrea tridentata*)

HEDIONDILLA "little stinker," *Larrea tridentata* (Roberts 1989)

HERBAJE herbage, herbaceous plants, not "grass" as translated by Brown (2001) and Web de Anza Archives

HIERBAS, ERBAS, ERVAS herbaceous plant, forb

HINOJO samphire (*Crithmum maritimum*), perhaps fennel

HORTIGAS nettles (*Urtica holosericea*)

JOJOBA (COCOBA) *Simmondsia chinensis*

JUCAROS shrub or tree with a fruit resembling the spiny black olive (*Bucida buseras*) of Cuba and southern Florida, possibly *Heteromeles arbutifolia* or *Rhus ovata* (Andrew Sanders, pers. comm.), in Bautista Canyon in the San Jacinto Mountains

JUNCOS reeds

JUNÍPEROS juniper, exclusively used in reference to juniper-like leaves of chamise (*Adenostoma fasciculatum*)

LAURELES California laurel tree, bay tree (*Umbellularia californica*)

LIRIOS lilies (*Lilium* spp.)

MADROÑO literally, "madrone," but more generally members of Ericaceae; refers to *Arbutus menziesii* at a few localities near San Francisco, and to *Arctostaphylos* spp. in Baja California (see Minnich and Franco-Vizcaíno 1998)

MALVAS mallow, probably the California poppy (*Eschscholzia californica*)

MANZANILLA manzanita (*Arctostaphylos* spp.)

MARISCO salt grass

MATORAL shrubland, weeds

MATORALLES ESPINOSOS thorny bushes or shrubs, low shrubs, e.g. mulefat along dry washes

MESCAL agave, most likely desert agave (*Agave deserti*); mescal heads made from *Yucca whipplei*

MESQUITE *Prosopis glandulosa*

MIRASOLES sunflowers, used to describe *Encelia farinosa* in coastal sage scrub near Riverside

MONTECILLO, MONTECITO low scrub

MONTUOSA thickets, in reference to chaparral

NOGALES walnut, *Juglans californica* in southern California and *J. hindsii* near Concord; traditionally, the pecan tree

NOPAL prickly pear thicket (a pad *Opuntia* ssp.)

NOPALIS prickly pear (*Opuntia* spp.)

PALO COLORADO coast redwood (*Sequoia sempervirens*)

PAJON, SETACIA a grass, coarse straw; in northern Mexico, a "tall grass"; in California, apparently refers to *Sporobolus* spp.

PAJONALES dry grass

PANTÁÑOS marshes

PARRAS grapes (*Vitus girdiana*)

PARVILLA burrobush (*Ambrosia dumosa*)

PELADOS bare or bald, barren, treeless

PALMA palm, in Baja California *Washingtonia filifera, W. robusta,* and *Brahea armata*

PALO ADAN ocotillo (*Fouquieria splendens*)

PALMITO little palm, Mojave yucca (*Yucca schidigera*)

PARROS   wild grape (*Vitus girdiana, V. californica*)
PARVILLA   burrobush (*Ambrosia dumosa*)
PASITA   raisin, possibly made from elderberry (*Sambucus* spp.)
PASTO   pasture, forage sufficient for livestock, "grass"
PENSAMIENTOS   heartsease, perhaps *Layia platyglossa*
PIMPAJAROS   sucker buds
PINABETE   literally, "pine fir," seen in the Santa Cruz Mountains, either coast redwood (*Sequoia sempervirens*) or Douglas fir (*Pseudotsuga menziesii, P. macrocarpa*)
PINO   pine, pines of Sierran mixed conifer forest, and serotinous pines of *Pinus. sect. sabinianae* and *P. sect. oorcarpae*, and in reference to *Pseudotsuga macrocarpa* in southern California (usually seen at a distance); pinos are invariably identified to species because modern stands occur in monospecific forests
PIÑON   pinyon pine (*Pinus monophylla, P. quadrifolia*), seen mostly in northern Baja California
PRADO   meadow
QUERCUS ILIX   oak with similar morphology to Mediterranean pin oak, perhaps *Q. douglasii*, or some member of the *Q. berberidifolia* complex (term used by Pedro Fages in Priestly 1937)
QUERCUS ROBUR   oak with similar morphology to deciduous English oak, very likely *Q. lobata* (term used by Pedro Fages in Priestly 1937)
QUERCUS SUBER   oak with a similar morphology to coast live oak, very likely *Quercus agrifolia* (term used by Pedro Fages in Priestly 1937)
RAIZES   onions, called "amole" by Native Californians
RAMAJOS   brush
ROBLE   deciduous oak tree, invariably in reference to valley oak (*Quercus lobata*); doubtfully attributed to *Q. douglasii* due to its smaller size to *Q. lobata* (i.e., would be poor timber)
ROMERILLO   sagebrush
ROMERO   rosemary, most likely *Eriogonum fasciculatum* or *Artemisia californica*
ROSA DEL CASTILLA, ROSALES   rose of Castille (*Rosa californica*), seen near streams in California; *R. minutifolia* in maritime desert scrub of Baja California
ROSEMARY   rosemary, shrub with rosemary-like leaves (needle-like leaves with fascicules), chamise (*Adenostoma fasciculatum*), or California buckwheat (*Eriogonum fasciculatum*)
SABINOS, SAVINOS   coast redwood (*Sequoia sempervirens*) in the Santa Cruz and Santa Lucia mountains, incense cedar (*Calocedrus decurrens*) in Sierra Nevada and San Bernardino Mountains
SALVIA   sage (*Salvia apiana, S. mellifera, S. leucophylla*), or possibly *Artemisia californica*
SAUCE   willow (*Salix* spp.)
SOTOLE   sour cane of an agave or yucca, possibly *Yucca whipplei* inflorescences
TASCALE   juniper, *Juniperus californica* in Baja California
TEXOCOTE   perhaps *Rhus integrifolia* or *R. ovata*; fruit of this plant gives a very refreshing drink, somewhat acid like the tamarind, made by soaking pulp in water (Priestly 1937)
TORNILLO   screwbean mesquite (*Prosopis pubescens*)
TRIGO   wheat, in reference to wild rye (*Elymus condensatus*)

TULAR   tule (*Scirpus* spp.)
TULIPANES   tulips, probably *Eschscholzia californica*
TUNA   pad opuntia (*Opuntia* spp.)
VEREJÓN   brush, tall sticklike shrub (*Baccharis glutinosa*), *Fraxinus trifoliolata* in Baja California
VIOLETA   violets (*Viola* spp.), or less likely *Nemophila* spp.
YERBAS   literally "herbs," forbs, green cover
ZACATALES   pasture patches, green cover, "grass"
ZACATE   pasture, green cover
ZACATE SECO   dry grass, dry pasture, dry cover; grandes zacatónes—large grass clumps
ZACATÓN SALADO   salt-tolerant grass, probably *Distichlis* spp.
ZARZA   bramble

Selected earliest botanical collections of exotic annual species in California

Species/ herbarium accession ID	collector	year	location
Spanish/Mexican Period exotics			
Avena fatua			
UC815842	Marcus Jones	1882	San Luis Obispo
JEPS78029	S. B. Parish	1887	San Bernardino
JEPS78028	S. B. Parish	1888	San Bernardino mesas
UC121634	T. S. Brandegee	1888	Santa Barbara County
UC37423	H. A. Brainard	1896	Santa Clara County, San Jose
UC53630	F. W. Hubby	1896	Santa Barbara County, Ojai
UC37424	J. Burtt Davy	1896	Antelope Valley
UC34038	J. Burtt Davy	1897	San Joaquin Valley
UC37441	J. Burtt Davy	1898	Mount Diablo
UC53691	J. Burtt Davy	1898	Alameda County, Livermore
UC69356	S. B. Parish	1899	San Bernardino
UC70817	J. Burtt Davy	1899	Long Valley, N Coast Range
UC188779	J. Burtt Davy, Walter C. Blasdell	1899	Mendocino, Ukiah
UC53688	J. Burtt Davy	1899	Monterey County, S Coast Range
UC53689	J. Burtt Davy	1899	Santa Clara County, Mount Hamilton

Species/ herbarium accession ID	collector	year	location
Spanish/Mexican Period exotics			
Brassica nigra			
UC10281	W. H. Brewer	1861	San Luis Obispo
JEPS53245	W. L. Jepson	1894	Napa County
UC10284	J. Burtt Davy	1898	Alameda County, San Leandro
UC10283	J. Burtt Davy	1900	Berryessa
UC10285	H. M. Hall	1901	San Jacinto Mountains, Idyllwild
UC56982, 56983	H. M. Hall	1902	Santa Barbara
UV55209	J. Burtt Davy	1902	Ventura County, Santa Paula
UCD42530	Lloyd Tevis	1902	Lake Tahoe
Erodium cicutarium			
UC109459	R. M. Austin	1874	Modoc County, Goose Valley
UC35201	unknown	1882	Sacramento County, Elk Grove
UC17082	H. P. Fitch	1883	Stockton
UC109456	T. S. Brandegee	1888	Santa Rosa Island
UC17081	J. A. Sanford	1890–91	San Joaquin County, Waverly
JEPS61389	W. L. Jepson	1891	Berkeley
UC17077	J. Burtt Davy	1893	Berkeley
UC204476	W. R. Dudley	1895	Stanford University
UC17098	J. Burtt Davy	1896	West Palmdale
UC17097	J. Burtt Davy	1896	Bakersfield
UC17096	F. W. Gunnison	1896	San Joaquin County, Linden
UC17097	J. Burtt Davy	1896	Bakersfield
UC17098	J. Burtt Davy	1896	West Palmdale, Antelope Valley
JEPS61391	W. L. Jepson	1897	Marin County, Olema
UC17078	Geo. F. Reinhardt	1897	San Jacinto Valley
UC17099	J. Burtt Davy	1897	Tulare
UC17100	J. Burtt Davy	1897	University of California Experiment Station
UC40970	J. Burtt Davy	1898	San Francisco
UC35202	J. Burtt Davy	1898	Colusa County, near Willows
UC70189	J. Burtt Davy	1898	Livermore Valley

UC17094	P. S. Woolsey	1898	Tulare County, Visalia
UC40968	J. Burtt Davy	1899.	Mendocino County, Rattlesnake Mountian
UC17088	J. Burtt Davy, Walter C. Blasdale	1899	Mendocino County, Sherwood Valley
UC17089	J. Burtt Davy	1899	Mendocino County, near Ukiah
UC17090	J. Burtt Davy	1899	Mendocino County, Humboldt, Bell Springs

Erodium moschatum

UC17104	state survey	1861	Santa Barbara County, Gaviota Pass
UC367130	J. G. Lemmon and wife	1884	Mojave River
JEPS61402	S. B. Parish	1888	San Bernardino
UC17108, 17109	J. A. Sanford	1890–91	Stockton, Waverly
UC17106	J. Burtt Davy	1893	Berkeley
UC204413	W.L. Dudley	1895	Stanford University
UC17107	J. Burtt Davy	1896	Kern County, head of San Joaquin Valley
UC17101	J. H. Barber	1897	Los Angeles County, Santa Monica Forestry Station
UC56279	Harley P. Chandler	1897	Los Angeles County, Claremont
UC70190	J. Burtt Davy	1898	Livermore
UC17102	J. Burtt Davy	1898	Marin County, near Olema
UC17103	J. Burtt Davy	1898	Colusa County, near Willows
UC17110	J. P. Tracy	1899	Humboldt County, Eureka

Hordeum murinum

POM112840	Marcus E. Jones	1882	Los Angeles
UC185567	T. S. Brandegee	1883	Santa Cruz Island
JEPS78228	S. B. and W. F. Parish	1885	San Bernardino Valley
UC120595	T. S. Brandegee	1890	Santa Catalina Island
UC73791	Mr. Sanford	1890	San Joaquin County, Waverly
UC50644	J. Burtt Davy	1893	Marin County, Mount Tamalpais
UC38656	McLean	1893	Marin County, Saucilito
UC120597	T. S. Brandegee	1895	San Diego County, El Cajon
UC1114137	W. L. Jepson	1896	Berkeley, Strawberry Canyon
UC38660	Geo. Hansen	1896	Amador County, Sequoia National Park

Species/ herbarium accession ID	collector	year	location
Spanish/Mexican Period exotics			

Hordeum murinum (continued)

UC50570	J. Burtt Davy	1896	Between Gorman and Fort Tejon
UC50571	J. Burtt Davy	1896	Modesto, railroad track
UC38651	J. Burtt Davy	1898	Alameda County
UC38657	J. Burtt Davy	1898	Berkeley
UC38658	Dr. Loughridge	1898	Tulare Experiment Station
UC38659	J. Burtt Davy	1898	Glenn, Norman
UC38647	H. M. Hall	1899	San Diego County, Warner Ranch
UC50665	J. Burtt Davy	1899	Monterey

Malva parviflora

UC18724	D. G. Cooper	1862	San Diego
UC35241	E. R. Drew	1882	Sacramento, Elk Grove
UC18738	E. L. Greene	1886	Berkeley
UC52298	W. R. Summers	1886	San Luis Obispo
UC188098	C. F. Sonne	1888	Truckee
JEPS68652	S. B. Parish	1888	San Bernardino
UC173650	T. S. Brandegee	1888	Santa Cruz Island
JEPS68650	W. L. Jepson	1891	Sutter County, Marysville Buttes
JEPS68649	W. L. Jepson	1892	Solano County
UC50666	J. Burtt Davy	1899	Mendocino County, Ukiah
UC50668	J. Burtt Davy	1899	Contra Costa County, Moraga Valley

Medicago polymorpha

UC16101	W. H. Brewer	1861	Santa Barbara
JEPS65513	W. H. Shockey	1886	Placer County, Auburn
UC16094	J. A. Sanford	1890–91	Stockton
UC16088	F. C. Bioletti	1891	San Francisco
UC16083, 16093	J. Burtt Davy	1893	Santa Clara, New Almaden
UC16089	M. S. Baker, Frank Nutting	1894 1895	Redding
UC16085	F. W. Bancroft	1897	Knight's Ferry

UC16087	J. Burtt Davy	1897	Contra Costa County, near Point Isabel
UC16100	J. Burtt Davy	1897	Tulare County
UC16092	Harley P. Chandler	1898	Los Angeles County, Claremont
UC16088	J. Burtt Davy	1899	Glenn County, near Norman
UC16086	J. Burtt Davy	1899	Tulare County
UC16091	Harley P. Chandler	1899	Berkeley
UC16095	J. Burtt Davy	1899	Humboldt County, around Scotia

Melilotus indica

UC16161	State Survey	1861	San Luis Obispo
UC16164	W. H. Brewer	1861	San Luis Obispo
UC186824	S. B. and W. F. Parish	1882	San Bernardino Valley
UC16160	W. L. Jepson	1885	Solano County, Alamo Creek
UC187203	C. F. Sonne	1887	Marin County, Tamalpais
UC16169	J. Burtt Davy	1893	West Berkeley Beach
UC16157, 16158	Ivar Tidestrom	1893	Monterey County, Pacific Grove
UC69576	J. Burtt Davy	1896	Bakersfield
UC16162	J. Burtt Davy	1896	Pomona
UC16164	J. H. Barber	1897	Los Angeles
UC16178, 16168	J. Burtt Davy	1897	Tulare County, Alkali Plain
UC55390	J. H. Barber	1898	San Luis Obispo County, Santa Ynez Canyon
UC16159	J. Burtt Davy	1898	Glenn, Norman
UC16163	W. L. Jepson	1898	Berkeley
UC16165	J. Burtt Davy	1898	Alameda County, Bay Farm Island

Trifolium gracilentum

UC16557	W. H. Brewer	1862	Contra Costa County, Kirker's Pass
UC16556	H. N. Bolander	1864	Sonoma County, Santa Rosa Creek
UC33531	E. R. Drew	1882	Sacramento
UC866785, 866786	Marcus E. Jones	1882	Los Angeles
UC52183, 73176	Mrs. R. W. Summers	1882	San Luis Obispo County
JEPS65904	W. L. Jepson	1884	Solano County, Vacaville

Species/ herbarium accession ID	collector	year	location

Trifolium gracilentum (continued)

JEPS65903	W. L. Jepson	1885	Solano County, Little Oak, Vacaville
UC80319	Nevin and Lyon	1885	San Clemente Island
UC80313	T. S. Brandegee	1888	Santa Cruz Island
UC80532	Katharine Brandegee	1889	Colusa County, Leesville
UC80533	Katharine Brandegee	1889	Tracy
JEPS95902	W. L. Jepson	1891	Sutter County
JEPS65901	W. L. Jepson	1891	Contra Costa County, Mount Diablo
JEPS65912	S. B. Parish	1891	San Bernardino
UC16551, 165586	J. Burtt Davy	1893	Alameda, Berkeley
UC1394590	J. Burtt Davy	1893	San Francisco
RSA 349115	Anstruther Davidson	1893	San Gabriel Mountains
UC16554	J. Burtt Davy	1893	San Francisco
UC16555	Mrs. Blockman	1893	San Luis Obispo County
UC 16508, 16509	J. Burtt Davy	1894	Santa Clara
UC80531	T. S. Brandegee	1895	San Diego
UC631032	Ivar Tidestrom	1895	Contra Costa County, near Lafayette
UC165628	J. Burtt Davy	1895	San Mateo
UC165629	J. Burtt Davy	1896	Kern County, Rosedale
UC165630	J. Burtt Davy	1896	Los Angeles County, Armagoja Creek
UC485392	J. Burtt Davy	1897	Merced
UC16550	J. H. Barber	1897	Pomona Experiment Station (Santa Monica)
UC71097	J. Burtt Davy	1898	Contra Costa County, Danville
UC1197195	J. P. Tracy	1899	Humboldt County, Bucksport

Trifolium willdenovii

UC16756	W. H. Brewer	1861	San Luis Obispo, Nipomo
UC16750, 16751, 16752,	W. H. Brewer	1862	Contra Costa County

16757	W. H. Brewer	1862	Marin County
UC16755	W. H. Brewer	1862	Sonoma County
UC16754	H. N. Bolander	1866	Marin County
UC16758	M. H. Gates	1879	Placer County
UC16806	M. E. Jones	1881	Santa Cruz County
UC380964	unknown	1882	Sacramento
UC33533	W. H. Shockey	1886	Placer County, Auburn
JEPS4358	Edward L. Greene	1887	Alameda County
UC16733	Mrs. R. W. Summers	1887	Berkeley
UC55701	unknown		San Luis Obispo
UC338268	S. B. and W. F. Parish	1887–88	San Bernardino
JEPS4138	T. S. Brandegee	1888	Santa Barbara
UC80494	Mrs. R. W. Sommers	1888	San Luis Obispo County, Chalcedon Hill
UC73178	Edw. L. Greene	1889	Berkeley
JEPS4156	Chestnut and Drew	1889	Tuolumne County, Lake Eleanor
UC33522	Katharine Brandegee	1889	Kern County
UC80499	T. S. Brandegee	1889	San Joaquin County
UC80498	Chestnut and Drew	1890	Placer County, Lake Tahoe
UC147859	W. L. Jepson	1891	Solano County, Vaca Mountains
JEPS4156	W. L. Jepson	1891	Sutter County, South Peak
JEPS4143	W. L. Jepson	1891	Solano County, Araquipa Rancho
UC16736	Geo. Hansen	1891	Amador County
UC16742, 194322	S. B. Parish	1891	Foothills, San Bernardino Mountains
UC80500	F. T. Bioletti	1892	Sonoma County, south los Guilicos
UC16731	Micherner and Bioletti	1892	Sonoma County
UC853222	Milo S. Baker	1893	Modoc County
UC16765	Mrs. Blockman	1893	San Luis Obispo
UC16410	W. L. Jepson	1893	Napa County (many collections)
JEPS4401	Edw. L. Greene	1893	San Francisco
UC16812	Milo S. Baker	1893	Modoc County, Forestvale
UC16715	J. Burtt Davy	1893	Santa Clara County
UC16743	R. D. Alderson	1893	San Diego
UC16408, 16817	Geo. Hansen	1893	Sequoia National Park region
UC194322	Mrs. Blockman	1893	San Luis Obispo

Species/ herbarium accession ID	collector	year	location
Spanish/Mexican Period exotics			

Trifolium willdenovii (continued)

UC194322	Mrs. Blockman	1893	San Luis Obispo
UC33314	R. D. Alderson	1894	San Diego County, Witch Creek
UC16741	T. S. Brandegee	1894	San Diego County, Lakeside
UC80502	J. Burtt Davy	1895	Contra Costa County, Antioch
UC165503	J. Burtt Davy	1895	Calavaras County
UC165593	J. Burtt Davy	1895	San Francisco, Lake Merced
UC16726	J. Burtt Davy	1895	Alameda County, near Newark
UC16727	Ivar Tidestrom	1895	Alameda County
UC16734	Mary Pulsifer	1895	Placer County, Auburn
UC397182	J. Burtt Davy	1896	Kern County, Rosedale
JEPS4158	S. B. Parish	1896	San Bernardino
UC16409	F. W. Gunnison	1896	Calavaras County
UC16725	J. Burtt Davy	1896	Kern County, Bakersfield
UC165495	J. Burtt Davy	1896	Kern County, near Junetts
UC630196	E. Stilson	1897	Butte County
CHSC993	J. Burtt Davy	1897	Tulare County
UC631063	J. Burtt Davy	1897	San Joaquin County, head of San Joaquin Valley
UC165496	J. Burtt Davy	1897	San Francisco, Lake Merced
UC485394	J. Burtt Davy	1897	Tulare County
UC631154	C. A. Purpus	1897	Mendocino County, Potter Valley
UC80458	M. S. Baker	1898	Sonoma County, Geyser Canyon
JEPS4162 canyon	M. S. Baker	1898	Sonoma County, Santa Rosa Creek
UC165494	P. S. Woolsey	1898	Tulare County, Kaweah, Visalia
UC16745	J. Burtt Davy and Walter Blasdale	1899	Humboldt County
UC16761, 16730, 33316		1899	Tehama County
JEPS4378	W. L. Jepson	1899	Marin County, Mount Tamalpais
UC16746	Harley P. Chandler	1899	

Late 19th Century Exotics

Avena barbata

UC121632, 121635	T. S. Brandegee	1888	Santa Cruz Island
JEPS68452	A. J. McClatchie	1895	Pasadena
UC 121636	T. S. Brandegee	1896	San Diego
UC37421	Geo. Hansen	1896	Amador County
UC37420	H. A. Brainard	1896	San Jose
UC69355	S. B. Parish	1897	San Bernardino
UC34036	J. Burtt Davy	1897	Tulare County
UC53768	J. Burtt Davy	1898	Contra Costa County, Point Isabel
POM115197	S. B. Parish	1898	San Bernardino Mountains
UC53767	J. Burtt Davy	1898	Livermore
UC53769	J. Burtt Davy	1898	Marin County, Olema
UC53770	J. Burtt Davy	1898	Colusa County, Princeton
UC53766	J. Burtt Davy	1899	near Ukiah
UC53772	J. Burtt Davy	1899	Mendocino County
UC53773	J. Burtt Davy	1899	Contra Costa County, Clairmont Canyon
UC53771	J. Burtt Davy	1899	Mendocino County/Ukiah
UC53772	J. Burtt Davy	1899	Mendocino County/ Walker Mountain
UC70938	J. Burtt Davy	1899	Mendocino County/ Sherwood Valley
UC1103327	J. P. Tracy	1900	Berkeley
UC53764	J. Burtt Davy	1901	Carmel

Bromus diandrus

POM112745	Marcus Jones	1884	Mojave Desert, Needles
JEPS78034	S. B. Parish	1888	Waterman Canyon, San Bernardino Mountains
UC60186	J. G. Lemmon	1889	Lake Tahoe
UC816288	F. E. Blaisdell, MD	1891	San Diego, Mokelumne Hill
UC50586	A. Davidson	1892	Los Angeles
UC37676	Geo. Hansen	1892	Amador County, Jackson
UC60217	J. Burtt Davy	1895	Calavaras County
UC37675	Geo. Hansen	1896	Sequoia National Park region

Species/ herbarium accession ID	collector	year	location

Late 19th Century Exotics

Bromus diandrus (continued)

UC50586	J. Burtt Davy	1896	San Emigdio Canyon
UC60220	W. L Jepson	1896	Berkeley
UC37637	H. A. Brainard	1896	San Jose
UC50599	J. Burtt Davy	1897	Contra Costa County, Point Isabel
UC50598, 60216	J. Burtt Davy	1897	Lake Merced, San Francisco
POM1264	Minnie Reed	1897	San Diego
UC50600, 50601, 50602, 53624	J. Burtt Davy	1898	Berkeley
UC60184	J. Burtt Davy	1899	Contra Costa County, Moraga Valley
UC60186	J. G. Lemmon	1899	Lake Tahoe
UC76983–86	J. Burtt Davy	1899	Mendocino County, Willits, Ukiah
UC187100	J. Burt Davy, Walter Blasdale	1899	Mendocino County
UC37674	H. M. Hall	1899	San Diego County, Cuyamaca Mountains
UC70723	F.C. Bioletti	1900	Yosemite
UC53619, 53620, 53622	J. Burtt Davy	1901	Big Sur, Monterey, Santa Lucia Mountains
UC50587	J. Burtt Davy	1901	Oxnard

Bromus madritensis

POM206139	R. F. Bingham	1882	Santa Barbara
UC337748	unknown	1885	San Luis Obispo County
JEPS78224	S. B. Parish	1887	Fort Tejon
UC37652	S. B. Parish	1889	San Bernardino, Reche Canyon
UC185562	T. S. Brandegee	1890	Santa Catalina Island
UC37649	S. B. Parish	1891	San Bernardino
POM363952	Merritt	1893	Orange County
UC60776	Davidson	1890s	Pasadena
UC37706	J. Burtt Davy	1895	Contra Costa County, Antioch
UC50604	J. Burtt Davy	1895	Calavaras County, Copperopolis

UC30431	J. Burtt Davy	1896	Mouth, San Emigdio Canyon
UC50574, 50575, 50577	J. Burtt Davy	1896	Kern County, Caliente, Kern Lake, Tulare
UC30432, 50576	J. Burtt Davy	1896	near Fort Tejon
UC37653	H. A. Brainard	1896	Santa Clara County, San Jose
UC53745	J. Burtt Davy	1897	Tulare Lake
UC50603	W. L. Jepson	1897	Solano County, Miller Canyon
JEPS69208	S. B. Parish	1898	San Bernardino County, Colton
UC37650	H. M. Hall	1899	San Bernardino County, Reche Canyon
UC34136	J. Burtt Davy, Walter Blasdale	1899	Mendocino County, Ten-Mile House
UC53749	F. C. Bioletti	1900	Yosemite Valley
UC37705, UC37651	H. M. Hall, H. P. Chandler	1900	Fresno County, Pine Ridge, Collins Meadow
UC60197	Benj. Cobb	1900	San Joaquin County, Tracy
UC188938	Harley P. Chandler	1901	Santa Clara County, San Martin
UC53748	J. Burtt Davy	1901	Big Sur
UC34040	H. M. Hall	1901	Chalk Hill, San Jacinto
POM50862	A. R. Abrams	1901	Los Angeles County, Ballona Harbor
POM50861	S. B. Parish	1901	San Bernardino
UC220877	A. D. E. Elmer	1902	Santa Barbara, Surf
UC175630	J. P. Tracy	1902	Lake County, Lakeport
UC650655	I. R. Wiggins	1902	San Diego County, Campo
UC149788	H. M. Hall	1902	Orange County, Santa Ana River
UC34029	J. J. Chapell	1902	Shasta County
POM50682	A. D. E. Elmer	1902	Santa Barbara County, Santa Ynez Mountains
POM112726	Marcus E. Jones	1903	San Bernardino Mountains
UC144814	A. A. Heller	1904	Santa Clara County, Los Gatos
UC127469	H. M. Hall	1905	Mount Pinos
UCR2244,	F. M. Reed	1906	Riverside
UCR2245–2247	F. M. Reed	1907	Riverside
JEPS69212	W. L. Jepson	1907	San Benino County, San Carlos Creek
POM1278	C. F. Baker	1909	Los Angeles County, Claremont

Species/ herbarium accession ID	collector	year	location

Late 19th Century Exotics

Bromus madritensis (continued)

UC816378	Agnes Chase	1910	Santa Barbara, Santa Inez Forest Reserve
UC204777	Geo. D. Butler	1910	Yreka

Bromus mollis

UC816299	Marcus Jones	1882	Oakland
UC336788	unknown	1889	Lake Tahoe
UC60205, POM363863	A. Davidson	1890	Los Angeles
UC816186	F. E. Blaisdell	1891	San Diego
UC60468	Dr. Edw. Palmer	1892	Marin County
UC50597	W. L. Jepson	1892	Solano County, Davis Hills
UC76889	Geo. Hansen	1893	Amador County
UC60203	W. L. Jepson	1893	Napa Valley
UC60206	W. L. Jepson	1896	Solano County, Little Oak
UC60192, 60195, 60204	J. Burtt Davy	1896	San Francisco Precidio
UC37627	Geo. Hansen	1896	Sequoia National Park region
UC60199	A. B. L.	1896	Petaluma
UC60207	W. A. Setchell	1897	Hollister
POM1255	Minnie Reed	1897	San Diego
UC50596, 50605	J. Burtt Davy	1898	Berkeley
UC60208	J. Burtt Davy	1898	Glenn County, Norman
UC60206	J. Burtt Davy	1899	Contra Costa County, Las Trampas Creek
JEPS69202	W. L. Jepson	1899	Tehama County, Stivers Ranch
UC34106, 188957	J.Burtt Davy and Walter C. Tisdale	1899	Humboldt County, Bell Springs, Bear River Ridge
UC70742	T. H. Gilbert	1899	Siskiyou County, Scott River valley
UC37626	H. M Hall, H. P. Chandler	1900	Fresno County, Pine Ridge

UC70724, 53625	F. T. Bioletti	1900	Yosemite Valley
UC50589	H. P. Chandler	1900	Napa County, Rutherford
UC53750	J. Burtt Davy	1900	Marin County, Point Reyes
UC53752	J. Burtt Davy	1901	Santa Clara
UC60105	K. D. Jones	1901	San Luis Obispo
UC816190	Geo. B. Grant	1901	Los Angeles
UC53754, 53755, 53746	J. Burtt Davy	1901	Big Sur, Monterey, Santa Lucia Mountains
UC34065	J. P. Tracy	1901	Eureka
UC53679	H. M. Hall	1901	Mount Diablo
UC60205	Miss K. D. Jones	1901	San Luis Obispo
UC163650	Harvey P. Chandler	1901	Berkeley
UC53681	H. M. Hall	1902	Casitas Pass
UC34082, POM1256	H. M. Hall	1902	San Bernardino County, Santa Ana River
POM155842	L. R. Abrams	1903	San Diego
UC153157	K. Brandegee	1903	San Diego County, Ramona
UC153150	A. A. Heller	1904	Santa Clara, Los Gatos
JEPS69190	W. L. Jepson	1905	Mendocino County, Ukiah
UCR2232	F. M. Reed	1906	Riverside
UC167147	S. B. Parish	1907	San Bernardino
JEPS69197, 69198	W. L. Jepson	1907	Fresno County, San Carlos Range
UC454056	I. J. Condit	1908	Cal. Poly, San Luis Obispo
JEPS69187, 69188	Geo. D. Butler	1909	Siskiyou County, Yreka

Bromus tectorum

UC70736	T. H. Gilbert	1899	Siskiyou County, Scott River valley
UC70719	F. T. Bioletti	1900	Yosemite Valley
UC37724	Harley P. Chandler	1901	Humboldt County, Klamath River
JEPS73745	George D. Butler	1908	Yreka
JEPS69221, 69222	George D. Butler	1909	Yreka
JEPS69223	W. L. Jepson	1909	Nevada Falls, Yosemite
JEPS69224, 69225	W. L. Jepson	1911	Yosemite
JEPS73973	A. S. Hitchcock	1913	Truckee
POM1275	O. W. Robinson	1916	Los Angeles County, Clarement

Species/ herbarium accession ID	collector	year	location
Late 19th Century Exotics			
Bromus tectorum (continued)			
UC726535, 816448	A. A. Heller	1916	Siskyou, NE base of Mount Eddy
POM1276	I. M. Johnston	1917	San Gabriel Mtns, Evy Canyon
UC205825	S. B. Parish	1917	San Bernardino Mountains, Diablo Canyon
POM363987	Thekla Mohr	1917	San Bernardino Mountains
JEPS69220	L. S. Smith	1917	Modoc County, Joseph Creek
UC726398	A. A. Heller	1919	Butte County, Sterling near Feather River
POM363980	Anstruther Davidson	1919	Los Angeles
POM8328	P. A. Munz	1919	San Bernardino Mountains, Devils Canyon
Brassica geniculata			
UC160451	A. Davidson	1911	Los Angeles
UC183820	S. B. Parish	1914	San Bernardino, street weed
UC177277	S. B. Parish	1914	San Bernardino, vacant lot
JEPS53261	Ehlers	1917	Berkeley
UC855640	Abrams	1917	San Francisco
UC1751179	S. B. Parish	1917	San Bernardino
Late 20th Century Exotics			
Brassica tournefortii			
UC723321	P. A. Munz	1941	Niland
UCR1573	W. R. Bowen	1962	west of Niland
UCR5695	O. F. Clarke	1967	Whitewater
UCR16147	Grace Sprague	1968	north Palm Springs
UCR17088	James K. Ryan	1969	ten miles east of Brawley
UCR13261	O. F. Clarke	1973	San Bernardino County, Colton
UCR122301	June Latting	1973	Imperial County, Calexico
UCR60708	R. F. Thorne	1973	Desert Hot Springs
SBBG39958, 39960	C. F. Smith	1975	Santa Barbara

UC1422528	C. F. Smith	1977	Imperial County, Glamis
UCR69989	Maureen Hales	1977	San Bernardino County, Cadiz Valley
UCRT20832	A. C. Sanders	1980	5 miles ESE of Thousand Palms
UCR28302	A. C. Sanders	1982	San Diego County, Otay Mountain
UCR41335	N. M. Fellows	1985	Snow Creek, Cabazon
UCR41733	G. R. Ballmer	1986	Los Angeles County, El Segundo dunes
UC125298	F. M. Roberts	1987	upper Newport Bay
UCR50592	A. C. Sanders	1988	Redlands
UCR51366	A. C. Sanders	1988	Riverside

Schismus barbatis

UC766963	R. F. Hoover	1936	Kern County, Bena
UC1132930	H. S. Yates	1937	Kern County, Lockern
UC1132931	H. S. Yates	1937	Whitewater
UCR23626	J. C. Roos	1940	Palm Springs
UC1108987	H. E. and S. T. Parks	1940	Indio
UC1117713	Alan A. Beetle	1946	San Bernardino County, Devore
UC907586	Curtis Bowsey	1946	Thermal
UC750633	Lewis S. Rose	1947	Baker
UCR26340	J. C. Roos	1947	Riverside County, S face of Eagle Mountains
UC905066	John Thomas Howell	1948	Kern County, Tehachapi Pass

SOURCE: SMASCH database, Jepson Herbarium, University of California, Berkeley, http://ucjeps.berkeley.edu/active.html.

References to wildflowers in the *Los Angeles Times,* *The Desert Magazine,* and the *Riverside Press Enterprise*
(year ending, July to June)

LOS ANGELES TIMES

1919	May 4
1920	January 1, March 4, May 19
1921	April 4, 17, May 15
1922	April 9, May 7, May 14
1923	May 10
1924	April 13
1925	April 5
1926	February 21, March 21
1927	March 10, 12, 13, 14, 15, April 3, 17
1928	January 3, 25, February 12, March 18, 24, April 4, 22
1929	March 10, 17, 24, April 2, 6, 7, 21
1930	March 28, April 6, 11, 13, May 1
1931	February 22, March 8, 24, April 12
1932	February 8, 28, March 2, 13, 19, 20, 21, 25, 27, 28, April 3, 7, 9, 10, 14, June 17
1933	February 13, 26, March 5, 19, April 2, 16, May 7
1934	March 18, March 25, April 1, 8
1935	February 17, 18, 24, March 3, 7, 10, 11, 14, 17, 20, 24, 25, 31, April 7, 17, 25
1936	March 19, 29
1937	January 12, March 7, 11, 13, 14, 18, 27, 28, April 4, 11, 18

1938	March 7, 18, 21, April 2, 3, 10, 17, 24, May 1
1939	March 26, April 2, 9, 16, 18
1940	February 13, 26, March 21, 24, 25, 30, 31, April 7, 8, 14
1941	March 2, 3, 15, 16, 30, April 10, 20, 27, May 11
1942	February 13, April 14, 29
1943	March 17
1947	March 15, 22
1949	February 23, March 3, 15, 16, 20, 26, April 17
1950	March 1, 17, 19, 26, April 7
1951	January 2, March 25, April 15
1952	February 20, March 2, 20, 27, April 3, 13, 20
1953	February 22, March 5, 7, 8, 15, April 1, 19
1954	March 5, 11, 24, April 4, 16, 20
1955	March 19, 20, 26, April 3
1956	January 25, March 11, April 1
1957	February 22, March 10, 23, 24
1958	March 1, 22, 28, April 4, 6, 12
1959	February 26, March 1, April 22, 26
1960	March 22
1961	February 17, 20, March 7, 23
1962	March 17, 19, April 1, 8, 19, 29
1963	March 29. April 22, 25, May 10, 13
1964	April 20, June 7
1965	March 12, 24, April 6, 12, 25, May 2
1966	December 10, March 15, April 10, 13, 23
1967	March 20
1969	March 2, 27, 30, April 20, May 11
1970	March 8
1971	April 11
1972	March 12
1973	March 8, 18, 27, April 15, 26
1974	February 3, March 31, April 11
1975	April 21, 27
1976	May 2
1977	March 13, 20
1978	March 27, April 13, 16, 25
1979	March 25
1980	March 20, April 6,

1981   April 5
1983   March 13
1985   March 31
1986   April 6
1987   March 28, April 11
1988   March 27, April 2
1990   April 9, 10, 15
1991   April 4, 18, 24
1992   March 5, 22, 29, April 8, 19, 29
1993   March 28, April 11
1994   March 19
1995   March 17, April 2, 6, 21
1996   April 18
1997   March 14, 20
1998   March 1, 10, 16, 22, April 13, 17
1999   April 9
2000   May 4
2001   March 18, 31, April 30, May 9
2002   June 23
2003   March 12, 30
2004   March 14
2005   January 30, March 1, 8, 13, 31, April 1, 12

THE DESERT MAGAZINE

1939   April, p. 28; May, p. 36
1940   April, p. 28
1941   April, p. 28; May p. 5
1942   April, p. 16
1946   April, p. 28; May, p. 28
1947   April, p. 12; May, p. 28
1948   April, p. 30; May, p. 29
1949   March, p. 26; April, p. 8; May, p. 28
1950   March, p. 26; April, p. 26; May, p. 28
1951   February, p. 29; April, p. 39; May, p. 2
1952   March, p. 25; April, p. 16; May, p. 27
1953   March, p. 26; April, p. 16; May, p. 20
1954   April, p. 6; May, p. 10

1955   March, p. 26; April, p. 30; May, p. 15

1956   April, p. 9

1957   April, p. 10; May, p. 6

1958   April, p. 10; May, p. 8

1959   March, p. 12

1960   April, p. 30

1961   April, p. 28

1962   May, p. 20

RIVERSIDE PRESS ENTERPRISE (INCOMPLETE BEFORE 1993)

1949   March 18

1952   April 4

1973   March 17

1993   January 6, March 12, April 16

1994   April 1

1995   March 10, 25, 29, April 2

1996   April 12

1997   March 5, 21, 29

1998   December 5, February 12, 20, March 6, 12, 13, 19, 20 April 9

1999   March 17, 19

2000   March 31, April 16

2001   March 30, April 18, 26

2002   March 29, April 14

2003   March 18, 21 April 2, 9, 11

2004   February 11, March 8, 10, 17, 30, April 2, 10

2005   February 11, March 9, 10, 13, 24

References

Abrams, L. 1904. Flora of Los Angeles and Vicinity. Stanford University Press, Stanford, CA. 474 pp.

———. 1940. Illustrated Flora of the Pacific States: Washington, Oregon, and California. 2 vols. Stanford University Press, Stanford, CA.

Anderson, M. K. 2005. Tending the Wild: Native American Knowledge and the Management of California's Natural Resources. University of California Press. Berkeley, 526 pp.

Anonymous. 1857. Letter from California, San Francisco, January 20, 1857. New York Evangelist 28 (February 19): 8. Signed by "W."

Anonymous. 1880. Brilliant tints of Californian flowers. Scientific American 43, no. 12 (September 8): 185.

Aschmann, H. H. 1959a. Evolution of a landscape and its persistence. Annals of the Association of American Geographers 9:35–56.

———. 1959b. The Central Desert of Baja California: Demography and Ecology. University of California Press, Berkeley. 282 pp.

———. (trans. and ed.). 1966. The Natural and Human History of Baja California from Manuscripts by Jesuit Missionaries. Baja California Book Series no. 7. Dawson's Book Shop, Los Angeles. 100 pp.

———. 1976. Man's impact on the southern California flora. In Symposium: Plant Communities of Southern California, J. Latting (ed.), 40–48. California Native Plant Society Special Publication no. 2. California Native Plant Society, Berkeley.

Baker, H. G., and G. L. Stebbins (eds). 1965. The Genetics of Colonizing Species. Academic Press, New York. 588 pp.

Bancroft, H. H. 1888. California Pastoral. The History Co., San Francisco. 808 pp.

Barbour, M. G., B. Pavlik, F. Drysdale, and S. Lindstrom. 1993. California's Changing Landscapes. California Native Plant Society, Sacramento. 244 pp.

Barry, W. J. 1972. The Central Valley Prairie. California Department of Parks and Recreation, Sacramento. 82. p.

Bartolome, J. W. 1979. Germination and seedling establishment in California annual grassland. Journal of Ecology 67:273–81.

Bartolome, J. W., and B. Gemmill. 1981. The ecological status of *Stipa pulchra* (Poaceae) in California. Madroño 28:172–84.

Bartolome, J. W., S. E. Klukkert, and W. J. Barry. 1986. Opal phytoliths as evidence for displacement of native Californian grassland. Madroño 33:217–22.

Baumhoff, M. A. 1963. Ecological determinants of California populations.University of California Publications in American Archaeology and Ethnology 49:155–235.

Bean, L. J., and H. W. Lawton. 1973. Some explanations for the rise of cultural complexity in native California, with comments on protoagriculture. Ballena Press Anthropological Papers, no. 1:v–xlvii.

Beck, W. A., and Y. D. Haase. 1974. Historical Atlas of California. University of Oklahoma Press, Norman. 202 pp.

Becker, R. H. 1964. Designs on the Land: Diseños of California Ranchos; Maps of Thirty-seven Land Grants [1822–46] from the Records of the United Stated District Court, San Francisco. Book Club of California, San Francisco. 151 pp.

Beechey, F. W. 1831. Narrative of a Voyage to the Pacific and Beering Strait to Co-operate with the Polar Expeditions; Performed on His Majesty's Ship Blossom under the Command of F. W. Beechey in the Years 1825, 1826, 1827, 1828. 2 vols. H. Colburn and R. Bentley, London.

Beetle A. A. 1947. Distribution of native grasses of California. Hilgardia 17:309–57.

Bell, C. J., and E. J. Lundelius Jr. (co-chairman), A. D. Barnosky, R. W. Graham, E. H. Lindsay, D. R. Ruez Jr., H. A. Semken Jr., S. D. Webb, and R. J. Zakrzewski. 2004. The Blancan, Irvintonian, and Rancholabrean mammal ages. In Late Cretaceous and Cenozoic mammals of North America, M. O. Woodburne (ed.), 232–314. Columbia University Press, New York.

Bidwell, J. 1904. Early California reminiscences. Out West 20 (March): 182–88.

———. 1928. Echoes of the Past about California. The Lakeside Press, R.R. Donnelley & Sons Co., Chicago.

———. 1937. A Journey to California. John Henry Nash, San Francisco. 48 pp.

———. 1948. In California before the Gold Rush. Ward Ritchie Press, Los Angeles. 111 pp.

———. 1966. Life in California before the Gold Discovery. Lewis Osborne, Palo Alto, CA. 76 pp.

Bigelow, J. M. 1856. General description of the botanical character of the country. Pp. 1–61 in vol. 3, pt. 5 of U.S. Department of War, Reports of Explorations and Surveys [Pacific Railroad survey].

Biswell, H. H. 1956. Ecology of California grasslands. Journal of Range Management 9, 19–24.

Black, E. B. 1975. Rancho Cucamonga and Doña Merced. San Bernardino County Museum Association, Redlands, CA. 322 pp.

Blackburn, T. C. 1975. December's Child: A Book of Chumash Oral Narratives. University of California Press, Berkeley. 359 pp.

Blake, W. P. 1856. Routes in California, to Connect with the Routes near the Thirty-fifth and Thirty-second Parallels. Geological Report. Vol. 5, pt. 2 of U.S. Department of War, Reports of Explorations and Surveys [Pacific Railroad survey].

Blumler, M. A. 1992. Some myths about California grasses and graziers. Fremontia 20:22–27.

———. 1995. Invasion and transformation of California's valley grassland, a Mediterranean analogue ecosystem. In Ecological Relations in Historical Times: Human Impact and Adaptation, R. A. Butlin and N. Roberts (eds.), 308–32. Blackwell, Oxford.

Blumler, M. A., and R. Byrne 1991. The ecological genetics of domestication and the origins of agriculture. Current Anthropology 32:23–54.

Bolton, H. E. 1916. Spanish Exploration in the Southwest, 1542–1706. Barnes and Noble, New York.

——— (trans.). 1926. Historical Memoirs of New California, by Fray Francisco Palóu. 4 vols. University of California Press, Berkeley.

———. 1927. Fray Juan Crespí, Missionary Explorer on the Pacific Coast, 1769–1774. University of California Press, Berkeley. 402 pp.

———. 1930a. Anza's California Expeditions. Vol. 2, Opening a Land Route to California [diaries of Anza, Díaz, Garcés, and Palóu]. University of California Press, Berkeley. 473 pp.

———. 1930b. Anza's California Expeditions. Vol. 4, Font's Complete Diary. University of California Press, Berkeley. 552 pp.

———. 1931. Outpost of empire: The Story of the Founding of San Francisco. A. A. Knopf, New York. 334 pp.

———. 1933. Font's Complete Diary: A Chronicle of the Founding of San Francisco. University of California Press, Berkeley. 552 pp.

Bossard, C. C., J. M. Randall, and M. C. Hoshovsky (eds.). 2000. Invasive Plants of California's Wildlands. University of California Press, Berkeley. 360 pp.

Boudet R. V. 1880. The wild-flower season. Californian March 1 (3): 291.no.

Brandegee K. 1892. Catalogue of flowering plants and ferns growing spontaneously in the streets of San Francisco. Zoe 2:334–86.

Brewer, W. H. 1883. Pasture and forage plants. Pp. 959–64 in U.S. Tenth Census, vol. 3, Report on the Productions and Agriculture as Returned at the Tenth Census. Government Printing Office, Washington, DC.

———. 1966. Up and Down California, 1860–1864. University of California Press, Berkeley. 583 pp.

Brewer, W. H., and S. Watson. 1876–80. Botany. Vol. 1, Polypetalae, by W. H. Brewer and S. Watson. Vol. 2, Apetalceae, Gymnospermae Monoctyledonous or endogenous plants, Cryptogamous plants, by S. Watson. Prepared by the Geological Survey of California, J. D. Whitney, state geologist. Welch, Bigelow, and Co., University Press, Cambridge, MA.

Brooks, M. L. 1999. Alien annual grasses and fire in the Mojave Desert. Madrono 46:13–19.

Brooks, M. L. et al. 2004. Effects of invasive alien plants on fire regimes. Bioscience 54:678–88.

Brown, A. K. 1967. The aboriginal population of the Santa Barbara Channel. Reports of the University of California Archaeological Survey, no. 69:1–100.

————. 2001. A Description of Distant Roads: Original Journals of the First Expedition into California, 1769, 1770, by Juan Crespí. San Diego State Press, San Diego. 848 pp.

Brown, D. E., and R. A. Minnich. 1986. Fire and changes in creosote bush scrub of the western Sonoran Desert. American Midland Naturalist 116:411–22.

Brumbaugh, R. W. 1980. Recent geomorphic and vegetal dynamics on Santa Cruz Island. In The Channel Islands: Proceedings of a Multidisciplinary Symposium, D. M. Power (ed.), 139–65. Santa Barbara Museum of Natural History, Santa Barbara.

Bryant, E. 1848. What I Saw in California: Being the Journal of a Tour by the Emigrant Route and South Pass of the Rocky Mountains, across the Continent of North America, the Great Desert Basin and through California. Appleton, New York. 455 pp.

Burcham, H. T. 1957. California Range Land: An Historical-Ecological Study of the Range Resource of California. California Division of Forestry, Sacramento. 261 pp.

Burrus, E. J. (trans.). 1966. Wenceslaus Linck's Diary of his 1766 Expedition to Northern Baja California. Baja California Travel Series no. 5. Dawson's Book Shop. Los Angeles. 115 pp.

Burrus E. J. 1967. Diario del Capitán Comandante Fernando de Rivera y Moncada, vol. 1. Colección Chimalistac de Libros y Documentos acerca de la Nueva España no. 24. Ediciones José Porrúa Toranzas, Madrid.

Cabrera Bueno, J. G. 1734. Navegación Especulativa y Práctica. Manila. 303 pp.

California Ranchos by County. 2006. www.californiaweekly.com/ca_ranchos.htm.

California State Archives. U.S. Surveyor General for California. Spanish and Mexican Land Grant Maps, 1855–1875. www.sos.ca.gov./archives/level3_ussg3.html

Carlsen, T. M., J. M. Menke and B. M. Pavlik. 2000. Reducing competitive suppression of a rare annual forb by restoring native California perennial grassland. Restoration Ecology 22:39–68.

Carmen, E. A., H. A. Heath and J. Minto. 1892. Special Report on the History and Present Condition of the Sheep Industry of the United States. Government Printing Office, Washington, DC. 1000 pp.

Clar, C. R. 1959. California Government and Forestry. Part 1: From the Spanish Days until the Creation of the Department of Natural Resources in 1927. California Department of Natural Resources, Division of Forestry, Sacramento. 292 pp.

Cleland R. G. 1964. Cattle on a Thousand Hills. Huntington Library, San Marino, CA.

Clements, F. E. 1916. Plant Succession: An Analysis of the Development of Vegetation. Carnegie Institution of Washington Publication no. 242. Carnegie Institution of Washington, Washington, DC. 512 pp.

————. 1920. Plant indicators. Journal of Ecology 22:39–68.

————. 1934. The relict method in dynamic ecology. Journal of Ecology 22:39–68.

Clements, F. E. and V. E. Shelford. 1939. Bio-ecology. Wiley. New York. 394 pp.

Coates, P.A. 2006. American perceptions of immigrant and invasive species: Strangers on the land. University of California Press, 256 pp.

Coffin Jones, L. 1881. On a California ranch. Lippincott's Magazine of Popular Literature and Science (April): 366.

Cohen, D. 1966. Optimizing reproduction in a randomly varying environment. Journal of Theoretical Biology 12:119–29.

Cole, K. L. 1986. The lower Colorado River Valley: A Pleistocene desert. Quaternary Research 25:392–400.

Cole, K. L., and R. H. Webb 1985. Late Holocene vegetation changes in Green Water Valley, Mojave Desert, California. Quaternary Research 23:227–35.

Cook, S. F. 1960. Colonial expeditions to the interior of California, Central Valley, 1800–1820. Anthropological Records 16:238–92.

———. 1962. Expeditions to the interior of California, Central Valley, 1820–1840. Anthropological Records 20:151–214.

Cooledge, S. 1872. The Calaveras big trees. Friends' Intelligencer 29 (September 12): 459.

Cornhill Magazine. 1883. Early spring in California. The Eclectic Magazine of Foreign Literature 7 (6): 733.

Coues, E. 1900. On the Trail of a Spanish Pioneer: The Diary and Itinerary of Francisco Garcés in His Travels through Sonora, Arizona, and California. 2 vols. Frances P. Harper, New York. 608 pp.

Cronise, Titus F. 1868. The Natural Wealth of California. H. H. Bancroft & Company, San Francisco. 696 pp.

D'Antonio, C. M., and P. M. Vitousek. 1992. Biological invasions by exotic grasses, the grass/fire cycle, and global change. Annual Review of Ecology and Systematics 23:63–87.

Dana, R. H. 1911. Two Years Before the Mast. Houghton-Mifflin Co. Boston. 553 pp.

Davidson, A. 1891. British plants in southern California. The Transactions and Journal of Proceedings of the Dumfriesshire and Galloway Natural History and Antiquarian Society, Session 1890–91, no. 7:112–15. Courier and Herald Offices, Dumfries, Scotland. Reprinted in Crossosoma 23 (1997): 68–70.

———. 1893a. Immigrant plants in Los Angeles County, California.—I. Erythea 1:56–61.

———. 1893b. Immigrant plants in Los Angeles County, California.—II. Erythea 1:98–104.

———. 1895. Immigrants of Los Angeles County. Western American Scientist 4:66–86.

———. 1907. Brome expansion in Los Angeles. California Academy of Sciences 6:11.

Davidson, A., and G. L. Moxley. 1923. Flora of southern California. Times-Mirror Press, Los Angeles. 452 pp.

Davis, W. H. 1929. Seventy-five Years in California. John Howell, San Francisco. 422 pp.

Davy, J. B. 1902. Stock Ranges of Northwestern California: Notes on the Grasses and Forage Plants and Range Conditions. U.S. Department Agriculture Bureau of Plant Industry Bulletin 12.

Dawson, T. M. 1874. A trip to Los Angeles, California. New York Evangelist 45 (April 2): 14.

Duflot de Mofras, E. 1937. Duflot de Mofras' Travels on the Pacific Coast. 2 vols. Trans., ed., and annotated by M. E. Wilber. Fine Arts Press, Santa Ana, CA.

Duhaut-Cilly, A. B. 1929. Duhaut-Cilly's account of California in the years 1827–28. Trans. C. F. Carter. California Historical Society Quarterly 8 (2–4): 131–66, 214–50, 306–36.

Dyer, A. R., and K. J. Rice. 1997. Intraspecific and diffuse competition: The response of Nassella pulchra in a California grassland. Ecological Applications 7:484–92.

Eastwood, A. 1893. Field notes at San Emidio. Zoe 4:145–47.

Egan, F. 1977. Frémont: Explorer from a Restless Nation. Doubleday. Garden City, New York. 582 pp.

Ellner, S. 1985a. ESS germination strategies in randomly varying environments, I: Logistic-type models. Theoretical Population Biology 28:59–79.

———. 1985b. ESS germination strategies in randomly varying environments, II: Reciprocal yield-law. Theoretical Population Biology 28:80–116.

Emory, W. H. 1857–59. Report of the United States and Mexican Boundary Survey Made under the Direction of the Secretary of the Interior. 2 vols. A. O. P. Nicholson, Washington, DC.

Engstrand, I. H. W. 1981. Spanish Scientists in the New World: The Eighteenth-Century Expeditions. University of Washington Press, Seattle. 220 pp.

Evarts, H. G. 1958. Jedediah Smith, Trailblazer of the West. Putnam, New York. 192pp.

Farquhar, F. P. 1937. The topographic records of Lieutenant George H. Derby. California Historical Society Quarterly 11:365–82.

Forbes, A. 1839. California: A History of Upper and Lower California from Their First Discovery to the Present Time, Comprising the Account of the Climate, Soils, Natural Productions, Agriculture, Commerce, Etc. Smith Eldeers and Co., London. 352 pp.

Forestier, Auber. 1870. A drive down Marysville. Saturday Evening Post, May 14.

Frémont, J. C. 1845. Report of the Exploring Expedition to the Rocky Mountains in the Year 1842 and to Oregon and California in the Years 1843–44. 28th Cong., 2nd sess., Senate Doc. 174.

———. 1848. Geographical Memoir upon Upper California. 30th Cong., 1st Sess., Senate Misc. Doc. 148.

Frenkel, R. E. 1970. Ruderal Vegetation along Some California Roadsides. University of California Press, Berkeley. 163 pp.

Gillis, M. J., and M. F. Magliari. 2003. John Bidwell and California: The Life and Writings of a Pioneer, 1841–1900. Arthur H. Clark Co., Spokane, WA.

Gordon, C. 1883. Report on cattle, sheep, and swine supplementary to enumeration of livestock on farms in 1880. Pp. 951–1116 in U.S. Tenth Census, vol. 3, Report on the Productions and Agriculture as Returned at the Tenth Census. Government Printing Office, Washington, DC.

Graumlich, L. J. 1993. A 1000-year record of temperature and precipitation in the Sierra Nevada. Quaternary Research 39:49–55.

Grove, A. T., and O. Rackham. 2001. The Nature of Mediterranean Europe: An Ecological History. Yale University Press, New Haven, CT. 384 pp.

Hall, C. 1929. California, our lady of flowers. National Geographic 60:703–50.

Hamilton, J. G. 1997. Changing perceptions of pre-European grasslands in California. Madroño 44:311–33.

Hamilton, J. G., C. Holzapfel, and B. E. Mahall. 1999. Coexistence and interference between a native perennial grass and non-native annual grasses in California. Oecologia 121:518–26.

Haston, L., and J. Michaelson. 1994. Long-term central coastal California precipitation variability and relationships to El Niño/Southern Oscillation. Journal of Climate 7:1373–87.

Heady, H. F. 1958. Vegetational changes in the California annual type. Ecology 39:401–16.

———. 1977. Valley grassland. In Terrestrial Vegetation of California, M. G. Barbour and J. Major (eds.), 491–514. Wiley, New York.

Heady, H. F., J. W. Bartolome, M. D. Pitt, G. D. Savelle, and M. C. Stroud. 1992. California prairie. In Ecosystems of the World 8A: Natural Grasslands, R. T. Coupland (ed.), 313–36. Elsevier Scientific Publishing Company, Amsterdam.

Hendrickson, J. 1988. Valley Hunt Club. www.nationmaster.com/encyclopedia/Valley-Hunt-Club.

Hendrickson, J. 1989. Tournament of Roses: The First 100 Years. Knapp Press, Los Angeles. 276 pp.

Hendry, G. W. 1931. The adobe brick as a historical source. Agricultural History 5:110–27.

Hendry, G. W., and M. P. Kelley. 1925. The plant content of adobe bricks. California Historical Society Quarterly 4:361–73.

Hervey, D. F. 1949. Reaction of a California animal-plant community to fire. Journal of Range Management 2:116–21.

Hickman, J. C. 1993. The Jepson Manual: Higher plants of California. University of California Press. 1400 pp.

Hittell, J. S. 1874 [1863]. The Resources of California: Comprising the Society, Climate, Salubrity, Scenery, Commerce and Industry of the State. A. Roman & Co., San Francisco. 464 pp.

Holder, C. F. 1889. All about Pasadena and Its Vicinity: Its Climate, Missions, Trails, and Cañons, Fruits, Flowers and Game. Lee and Shepard, Boston. 141 pp.

Hornbeck D. 1983. California Patterns: A Geographical and Historical Atlas. Mayfield Publication Co., Palo Alto, CA. 117 pp.

Howell, J. T. 1937. A Russian collection of California plants. Leaflets of Western Botany 2:17–20.

Huenneke, L. F. 1989. Distribution and regional patterns of California grasslands. In Grassland Structure and Function: California Annual Grassland, L. F. Huenneke and H. Mooney (eds.), 1–12. Kluwer Academic Publishers, Dordrecht and Boston.

Huenneke, L. F., and H. A. Mooney (eds.). 1989. Grassland Structure and Function: California Annual Grassland. Kluwer Academic Publishers, Dordrecht and Boston. 220 pp.

Huenneke, L. F., S. P. Hamburg, R. Koide, H. A. Mooney, and P. M. Vitousek. 1990. Effects of soil resources on plant invasion and community structure in Californian serpentine grassland. Ecology 71:478–91.

Hunt, T. W. 1859. California, Oregon and Washington. Journal of the American Geographical and Statistical Society 1 (May): 5.

Hunter, R. 1991. Bromus invasions of the Nevada Test Site: Present status of *B. Rubens* and *B. tectorum* with notes on their relationship to disturbance and altitude. Great Basin Naturalist 51:176–82.

Jackson, J. B. C., et al. 2001. Historical overfishing and the recent collapse of coastal ecosystems. Science 293:629–37.

Jackson, L. E. 1985. Ecological origins of California's Mediterranean grasses. Journal of Biogeography 121:349–61.

Jaeger, E. C. 1941. Desert wildflowers. Stanford University Press, Stanford, CA. 322 pp.

James, J. F. 1880. A botanist in Southern California. American Naturalist 13: 492–498.

Jepson, W. L. 1925. A Manual of the Flowering Plants of California. Sather Gate Bookshop, Berkeley, CA. 1238 pp.

Kahrl, W. L., W. A. Bowen, S. Brand, M. L. Shelton, D. L. Fuller, and D. A. Ryan. 1979. The California Water Atlas. State of California, General Services, Publication Section, North Highlands, CA.

Keeley, J. E. 1989. California valley grassland. In Endangered Plant Communities of Southern California, A. A. Schoenherr (ed.), 3–23. Southern California Botanists Special Publication No. 3. Southern California Botanists, Claremont, CA.

———. 1993. Native grassland restoration: The initial state—assessing suitable sites. In Interface between Ecology and Land Development in California. J. E. Keeley (ed.), 271–81. Southern California Academy of Sciences, Los Angeles.

———. 2000. Chaparral. In North American Terrestrial Vegetation, M. G. Barbour and W. D. Billings (eds.), 203–53. Cambridge University Press, Cambridge.

Keeley, J. E., and C. J. Fotheringham 1997. Trace gas emissions and smoke-induced germination. Science 276:1248–50.

———. 1998. Mechanism of smoke-induced germination in a postfire annual. Journal of Ecology 86:27–36.

Kimball, S., and P. M. Schiffman. 2003. Differing effects of cattle grazing on native and alien plants. Conservation Biology 17:1681–93.

King, C. 1915. Mountaineering in the Sierra Nevada. Charles Scribner's Sons, New York. 378 pp.

King, T. J. 1976. Late Pleistocene-early Holocene history of coniferous woodlands in the Lucerne Valley region, Mohave Desert, California. Great Basin Naturalist 36:227–338.

King, T. J., and J. R. Van Devender. 1977. Pollen analysis of fossil packrat middens from the Sonoran Desert. Quaternary Research 8:191–204.

Kip, W. I. 1954. Early Days of My Episcopate. Biobooks, Oakland, CA. 263 pp.

Kroeber, A. L. 1953. Handbook of the Indians of California. California Book Co., Berkeley. 995 pp.

L'Enfant. 1848. A hunt in the mountains of California. Spirit of the Times: A Chronicle of the Turf; Agriculture, Field Sports, Lite 18 (July 1): 19.

Leonard, Z. 1959. Adventures of Zenas Leonard, Fur Trader, 1833–34. Ed. J. C. Ewers. University of Oklahoma Press, Norman. 172 pp.

Lewis, H. T. 1973. Patterns of Indian burning in California: Ecology and ethnohistory. Ballena Press Anthropological Papers, no. 1:3–101.

Lindley, W., and J. P. Widney. 1888. California of the South: Its Physical Geography, Climate, Resources, Routes of Travel, and Health Resorts. Appleton and Co., New York. 377 pp.

Lockmann, R. F. 1981. Guarding the Forests of Southern California: Evolving Attitudes toward Conservation and Watershed, Woodlands and Wilderness. Arthur H. Clark Co., Glendale, CA.

Lougheed, H. E. 1951. Harvesting wild mustard. Westways 43:14–15.

Lymau, C. S. 1850. Art. VI.—California. El Dorado, or Adventures in the Path of Empire, by Bayard Taylor, Author of "Views a-foot"; and "Three Years in California," by Rev. Walter Colton, U.S.N., &c. New Englander 8 (November): 32.

Mack, R. N. 1981. Invasion of *Bromus tectorum* L. into western North America: An ecological chronicle. Agro-Ecosystems 7:145–65.

———. 1989. Temperate grasslands vulnerable to plant invasion: Characteristics and consequences. In Biological Invasions: A Global Perspective, J. A. Drake et al. (eds.), 155–79. Wiley, New York. 525 pp.

Mack, R. N., and J. N. Thompson, J. N. 1982. Evolution in steppe with few large, hooved mammals. American Naturalist 119:757–73.

Major, J., and W. T. Pyott. 1966. Buried viable seeds in two California bunch grass sites and their bearing on definition of a flora. Vegetatio 13:253–82.

Martin, P. S. 2005. Twilight of the Mammoths: Ice Age Extinctions and the Rewilding of America. University of California Press, Berkeley. 250 pp.

Martínez, M. 1947. Baja California: Reseña a Histórica del Tieritorio y Su Flora. Ediciónes Botas, Mexico. 154 pp.

Mattoni, R., and T. R. Longcore. 1997. The Los Angeles coastal prairie, a vanished community. Crossosoma 23:71–102.

Mayfield, T. J. 1929. San Joaquin Primeval. Uncle Jeff's Story. Arranged by F. F. Latta. Tulare Times Press, Tulare, CA.

McGuire, B. 1935. Distributional notes on plants of the Great Basin Region-I. Leaflets of Western Botany 1:185–88.

McNaughton, S. J. 1968. Structure and function of California grassland. Ecology 49:962–72.

McWilliams, C. 1946. Southern California: An Island on the Land. Duel, Sloan & Pearce, New York. 387 pp.

Mead, J. I., and A. M. Phillips III. 1981. The late Pleistocene and Holocene fauna of Vulture Cave. Southwestern Naturalist 26:257–88.

Meigs, P. 1935. The Dominican Mission Frontier of Lower California. University of California Publications in Geography no. 7. University of California Press, Berkeley. 231 pp.

Mensing, S., and R. Byrne. 2000. Invasion of Mediterranean weeds into California before 1769. Fremontia 27 (3): 6–9.

Merrifeld, G. C. 1851. California wildflowers. Prairie Farmer 11:1.

Meyer, M. D., and P. M. Schiffman 1999. Fire season and mulch reduction in a California grassland: A comparison of restoration strategies. Madroño 46:25–37.

Millspaugh, C. F., and L. W. Nuttall. 1923. Flora of Santa Catalina Island. Field Museum of Natural History Pub. 212, Botanical Series vol. 5, Chicago. 413 pp.

Minnich, R. A. 1980. Vegetation of Santa Cruz and Santa Catalina Islands. In
The Channel Islands: Proceedings of a Multidisciplinary Symposium, D. M.
Power (ed.), 123–37. Santa Barbara Museum of Natural History, Santa Barbara, CA.
———. 1983. Fire mosaics in southern California and northern Baja California.
Science 219:1287–94.
———. 1988. The Biogeography of Fire in the San Bernardino Mountains of
California. University of California Publications in Botany vol. 28. University of California Press, Berkeley. 120 pp.
———. 2003. Fire and Dynamics of Temperate Desert Woodlands in Joshua Tree
National Park. U.S. Department of Interior, National Park Service Contract
P8337000034/0001.
———. 2006. California climate and fire weather. In Fire in California's Ecosystems, N. Sugihara, G. J. W. Van Wagtendonk, K. E. Shaffer, J. Fites-Kaufman,
and A. E. Thode (eds.), 13–37. University of California Press, Berkeley.
Minnich, R. A., and E. Franco-Vizcaíno. 1998. Land of Chamise and Pines. Uni-
versity of California Publications in Botany vol. 80. University of California
Press, Berkeley. 166 pp.
Minnich, R. A., and A. C. Sanders 2000. *Brassica tournefortii* Gouan. In Inva-
sive Plants of California's Wildlands, C. C. Bossard, J. M. Randall, and M. C.
Hoshovsky (eds.), 68–72. University of California Press, Berkeley. 360 p.
Montané Martí, J. C. M. 1989. Juan Bautista de Anza: Diario del Primer Viaje a
la California, 1774. Sociedad Sonorense de Historia, Hermosillo, Sonora,
  Mexico. 121 pp.
Mooney, H. A., and J. A. Drake (eds.). 1986. Ecology of Biological Invasions of
North America and Hawaii. Springer-Verlag, New York. 321 pp.
Mooney, H. A. Hamberg, S. P., and J. A.. Drake. 1986. The invasions of plants
and animals into California. In Ecology of Biological Invasions of North Amer-
ica and Hawaii, H. A. Mooney and J. A. Drake (eds.), 250–72. Springer-Verlag,
New York.
Morgan, D. L., and J. R. Scobie. 1964. Three Years in California: William Perkins'
Journal of Life at Sonora, 1849–1852. University of California Press. Berke-
ley, California. 424 p.
Morris, E. 2001. Theodore Rex: The Rise of Teddy Roosevelt. Random House,
New York. 772 pp.
Muir, J. 1874. Summering in the Sierra. Overland Monthly and Out West Magazine 12, no. 1 (January): 79.
———. 1883. The Sierra Madre Mountains—by John Muir. In Southern California Paradise—Being a Historic and Descriptive Account of Pasadena, San
Gabriel, Sierra Madre, and La Canada. Edited and published by R. W. C.
Farnsworth, Pasadena, CA.
———. 1904. The Mountains of California. The Century Co., New York. 381 pp.
———. 1948. Yosemite and the Sierra Nevada. Houghton Mifflin, Boston. 132
pp.
———. 1974. Rambles of a Botanist among the Plants and Climates of California. Dawson's Book Shop, Los Angeles. 43 pp.
Munz, P. A. 1974. A Flora of Southern California. University of California Press,
Berkeley. 1086 pp.

Munz, P., and D. Keck 1949. California plant communities. El Aliso 2:87–105.
————. 1959. A California Flora. Rancho Santa Ana Botanical Garden, with the University of California Press, Berkeley. 1681 pp.

Naveh, Z. 1967. Mediterranean ecosystems and vegetation types in California and Israel. Ecology 48:445–59.

Nelson L. L., and E. B. Allen. 1993. Restoration of *Stipa pulchra* grasslands: Effects of mycorrhizae and competition from *Avena barbata*. Restoration Ecology 1:40–50.

Newberry, J. S. 1857. Routes in California, to connect with the Routes near the Thirty-fifty and Thirty Second Parallels. Geological Report, vol. 6 pt. 2, of the U.S. Department of War, Reports of the Explorations and Surveys [Pacific Railroad Survey].

O'Leary, J. F., and Minnich, R. A. 1981. Postfire recovery of creosote bush scrub vegetation in the western Colorado Desert. Madroño 28:61–65.

Online Archive of California. Bound Manuscripts Collection, Diseños, http://www.oac.cdlib.org.

Ornduff, R. 2003. Introduction to California Plant Life. University of California Press, Berkeley. 341 pp.

The Paleobiology Database. http://paleodb.org.

Parish, S. B. 1890a. Notes on the naturalized plants of southern California, IV. Zoe 1:182–88.

————. 1890b. Notes on the naturalized plants of southern California. Zoe 1:7–10.

————. 1909. Notes on some introduced plants of southern California—I. Muhlenbergia 5:109–15.

————. 1920. The immigrant plants of southern California. Bulletin of the Southern California Academy of Sciences 19:3–30.

Parke, J. G. 1857. Routes in California, to connect with Routes near the Thirty-fifth and Thirty-Second Parallels. Vol. 7, pt. 1. of the U.S. Department of War, Reports of the Explorations and Surveys [Pacific Railroad Survey].

Parry, C. C. 1859. Part 1, Botany of the boundary, Introduction. Vol. 2 of Emory, Report of the United States and Mexican Boundary Survey Made under the Direction of the Secretary of the Interior. A. O. P. Nicholson, Washington, DC.

Parsons, D. J., and T. J. Stohlgren. 1989. Effects of varying fire regimes on annual grasslands in the Sierra Nevada of California. Madroño 36:154–68.

Pavlik, B. M., P. C. Muick, S. Johnson, and M. Popper. 1991. Oaks of California. Cachuma Press, Inc. Los Olivos, CA .184 pp.

Perkins, J. E. 1863. Sheep husbandry in California. Transactions of the California State Agricultural Society: 134–45.

Phillips, G. H. 1993. Indians and intruders in central California, 1769–1849. University of Oklahoma Press, Norman and London. 223 pp.

Priestly, H. I. (trans.). 1937. A Historical, Political and Natural Description of California by Pedro Fages, Soldier of Spain. University of California Press, Berkeley. 83 pp.

Raven, P. H. 1963. Amphitropical relationships in the floras of North and South America. Quarterly Review of Biology 38:151–77.

Reed, F. M 1916. Catalogue of the plants of Riverside and vicinity. Unpublished. 79 pp.

Revere, J. W. 1849a. A tour of duty in California. The Literary World 4 (February 24):108.

———. 1849b. A tour of duty in California. Christian Inquirer 3 (February 17): 19.

———. 1849c. A tour of duty in California. C. S. Francis and Co., New York. 305 pp.

Roberts, N. C. 1989. Baja California plant field guide. Natural History Publishing Company, La Jolla, CA. 309 pp.

Robbins, W. W. 1940. Alien Plants Growing without Cultivation in California. California Agricultural Experiment Station Bulletin 637.

Robbins, W. W., M. K. Bellue, and W. S. Ball 1951. Weeds of California. California Department of Agriculture, Sacramento.

Robinson, J. W. 1969. José Joaquin Arrillaga: Diary of His Surveys of the Frontier, 1796. Trans. Fray Tiscareño. Ed. and annotated by J. W. Robinson. Dawson's Bookshop, Los Angeles. 103 pp.

Rundel, P. W. 1983. Impact of fire on nutrient cycles in Mediterranean-type ecosystems with reference to chaparral. In Mediterranean-Type Ecosystems: The Role of Nutrients, F.J. Kruger, D.T. Mitchell, and J. Jarvis (eds), 192–207. Springer-Verlag, New York. 552 pp.

Salo, L. F. 2004. Population dynamics of red brome (*Bromus madritensis* subsp. *rubens*): Times for concern, opportunities for management. Journal of Arid Environments 57:291–96.

———. 2005. Red brome (*Bromus rubens* subsp. *madritensis*) in North America: Possible modes for early introductions, subsequent spread. Biological Invasions 7:165–80.

Salvator, L. L. [Archduke]. 1929. Los Angeles in the Sunny Seventies: A Flower from the Golden Land. Translated from German by M. E. Wilber. Bruce McCallister Jake Zeitlin, Los Angeles. 188 pp.

Sauer, J. D. 1988. Plant Migration: The Dynamics of Geographic Patterning in Seed Plant Species. University of California Press, Berkeley. 282 pp.

Saunders, C. F. 1914. With the Flowers and Trees in California. McBride, Nast, and Co., New York. 286 pp.

———. 1931 [1913]. Under the Sky in California. Robert M. McBride & Co., New York. 299 pp.

Schiffman, P. M. 2000. Mammal burrowing, erratic rainfall and the annual lifestyle in the California prairie: Is it time for a paradigm shift? In 2nd Interface between Ecology and Land Development in California, J. E. Keeley, M. Baer-Keeley, and C. J. Fotheringham (eds.), 75–78. U.S. Geological Survey Open File Report 00–62.

Simpson, L. B. 1938. California in 1792: The Expedition of José Longinos-Martínez. Huntington Library, San Marino, CA. 111 pp.

——— (trans. and ed.). 1961. Journal of Jose Longinos Martínez: Notes and Observations of the Botanical Expedition in Old and New California and the South Coast, 1791–1792. John Howell Books, San Francisco. 114 pp.

Sims, P. L., and P. G. Risser. 2000. Grasslands. In North American Terrestrial Vegetation, M. G. Barbour and J. Major (eds.), 323–56. Cambridge University Press, Cambridge.

Smiley, F. J. 1922. Weeds of California and methods of control. Monthly Bulletin of the Department of Agriculture 11. California State Printing Office, Sacramento.

Spaulding, W. G. 1983. Late Wisconsin macrofossil records of desert vegetation in the American Southwest. Quaternary Research 19:256–264.

———. 1990. Vegetational and climatic development of the Mojave Desert: The last glacial maximum to the present. In Packrat Middens: The Last 40,000 Years of Biotic Change, J. L. Betancourt, T. R. Van Devender, and P. S. Martin (eds.), 166–99. University of Arizona Press, Tucson.

Talbot, M. W., H. H. Biswell, and A. L. Hormay. 1939. Fluctuations in annual vegetation of California. Ecology 20:39–402

Teggart F. J. (trans.). 1911. The Portola Expedition of 1769–1770: Diary of Miguel Costanso. Publications of the Academy of Pacific Coast History, University of California Press, Berkeley. 167 pp.

Thornton, J. Q. 1848. Oregon and California in 1848. Harper and Brothers, New York. 379 pp.

Timbrook, J., J. R. Johnson, and D. D. Earle. 1982. Vegetation burning by the Chumash. Journal of California and Great Basin Anthropology 4:163–86.

Torrey, J. 1856. Description of the general botanical collections. Pp. 61–182 in vol. 4, pt. 1 of U.S. Department of War, Reports of Explorations and Surveys [Pacific Railroad survey].

Twisselmann E. C. 1967. A flora of Kern County, California. Wasmann Journal of Biology 25 (1–2): 1–395.

U.S. Department of War. 1855–61. Reports of Explorations and Surveys to Ascertain the Most Practicable and Economical Route for a Railroad from the Mississippi River to the Pacific Ocean, 1853–54 [Pacific Railroad survey]. 12 vols. U.S. Department of War, Washington, DC.

Van Devender, T. R. 1990. Late Quaternary vegetation and climate of the Sonoran Desert, United States and Mexico. In Packrat Middens: The Last 40,000 Years of Biotic Change, J. L. Betancourt, T. R. Van Devender, and P. S. Martin (eds.), 134–65. University of Arizona Press, Tucson.

Van Dyke, T. S. 1886. Southern California: Its Valleys, Hills, and Streams, Animals, Birds, and Fishes, Gardens, Farms, and Climate. Fords, Howard, and Hulbert, New York, 233 pp.

Vasey, G. 1885. A Descriptive Catalogue of the Grasses of the United States Including Especially the Grass Collections at the New Orleans Exposition Made by the U.S. Department of Agriculture. Gibson Brothers. Washington, DC.

Venable, D. L., and J. S. Brown. 1988. The selective interactions of dispersal, dormancy, and seed size as adaptations for reducing risk in variable environments. American Naturalist 131:360–84.

Vizetelly, H. 1848. Four Months among the Gold Finders in Alta California: Being a Diary of an Expedition from San Francisco to the Gold Districts. David, Bogue, London. 207 pp.

Weaver J. E. and F. E. Clements, 1938. Plant ecology. McGraw-Hill, NY. 601p.

Web de Anza Archives. Journals of Miguel Costansó, July 14, 1769–February 7, 1770 [Portolá expedition]; Juan Bautista de Anza, January 8–May 27, 1774 [first Anza expedition]; Juan Bautista de Anza, October 23,1775–June 1, 1776 [second Anza expedition]; Pedro Font, September 28, 1775–June 2, 1776 [second Anza expedition]; and Pedro Font expanded, September 28, 1775–June

2, 1776 [second Anza expedition]. Center for Advanced Technology in Education, University of Oregon. http://anza.uoregon.edu/archives.html.

Weislander, A. E. 1938. A vegetation type map of California. Madroño 3:140–44.

Wells, P. V. 1962. Vegetation in relation to geological substratum and fire in the San Luis Obispo Quadrangle. Ecological Monographs 32:79–103.

Wester, L. 1981. Composition of native grasslands in the San Joaquin Valley, California. Madroño 28:231–41.

Whipple, A. W. 1856a. Report. Explorations for a Railway Route, near the Thirty-fifth Parallel of North Latitude, from the Mississippi River to the Pacific Ocean. Vol. 3, pt. 1 of U.S. Department of War, Reports of Explorations and Surveys [Pacific Railroad survey].

———. 1856b. Report of the Topographic Features and Character of the Country. Vol. 3, pt. 2 of U.S. Department of War, Reports of Explorations and Surveys [Pacific Railroad survey].

———. 1856c. Route of the Thirty-fifth Parallel of North Latitude. Report of the Zoology of the Expedition. Vol. 4, pt. 6 of U.S. Department of War, Reports of Explorations and Surveys [Pacific Railroad survey].

———. 1961. The Whipple Report: Journal of an Expedition from San Diego, California, to the Rio Colorado, from September 11 to December 11, 1849, by A. W. Whipple, Lieutenant, United States Topographical Engineers. Westernlore Press, Los Angeles. 100 pp.

———. 1967. Native bunchgrass (*Stipa pulchra*) on Hastings Reservation, California. Ecology 48:949–55.

Wilkes, C. 1844. Narrative of the United States Exploring Expedition. Vol. 5, pp. 77–214. Entered according to the Act of Congress, Clerk's office of the District Court for the District of Columbia.

———. 1849. Western America, Including California and Oregon with Maps of Those Regions and the Sacramento Valley. Lee & Blanchard, Philadelphia. 130 pp.

Williamson, R. S. 1856. Report. Explorations in California for Railroad Routes, to Connect with the Routes near the Thirty-fifth and Thirty-second Parallels of North Latitude. Vol. 5, pt. 1 of U.S. Department of War, Reports of Explorations and Surveys [Pacific Railroad survey].

———. 1857. Report of Lieutenant Henry L. Abbot, Corps of Topographical Engineers upon Explorations for a Railroad Route, the Sacramento Valley to the Columbia River. Vol. 6, pt 1 of U.S. Department of War, Reports of Explorations and Surveys [Pacific Railroad survey].

Wills, R. 2000. Effective fire planning for California native grasslands. In 2nd Interface between Ecology and Land Development in California, J. E. Keeley, M. Baer-Keeley, and C. J. Fotheringham (eds.), 75–78. U.S. Geological Survey Open File Report 00–62.

Woodburne, M. O. 2004. Global events and the North American mammalian biochronology. In Late Cretaceous and Cenozoic Mammals of North America, M. O. Woodburne (ed.), 315–43. Columbia University Press, New York. 391 pp.

Zohary, M. 1962. Plant Life of Palestine. Ronald Press, New York. 262 pp.

———. 1973. Geobotanical Foundations of the Middle East. 2 vols. Gustav Fischer Verlag, Stuttgart.

Index

Text and Display: Sabon
Compositor: Integrated Composition Systems
Indexer: Andrew L. Christenson
Cartographer: Bill Nelson
Printer and binder: Thomson-Shore, Inc.